Y0-BTC-424

CONSTRUCTION OF LININGS FOR RESERVOIRS, TANKS, AND POLLUTION CONTROL FACILITIES

Wiley Series of Practical Construction Guides

M. D. Morris, P. E., Editor

Jacob Feld
CONSTRUCTION FAILURE

William G. Rapp
CONSTRUCTION OF STRUCTURAL STEEL BUILDING FRAMES

John Philip Cook
CONSTRUCTION SEALANTS AND ADHESIVES

Ben C. Gerwick, Jr.
CONSTRUCTION OF PRESTRESSED CONCRETE STRUCTURES

S. Peter Volpe
CONSTRUCTION MANAGEMENT PRACTICE

Robert Crimmins, Reuben Samuels, and Bernard Monahan
CONSTRUCTION ROCK WORK GUIDE

B. Austin Barry
CONSTRUCTION MEASUREMENTS

D. A. Day
CONSTRUCTION EQUIPMENT GUIDE

Harold J. Rosen
CONSTRUCTION SPECIFICATION WRITING

Gordon A. Fletcher and Vernon A. Smoots
CONSTRUCTION GUIDE FOR SOILS AND FOUNDATIONS

Don A. Halperin
CONSTRUCTION FUNDING: WHERE THE MONEY COMES FROM

Walter Podolny, Jr., and John B. Scalzi
CONSTRUCTION AND DESIGN OF CABLE-STAYED BRIDGES

John P. Cook
COMPOSITE CONSTRUCTION METHODS

William B. Kays
CONSTRUCTION OF LININGS FOR RESERVOIRS, TANKS, AND POLLUTION CONTROL FACILITIES

CONSTRUCTION OF LININGS FOR RESERVOIRS, TANKS, AND POLLUTION CONTROL FACILITIES

WILLIAM B. KAYS

A Wiley-Interscience Publication

JOHN WILEY & SONS

New York London Sydney Toronto

Library of Congress Cataloging in Publication Data:

Kays, William B 1919-
Construction of linings for reservoirs, tanks, and pollution control facilities.

(Wiley series of practical construction guides)
"A Wiley-Interscience publication."
Bibliography: p.
Includes index.
1. Reservoirs—Linings. 2. Tanks—Linings.
3. Pollution control equipment—Linings. I. Title.
TC167.K38 628.1'3 77-3944
ISBN 0-471-02110-5

Printed in the United States of America

10 9 8 7 6 5 4 3 2 1

TO
ARTHUR M. LOCKHART
HOWARD D. WEBB

Series Preface

The construction industry in the United States and other advanced nations continues to grow at a phenomenal rate. In the United States alone construction in the near future will exceed ninety billion dollars a year. With the population explosion and continued demand for new building of all kinds, the need will be for more professional practitioners.

In the past, before science and technology seriously affected the concepts, approaches, methods, and financing of structures, most practitioners developed their know-how by direct experience in the field. Now that the construction industry has become more complex there is a clear need for a more professional approach to new tools for learning and practice.

This series is intended to provide the construction practitioner with up-to-date guides which cover theory, design, and practice to help him approach his problems with more confidence. These books should be useful to all people working in construction: engineers, architects, specification experts, materials and equipment manufacturers, project superintendents, and all who contribute to the construction or engineering firm's success.

Although these books will offer a fuller explanation of the practical problems which face the construction industry, they will also serve the professional educator and student.

M. D. Morris, P.E.

Contents

CONSTRUCTION OF LININGS FOR RESERVOIRS, TANKS, AND POLLUTION CONTROL FACILITIES

Introduction

Within the past 15 years a new discipline has developed that is quickly expanding into a major consideration for designers of water storage and transmission facilities, for those who work in the wastewater treatment field, and for almost all engineers in search of better ways to control pollution. *Linings* is the name applied to this new technology, and the following pages are concerned with promoting a better understanding of this often misunderstood subject.

To be correct, we should not refer to linings as new. They have been around for many thousands of years. What is new is the rapid growth in their use as a means of controlling seepage from hydraulic facilities. Also new is a developing awareness by users and designers that there is a separate and important technology concerned with the use of linings and that this technology is unlike any other that now exists.

When the subject of linings is mentioned, the most common reaction is to think of the cut-and-fill reservoir, that scooped-out place in the ground built to store water or other liquids. It is linings for this type of facility with which this book is initially concerned. The technology discussed has wide application also to canals, which are close cousins to the cut-and-fill reservoir. In addition, the book serves the expanding interest in linings for concrete and steel tanks. Finally, since linings are one of the major weapons in the battle to control pollution, the special facilities used for this purpose are carefully analyzed. Included is the emerging problem of thermal pollution, the solution of which will undoubtedly involve seepage control membranes.

"Hydraulic linings," within the framework of this book, means any material laid down in a holding or conveyance facility to prevent the movement of water from one point where its presence is desirable or least objectionable to another point where its presence is undesirable. Although, by definition, the liquid involved is water, the term has by custom come to include liquids in general, such as oil, brines, sewage, and chemical solutions of all types. The definition of a holding facility is taken to include such things as concrete, steel, and wooden tanks, in addition to cut-and-fill reservoirs. The conveyance systems discussed herein are primarily open channels (canals).

A search of the literature today would reveal almost nothing about the technology of linings. What little information there is exists largely in the minds of a handful of people who have been working in the field during the past 20 years. A few magazine articles have discussed the subject from a general point of view, and specific information will be found on the various kinds of linings in manufacturers' bulletins. There are also write-ups from laboratories and government publications which discuss the results of field trials in water holding and conveyance systems. All of these discussions, however, give little or no information about the overall technology with which these barrier systems are associated. It is the total system, which includes the lining and its environment, that is the key to successful utilization.

A failure to look at the overall picture has led to a number of unfortunate occurrences in regard to linings. In almost all of these cases, at least initially, the lining was labeled as the weak link in the chain that triggered the failure. No doubt this was sometimes true, but much more often the final analysis showed that other factors were the basic cause of the malfunction. The lining failure, in most cases, was the result of an action rather than its cause.

It is the purpose of this book to examine hydraulic linings from the broad point of view of the overall system of design and utilization. An attempt will be made to include all of the various materials now in general use, their trade names and cost estimates, their advantages and disadvantages, and the ways in which they should be utilized for maximum effectiveness. An entire chapter is devoted to failure mechanisms, for much can be learned from the past. The future will be probed too, in an attempt to find out where the present trends may be taking us.

To complete the book, three appendices present some of the pertinent design information associated with this field, a list of trade names and sources of common lining materials, and (here published for the first time) a set of reservoir tables to aid the design engineer in making the most effective use of his environment. A glossary and a bibliography conclude the book.

The outline that will be followed for the analysis of lining systems in this book includes all materials that have actually enjoyed some substantial usage in the field. No attempt has been made to include experimental linings still in the developmental stage. There are a number of these, but data on them are limited and any information available is on a restricted basis. Some lining materials have been used in combination with each other, and advanced combination systems are discussed in chapter 5.

Linings may be classified in three ways: flexible or rigid, impervious or semiimpervious, continuous or noncontinuous. Table I.1 shows how the common primary linings are categorized in these three classification

TABLE I.1 Lining Classifications

Flexible	Rigid	Miscellaneous
Plastics	Gunite	Bentonite clays
Elastomers	Concrete	Chemical treatments
Asphalt panels	Steel	Waterborne treatments
Compacted soils	Asphalt concrete	Combinations
	Soil cement	

Impervious	Semiimpervious
Plastics	Compacted soils
Elastomers	Gunite
Asphalt panels	Concrete
Steel	Asphalt concrete
	Soil cement
	Bentonite clays
	Chemical treatments
	Waterborne treatments

Continuous	Noncontinuous
Plastics	Compacted soils
Elastomers	Gunite
Asphalt panels	Concrete
Steel	Asphalt concrete
	Soil cement
	Bentonite clays
	Chemical treatments
	Waterborne treatments

systems. The table deals with lining systems that, through general usage, have been more or less accepted by industry as proven materials when used in properly designed and operated facilities.

Every cut-and-fill reservoir requires a lining, primarily to control seepage losses that could cause the facility to fail or could otherwise subject surrounding property or life to undue risks. The lining also reduces embankment erosion due to wave action or rain, and it protects against the undermining of footings and foundations of the reservoir's rigid structural components.

When the proper type of clay is present, it may be used to construct the lining without need for any additional membrane materials. In these cases the lining thickness will usually run from 1 to 10 ft or more. Although compacted clay linings are basically not impervious, they do control seepage and are often used because of their low cost and flexible nature. Normally, they are protected from wave erosion and rain damage by a Gunite or asphalt concrete topping, although other protective systems have also been used with varying degrees of success.

If proper clays are not available, the engineer must look to other materials to meet his lining needs. He may choose one of the thin membranes, instead of on-site clay, because of their more impervious nature. Delivery and installation time is much better, and several membranes carry attractive long-term guarantees by the manufacturers.

The need for the lining must be evaluated, and a determination made as to how effective it should be. This will depend on several factors: reservoir depth, side slope, contents, foot and animal traffic expected, leakage tolerance, risks of failure, and budget considerations, to name a few.

The construction of a cut-and-fill reservoir involves a number of sequential steps. The site determination is dependent on such things as reservoir size, elevation (pressure) requirements, proximity to the service area, environmental impact studies, land availability, safety factors, soils engineering data, and financing.

On the basis of the available land, the engineer determines the measurements of the facility, either by direct calculation or by means of standard reservoir tables (Appendix III). The dimensional parameters of the reservoir are defined as five items:

1. Plan length at top of the slope.
2. Plan width at top of the slope.
3. Total vertical depth.
4. Slope (horizontal : vertical ratio) of side walls.

5. Freeboard (vertical distance from top of berm to normal high water level).

The slope ratio, generally called the slope, does not necessarily have to be uniform for all slopes, although it usually is for ease of construction. If it is not constant, its variation for each slope should be given.

After the engineer has defined the precise area for the reservoir, the land is cleared of trees, shrubs, and any interfering structures and the excavation is begun. In a true cut-and-fill reservoir, the excavated material is used to build up the walls of the reservoir for added capacity. The soils engineer is the one who determines whether the excavated soil can be used in this manner, or whether soil must be imported from another location. He is especially interested in eliminating soils that swell or shrink when wet, as well as those with soluble components, high organic content, or large rocks that will interfere with the attainment of proper compaction.

Excavation is usually done by means of a bulldozer, dragline, or clam shell, with bottom loading equipment required in the case of larger installations. Auxillary trucks and front-end loaders are often used, depending on the dictates of the job. Blasting is utilized when necessary. Although this operation is not very common on small facilities because of economics, it may be feasible on larger installations where blasting costs represent a much lower proportion of the total expense.

Usually the reservoir is overexcavated (dug deeper than the finished elevation), since naturally deposited soils do not normally occur at the densities required for this type of land use. Well-compacted select fill replaces the overexcavated material to give the supporting earth members greater stability. An experienced soils engineering firm determines the procedure, the overexcavation required, the makeup of the replacement soil (if different from the overexcavated material), and the compactive effort necessary to produce the required stability and properties.

Sometimes the compacted soil is built up to an elevation slightly higher than the final grade, and the entire area is "trimmed" back to proper elevation as shown on the drawings, using a motor patrol (road grader) or other appropriate piece of equipment. It can be seen that the slope of the reservoir walls affects the cost of construction. When steep slopes (2½ : 1 or steeper) are used, equipment must be winched up and down the slope—a slow and costly procedure. The type of finishing specifications and the operations and equipment required to meet them are dependent on the lining system to be used.

Before the lining is installed, underdrain trenches are dug and perforated pipes and/or small rock is put into them. Some sort of padding or protection (e.g., porous fine-graded earth or heavy filter cloth material) is then installed on top of the rock layer to insulate it from direct contact with any of the flexible membrane linings. If a continuous underdrain (porous concrete or asphalt concrete) is to be installed, this is done next. The design of the underdrain lines and the continuous underdrain should be under the supervision of an experienced soils engineer. If not properly designed, the system can contribute to serious reservoir malfunctions and even complete failure.

If a thin membrane lining is to be placed directly on the soil, fine grading is the next step. This is normally a hand operation requiring rake, shovel, and muscle, but rolling with a heavy steel roller or related equipment, in conjunction with minor manual touchup, may be enough in some cases. In any event the finishing operation should eliminate broken glass, clods, roots, holes, or other surface irregularities that might do damage to the thin membranes. If the surface is to be paved with concrete, asphalt concrete, or Gunite, fine-grading steps are often eliminated or modified. In these cases accurate, uniform grade is desirable to keep material costs to a minimum, but final surface texture may be rougher. If, on the other hand, a thin membrane is to be installed over these rigid linings, their top surface should be smooth: for concrete, troweling is recommended; for asphalt concrete, a dense, smooth finish coat with no voids is best; and for Gunite a smooth float finish to eliminate rebound particles is mandatory.

Should a structural roof be required, column footings are generally dug after earthwork, fine grading, and underdrains are completed, although some contractors alter the scheduling of this portion of the work to meet special circumstances. Usually the engineer wants the roof structure completed before the lining contractors move onto the job, so that objects which fall during the structural roof work will not damage the lining below. Although this is no doubt the best schedule, it means that artificial lighting is required during the lining installation.

In the case of vertical-wall concrete tanks, most lining work is done as a remedial measure on existing facilities. Here the tank leakage must be controlled to prevent economic loss, avoid unsightly conditions on or adjacent to the tank, or arrest possible corrosion of the reinforcing bars.

Normally, these linings are not bonded to the concrete. Unless patching materials formerly applied to the tank (usually liquid systems that dry or cure) are not compatible with the lining, they are left in place. If old patching materials have sharp edges or protrusions, however, these must be smoothed to prevent lining damage. Any oily deposits also have to be removed, and all cracks treated with nonshrink grout and concrete adhesives. Surface dirt, grime, depositions, and so on are removed from the substrate in areas where the lining is to be sealed around pipes or structures. Rough concrete or Gunite must be smoothed to accept the lining. This may be accomplished by scraping, grinding, or flood coating, or by placing a chafer pad of heavy membrane or filter cloth over the surface before the lining is installed.

Existing cut-and-fill reservoirs are often lined, either because an existing lining that has lost its effectiveness must be replaced or because the facility was never formally lined after it was constructed. Rigid linings are among the types of surfaces that are commonly relined. To reline a facility of this type usually requires preparation of varying degree. As with concrete tanks, cracks are filled with nonshrink grout, using adhesives to bond it to the old concrete. Construction joints are usually cleaned of old caulking compound and dirt and are recaulked before relining. Sharp changes in elevation due, for example, to slab settlement are "feathered" to give a smooth transition between the different elevations of adjacent slabs. Cement grouts, epoxy compounds, and special curing-type elastomerics are used for this work.

In lieu of grouting, a protective sheet (usually copper) may be placed over the crack and mechanically fastened to *one* side of it to protect the lining against bridging tendencies at such points. The copper serves as a slip sheet, allowing slab movement beneath it without detrimental effect on the lining.

If column footings are protruding from the bottom of the tank, four things can be done. The lining can be installed without removing the pedestals, in which case a lining seal is made on the floor around the base of each pedestal if the pedestals are rough and/or protrude excessively above the floor. If the pedestals are small with slight protrusion, the lining may go over the pedestal itself, completely sealing in each one separately, a more expensive procedure. A third possibility is to completely remove the pedestal. Finally, the reservoir may be given a sand blanket thick enough to cover and eliminate the surface irregularity caused by the footings.

1

History

In the beginning, man was not concerned with trying to keep water in any particular place. He merely built his various civilizations where water was present. His home was first a tree or a cave and later a small hut or shelter of varying design. When these were destroyed or damaged by high water, he quickly rebuilt or moved to higher ground. Since his home was simple in any event, he seldom experienced problems with wet or unstable foundations. When this contingency occurred, it was easier to move than to solve the problem by ingenuity.

At first, he kept close to the water supply, so that he could haul water by manpower or animal power, utilizing small containers. These served as man's first reservoirs. Seepage losses from these small vessels were only a minor problem, as replacement water was close at hand. As man progressed he saw fit to place civilizations at locations other than those blessed with water.

The exact time when man was first faced with the problem of moving water any appreciable distance is open to some debate. If we believe the writings of Plato, this may well have occurred some 12,500 years ago on the lost island of Atlantis.[1]* Of course, to accept such an account, it is

*New evidence indicates that an error in the translation of Plato's writings may have been made. In his dialogues *Timaeus* and *Critias,* the sinking of the island of Atlantis was said to have occurred 9000 years before Solon (638 to 558 B.C.) or about 9638 B.C. A scientific analysis by Galanopoulos and Bacon[23] suggests that, because of a translation error involving a factor of 10, the event occurred 900 years before Solon instead of 9000 years as previously believed. This would place the destruction of Atlantis at about 1538 B.C. If this view is correct, the irrigation exploits by these island inhabitants may not have deserved the early ranking given them by Mr. Huffman in his book.

necessary to believe in the existence of this island. At any rate, modern historians seem agreed on one thing: man early became concerned with water conveyance and storage because of his agricultural needs. That this water was used for drinking too is rather obvious, but it was man's farming pursuits that started his primitive undertakings in the field of hydraulics. Of course canals were important also in military campaigns, serving both as a means of transport and as a defense barrier. In fact, the development of canal systems for military purposes paralleled and, in some cases, may have preceded their use as instruments for agricultural development.

There is a gap between the Atlantis date and the next historical evidence of water transport and storage. The first public irrigation in Egypt apparently occurred before modern recorded history, probably in the period from 3000 to 2500 B.C..[2] King Menes, the first ruler of united Egypt, dammed the Nile, supposedly with clay, 5000 years ago. The details are obscure, but if he indeed used clay this is probably the first use of a hydraulic lining material for which the record may be matched reasonably well with a date. Such a conclusion is by inference, however, as we do not know whether the king took any type of clay available or a specific one particularly suited as a sealing membrane against seepage.[3]

But King Menes does not deserve the honor of being the first canal builder. Another king preceded him—a king without a name[4] or a history. So vague is the record of this event that we do not know where or when he built his canal, or what other deeds he did. But we know that he built one, that its purpose was for irrigation, and that this "no-name" king was so proud of the ceremonial opening of his project that he had the event portrayed on his mace head.[5] The carving on this power emblem is preserved in the Ashmolean Museum at Oxford.

By about 2000 B.C. written records indicate that irrigation waters were being proportioned to various districts in Egypt.[6] About the same time Hammurabi of the Babylonian Empire expressed pride in his accomplishments in "changing desert plains into well-watered lands." One of these accomplishments included a man-made lake 42 miles in circumference and 35 ft deep to store Euphrates River floodwaters for irrigation.[7]

Canal building was an art well developed by the Egyptians. To them, water was precious, water was power, water was sacred. Pharaohs distinguished themselves by constructing canals, and they attached great importance to the dedication ceremonies for these works. Generation after generation, the work of building canals went on because the Egyptians were obsessed with it.[8]

We can understand this feeling better when we look back over our own history a mere four or five generations ago. We too became obsessed with canal building. Indeed, the fever rose so high that in 1826 the governor of Pennsylvania signed "an Act to provide for the Commencement of a Canal" before anyone knew where it was going to be built or why it was needed.[9] So unappropriate was the terrain for this kind of adventure that in the canal's 394 official miles from Philadelphia to Pittsburgh there were only 276 miles of actual canal. This portion of the voyage was cluttered with 174 locks, or one about every 1.5 miles. For the first 90 miles the boat was hauled by rail, including a 2805 ft cable lift just outside Philadelphia. Midway in the journey it was necessary to lift the boat 1400 feet over the Allegheny Mountains and then gently lower it 1170 feet into Johnstown. This feat was accomplished using two 35-horsepower steam engines and 10 inclined planes, 5 up and 5 down. The canal took 8 years to build at a cost of $14 million. It was a white elephant from the very beginning, but Pennsylvania had a canal!

Other civilizations built canals too, for irrigation, for domestic use, and for military purposes. Sometimes it seems that the ancients spent all their time threading canals to and fro across the landscape. The Sumerians, the Assyrians, the Greeks, the Arabs, the Italians, the English, the Russians, and the Dutch—all of them built canals! On a percentage of participation basis, Caliph al-Mansur set the record in the 8th century in Baghdad, where it was estimated that practically everyone in the city was engaged in the building, financing, or provisioning of canal digging operations or in litigation over water rights.

Credit for the most colossal of all undertakings in this category belongs to the Chinese. They won the title in the 7th century, according to one historian, who recorded the facts at the opening of the Pien Canal in A.D. 609. For this project all able-bodied men between the ages of 15 and 50 were conscripted, along with one person from each household as a backup force. The total employed on this enterprise amounted to 5,430,000 persons.[10]

The Romans too built aqueducts and canals from the beginning of their history. Although their aqueduct program seems to have been well run, they had some sad experiences in the canal division. Under the mad Claudius, 30,000 workmen labored for 10 years to cut a 7 by 7 ft tunnel for 3 miles through solid rock, to drain Lake Fucinus. No one knows why Claudius wanted the tunnel built; perhaps it was just an obsession. In any event the gods were against him. At the dedication the canal was opened, but only a trickle of water appeared. Sometime later the tunnel was deepened, and this time, when the gate was opened, so much water rushed out that the spectators at the dedication ceremony were inundated. It is not recorded how many incompetent engineers died at Claudius' hand because of these two incidents, but no doubt many did. In those days engineers were seldom given the opportunity to make more than one mistake, a custom that obviously did not prevent all failures but certainly eliminated repetitive malpractice.

Later, Nero, last of the Roman canal builders, tried his hand at the art, with mostly disastrous results. This emperor had an overwhelming desire to write his name across the face of the earth, and in what better way could this be done than by digging tunnels and canals? With his coming, canal building entered the realm of fantasy. Some of his undertakings were colossal, if not well thought out. Nothing seemed to work out right for him in this field of endeavor, and his death by suicide appears linked to his deficiencies as a manager of canal digging projects.[11]

These very early feats in the waterworks field were accomplished without regard for linings as we know them today. Seepage must have been an important factor in their processes. Being agriculturists, these builders were obviously aware of the percolation of irrigation water into the soil. Thus they had the background for deducing the possibility of the same phenomenon occurring in the bottom of their canals and reservoirs. If they thought about it or took any preventive action in this regard, they said nothing about it. Their works were on a large scale for those times, and perhaps, like today, seepage losses from projects of such magnitude were ignored as being the only practical thing to do.

Discounting King Menes' questionable use, one of the earliest known applications of an impervious-type lining in a hydraulic facility occurred over 3200 years ago during the construction of the Tigris River embankment lining at Assur. Alternate layers of bitumen–sand–gravel–clay mixtures and burnt bricks in bitumen mortar were used as the impervious membrane.[12]

The digging of wells and cisterns is such an ancient art as to have been lost in the obscure past. No historian seems to have placed a time label on when the practice began. Logically, it should have occurred before canal digging developed, because, with this tool of technology, man could still live close to watercourses and provide himself with reserve supplies in time of low flow conditions in the stream or river. Small facilities could take care of domestic needs. Only when man began agricultural pursuits would vastly larger quantities of water be required, along with distribution networks.

Inscriptions dating back to about 2000 B.C. record the digging of wells, so we know the practice is at least that old. Cisterns of a sort may well have been utilized by prehistoric man many millenia ago. The first mention of linings in connection with wells shows that they were used about

865 B.C. at Nimrud. There, in the palace of Ashur-nasir-pal II, was a stone storage tank with a capacity of 100 gal. Nearby were other cisterns lined (waterproofed) with bitumen.[13]

Although linings had been employed in well digging before that time, they were structural in character, designed to prevent cave-in of the well. Early technology describes well digging as utilizing the diggings to build up an earthen embankment around the well to prevent animals from falling into the hole and to shield the excavation from blown sand and other debris. These wells, therefore, bore a close relationship to the cut-and-fill reservoir, even though the embankment served an entirely different purpose.

In most areas of Arabia there are remains of cement lined tanks. At Aden a series of 50 great cisterns with a capacity of over 30 million gal exists. Their age is given as antiquity.[14]

Early civilizations were familiar with a variety of mortars and related materials. Structural remains indicate the use of mortar binders, or what we would now call masonry construction. They also made extensive use of stuccos. The technology was well understood, as shown by the remarkable state of preservation of some of these structures today.[15]

Mortar binders may be divided into three classes: limes, cements, and an in-between category known as hydraulic limes. The hydraulic limes were largely used in western and central European civilizations in the construction of hydraulic facilities. Cements, likewise, may be divided into two classes: natural and portland. Both are kilned products later ground to fine powder, but the former are burned at a much lower temperature. Portland cement is a scientific blend of proper materials calcined to incipient fusion.[16]

In historical accounts of hydraulic structures little attempt is made to classify the construction in accordance with the foregoing breakdown. "Cement" is used interchangeably with "mortar" and further confused with such terms as "plaster" and "concrete." "Hydraulic concrete" would probably be a better technical description, although visually most of these aged products have similar textures and appearances.

From the ancient canal and reservoir construction days, the basic lining materials were expanded over the years to include various combinations of wood, plaster, rock set in mortar, and bituminous materials; there was also an increasing awareness that some clays exhibited excellent waterproofing capabilities. By the early 1940s concrete, airblown concrete, steel, asphalt, asphalt concrete, and a special type of clay known as bentonite were in use. At the outbreak of World War II these were the only materials available to the design engineer faced with the problems of the containment and transportation of water.

The apparent stagnation with respect to really new developments in linings from ancient times to nearly the midpoint of the twentieth century stemmed partly from lack of necessity and partly from economic and political considerations. Until 1940 little publicity had been given to the problems concerned with a dwindling water supply. Then, however, there began a slow realization that projected population growth rates would result in exhaustion of the relatively stable sources of supply, particularly in heavily urbanized areas. The alarm was sounded by an ever-increasing flood of articles on the need for the development of additional water supplies and for more stringent conservation of those already available. This trend was particularly noticeable in the western part of the country and was climaxed with the announcement of the California Water Plan in 1957. Three years later almost $2 billion of state bond revenue was voted for the first phase of this program, designed to bring excess water from the northern mountains to the water-deficient areas in the southern half of the state.

Water shortages were not confined to the western part of the country; the problem was being experienced in many locations. In 1965 in New York City low rainfall during the previous 4 years created the worst drought on record. The situation became so bad that water use restrictions even abolished the glass of water normally served to restaurant customers. During the last few years car washing and lawn watering restrictions have been imposed in several cities. Ironically enough, the majority of these instances have been in the East and Midwest, whose residents had never thought in terms of a water scarcity.

Another consideration that gave a sharp boost to the lining field was coming into notice about the same time as public awareness of disappearing water supplies. This problem has more serious overtones than any that mankind has faced since his existence began. The problem is *pollution.* It has been accompanied by complacency on the part of individuals and corporations alike. Although an unpopular political issue at first, it is gradually beginning to win the support of all people, even the ones most guilty of pollution.

Pollution control problems are drastically affecting the linings field. Linings are coming to be regarded as a prevention–defense against pollution. Originally, they were viewed almost entirely as a means of economic control against direct monetary loss.

Historically, pollution became a factor when life on earth began. In those times the land, the oceans, and the air were great absorbers of anything in the way of pollutants. This worked well because there was such a small amount of pollutants compared to the quantities of things doing the absorbing—in other words, the ratio was heavily in favor of the diluents. Today, however, the ratio is rapidly becoming unbalanced, and we are quickly learning the consequences.

This is not the first time that mankind has learned about water pollution. In fact, history tells that this lesson has had to be learned three separate times in the course of civilization. Turning back the pages, we find many instances plainly showing that our ancient ancestors had more basic expertise in handling pollution problems than we have today, and that is saying quite a bit, considering the technology gap that existed. In our time, standing knee deep in sewage, we launched a rocket to the moon. But listen to what they did.

Back in Mesopotamia during the Akkadian supremacy (3000 B.C.), and at the same time in the Indus valley, ample drains existed to handle domestic sewage. They ran beneath the streets and were constructed of burnt bricks jointed and lined with bitumen. The bathrooms had a pavement of bricks with an overlay lining of bitumen. These systems removed the sewage from the area of the homes and villages, and it was eventually flushed out into the Indian Ocean or the Mediterranean Sea.[17] In those days it was possible to use the ocean as a cesspool because the population was relatively small. Hence the extensive use of linings then was more for pollution control than for seepage control from reservoirs and canals—a situation analogous to the one that seems to be developing now.

In the 6th century B.C. Rome had built an elaborate combination of sewers leading into the Cloaca Maxima, an open canal that was later roofed. But with the fall of the Roman Empire a thousand years later, the world forgot the lessons of pollution control learned earlier and all pretense of sanitation was abandoned.[18] All through history from that time on, problems of contaminated water occurred with surprising regularity. But the civilizations involved, ignorant of chemistry or biology, did not connect their problems with the true cause; they blamed evil spirits, witchcraft, or other sources. Therefore mankind had to learn the lesson on pollution again, and this time the die was cast in London in 1849.

For many years in England, installation of water closets proceeded at a rapid rate. At first the authorities insisted that these be connected to cesspools, and they were. But the practice was slowly discontinued, eventually by governmental order, because the ground near the cesspools became soggy and foul smelling and was a breeding place for flies. The new edict, one of the most disastrous and far-reaching mistakes ever made by civilization, directed that the pipes be disconnected from the cesspools and tied in directly to the storm sewers.[19] Had this mistake not been made, perhaps the problem of interconnected sewage and storm drains in practically every U.S. city would not be with us, because, as our towns were laid out, we merely copied what the English had done.

But such was the practice in England at that time, and her rivers became nothing more than open sewers. Londoners were drinking polluted water from the Thames. One water intake line, which supplied 7000 families, was within a few yards of the Ranelagh sewer outlet.[20] This was the situation when cholera struck in 1849 and again in 1853. Over 20,000 people died in London—the greatest pollution disaster in the history of the western world.[21] The cause of the problem was finally diagnosed as sewage, so mankind had learned about pollution for the second time!

The government's first step was to lay large sewers running parallel to the rivers so that their sewage could be dumped downstream from London. This was a great solution for the city, and overnight it became a healthy place in which to live. There was, however, one small problem: the people on the downstream side of London got a much heavier and more concentrated dose of sewage, and they complained. Finally, in a brilliant move, the city constructed works to precipitate and chemically clarify the outfall water. Early in modern history, therefore, London achieved a sewage system much more efficient and sophisticated than the ones most American cities have today.[22]

Now, for the third time in history, we are learning the lesson of environmental pollution. Unfortunately, we have to solve three problems today—not only that of water pollution, but those of air and thermal pollution as well. It is now rather obvious that the use of linings for the control of water pollution is not going to be the long-term solution. Although they will play an important role and enjoy a tremendous increase in usage during the next several years, the eventual breakthrough that will really solve the basic problem will involve the treatment of pollution at the source. Retention and dilution, we know now, only forestall the problem and delay the inevitable.

No one yet knows the answer to this challenge, but the proper response will eventually be found. In the meantime, linings are essential as a means of plugging the gap until such time as the basic solution is developed and put into practical application.

References

1. Roy E. Huffman, *Irrigation Development and Public Water Supply,* Ronald Press, New York, 1953, p. 7.
2. Clessen S. Kinney, *A Treatise on the Law of Irrigation,* W. H. Lowdermilk & Co., Washington, D. C., 1894, p. 15.

3. Michael Overman, *Water,* Doubleday & Co., Garden City, N. Y., 1969, p. 69.
4. Robert Payne, *The Canal Builders,* Macmillan Co., New York, 1959, p. 10.
5. *Ibid.,* pp. 10–12.
6. W. L. Westerman, "The Development of the Irrigation System of Egypt," *Classical Philology,* Vol. XIV, No. 2, April 1919, pp. 158–164.
7. Charles F. Davis, *The Law of Irrigation,* Publishers Press Room & Bindery Co., Denver, Colo., 1915, p. 8.
8. Payne, *op. cit.,* pp. 16–26.
9. *Ibid.,* pp. 155–156.
10. *Ibid.,* p. 61.
11. *Ibid.,* pp. 43–47.
12. Asphalt Institute, *Asphalt in Hydraulics,* 5th printing, College Park, Md., March 1965, p. 3.
13. Charles Singer, E. J. Holmyard, and A. R. Hill, *A History of Technology,* Clarendon Press, Oxford, England, 1954, p. 827.
14. *Ibid.,* p. 530.
15. Robert W. Lesley, *History of Portland Cement Industry in the United States,* International Trade Press, Chicago, 1924, p. 1.
16. *Ibid.,* pp. 2–3.
17. Donald E. Carr, *Death of the Sweet Waters,* W. W. Norton Co., New York, 1966, p. 38.
18. *Ibid.,* p. 39.
19. *Ibid.,* p. 42.
20. J. Gordon Cook, *The World of Water,* Dial Press, New York, 1957, p. 107.
21. *Ibid.,* p. 43.
22. *Ibid.,* p. 44.
23. A. G. Galanopoulos and Edward Bacon, *Atlantis: The Truth Behind the Legend,* Bobbs-Merrill Co., Indianapolis/New York, 1969, p. 133.

2

Types of Reservoirs

Throughout history, man has been faced with the problem of building facilities for storing water and other liquids. As he learned more modern building methods, designs changed and other alterations were made as new materials became available. Although initial construction efforts were rather primitive by today's standards, many structures built by the ancients are still standing, and it is interesting to note that three out of four basic designs used today for water storage were developed thousands of years ago. Of the types listed below, only the steel tank is a relatively modern development. The introduction of reinforcement steel into concrete, however, did not change the basic concept of the concrete tank; it merely improved the design.

Despite many variations, only two general methods are used in designing large water holding facilities. As the following tabulation shows, there are four variations of tanks, and the last of these may be further classified in three ways:

1. Cut-and-fill reservoirs.
2. Tanks.
   A. Wooden.
   B. Steel.
   C. Masonry.
   D. Concrete
      1. Cast.
      2. Prestressed.
      3. Wire wrapped.

Cut-and Fill Reservoirs

When the term "reservoir linings" first appeared, these linings were confined to a certain type of design known as the cut-and-fill reservoir. In this construction method, a saucer-shaped depression is scooped out of the ground (Figure 2.1). The earth so removed is utilized to construct a fill or embankment around the top of the excavation to increase the effective height of the water and to reduce excavation costs. Part of the water is held below the original ground level (in cut) and part above this level (in fill). The slopes and bottom of the excavation are then lined with an appropriate material to prevent the loss of water by seepage and also to protect the earthen components of the facility. Even when the excavated material is not used in the actual reservoir construction, the design is still referred to as cut-and-fill, particularly in discussions involving reservoir linings. A dammed canyon is a special case of a cut-and-fill

FIGURE 2.1. Installation photograph of a PVC lining in a western irrigation reservoir.

reservoir. The "cut" has already been dug by the flowing water, while the dam embankment is the "fill" portion of the structure.

Although the plan view of a cut-and-fill reservoir may be of any shape, the usual practice is to make it in the form of a square, rectangle, or circle. Since the circle is more difficult to construct, more costly to line, and usually less efficient in land use, the square or rectangular shape predominates. Sometimes, in very large facilities, the lay of the land will dictate a polygon shape with five, six, or even more sides.

With the advent of heavy earth moving equipment and the efficient way in which this type of construction can be done, it is no wonder that the cut-and-fill reservoir has received such widespread acceptance. This type of construction was a natural boon to the development of lining systems, particularly since these reservoirs were being built with increasing frequency in the newly expanding areas in the western and southwestern parts of the country. It was here that a natural shortage of water, coupled with generally porous soil conditions, helped to focus attention on water loss by seepage. And it was here also that linings began their slow and steady rise to prominence.

In the past the thin membrane lining was often considered as the single element upon which rested the integrity of the reservoir. This discarding of the "team" philosophy of components is an erroneous concept, and adherence to it will usually lead to difficulties. Such a lining will not hold a reservoir together by any stretch of the imagination. Perhaps it may provide this type of performance in a very minor way, but its structural capabilities are indeed limited. These constitute the primary function of the earthen portion of the facility.

The most important advantage of the cut-and-fill reservoir is its basic low cost. This is accompanied by a low profile, an important consideration when the reservoir is to be constructed in newly developing areas where the accent is on beauty. The structural strength of the facility comes from the compacted earth itself, which can be formed into place more economically than steel, concrete, or other materials of construction. No other type of reservoir can be built as economically as this one.

Unfortunately, this seemingly ideal water holder is not without its disadvantages. Although the earth is really holding the reservoir together, it must do so without fail. The technical aspects associated with this requirement are not assembled in any one place for easy access by the engineer. If he has never built a cut-and-fill reservoir before, he has considerable homework ahead of him. Several disciplines are involved, and he must become familiar with all of them. In contrast, steel and concrete tanks are predesigned by various marketing organizations, and the engineer using this type of storage facility need only copy these standard designs.

Moreover, unlike the steel, wooden, or concrete tank, the cut-and-fill reservoir cannot hold water in a vertical configuration. It must utilize side supports that have slopes. In reservoir technology the "slope" refers to the ratio of the horizontal distance to the vertical distance. A side slope of zero is a vertical wall. Generally, a slope of 1½ : 1 is about as steep as any engineer will want to go with an earthen support member. Today, most designers are using slopes of 2½ : 1 or flatter, with the trend pointing toward 3 : 1, a ratio particularly favored on critical installations.* The economic disadvantage of the flatter slopes is partially offset by the fact that anyone falling into the reservoir should be able to get out unassisted, a safety advantage of obvious merit.

The soil should not contain soluble material, organic deposits, or expansive clays. If present, these items should be removed by overexcavation and replaced with properly compacted select material. Compaction requirements vary but for more critical reservoirs they usually range between 95 and 98% of standard maximum density. Except in very large installations, there should be no rock or substrate that would require blasting, as this would make the facility too expensive in comparison to other types of storage.

One very important requirement, often overlooked by the uninitiated, is that the substrate soil materials be stable *wet or dry*! The slopes must not sluff or lose their stability or load bearing qualities when wet. Collapse in these areas will damage or destroy the lining and has often been responsible for the destruction of the entire facility. The bottom of the reservoir must also retain good stable properties when wet. It should not become mushy or lose its bearing strength under saturated conditions.

Current practice is to protect the larger and deeper earthen reservoir with an underdrain system unless it lies entirely in cut. This may take several forms, but they all have one objective: to allow any water that gets behind the lining for any reason to be conducted through the reservoir embankments without taking any soil particles with it or otherwise contributing to instability.

A modern underdrain system consists of five essential parts:

*Critical structures are defined as those in which leakage must be held to absolute zero (as in the case of reflection pools built over office space or living quarters), or those in which leakage can cause settlement of columns or foundations or can contribute to instability of embankments. This term also covers facilities whose failure would endanger life or property in a catastrophic sense.

1. Interceptor.
2. Collector.
3. Filter.
4. Conveyor.
5. Disposal area.

Each of these parts has a definite function, and all are usually needed for proper operation of the system.

The interceptor, as its name implies, serves to intercept all water behind the lining at any location, regardless of how the water got there. One of the best methods of doing this makes use of an open-graded (popcorn) pervious asphalt concrete blanket about 4 in. thick, overlaid with an open-graded asphalt concrete smoothing course to prevent any tendency of the lining membrane to be squeezed into the interstices of the open-graded component. This continuous underdrain blanket covers the bottom and all slope areas. It is most often laid on a base of well-compacted impervious or semiimpervious earth substrate. Sometimes the earth is given a minimum two-spray coat of a cutback asphalt for increased imperviousness before the asphalt concrete is applied.

Another continuous underdrain design gaining some favor is based on a 6 in. blanket of small rock (usually pea gravel), overlaid with one of a number of materials to protect the membrane lining from damage by puncture or extrusion. The overlay protection materials have been carefully graded earth, lining membrane chafer blankets, or synthetic filter blankets. Most of the last type are made in 16 ft widths and weigh 400 to 500 g/sq m. Their composition may be nylon, polyester, polypropylene, or the like.

Regardless of which system is used, the idea is to ensure that all, or almost all, of the water intercepted by the underdrain will be moved to the collector pipes through a pervious layer. Shallow, noncritical facilities, or those entirely in cut, are often constructed without a formal underdrain at all, or the basic system of components is modified by eliminating certain features in the interest of saving cost.

The collector pipes are placed in trenches on the bottom of the reservoir just inside the toe of each slope. In large reservoirs they may also branch out through other areas of the bottom. In the more sophisticated designs various segments of the pipes are isolated from each other, with the different segments carrying their water to the collection area in separate pipes. With this plan leakage areas may be pinpointed in a specific part of the reservoir for more effective troubleshooting.

There are two disadvantages to segregating the pipes: (1) the cost is higher, (2) increased design and construction vigilance is needed to ensure that there will be no shortcutting of reservoir leakage water between pipe systems. If shortcutting does occur, the value of the segregrated system is considerably diminished.

After the perforated pipes have been placed in the ditch, usually on a thin layer of pervious material, the ditch is filled with *select rock*, and emphasis here is important. This ditch is filled, not merely with any kind of rock, but with select rock, based on soils engineering considerations. The size and gradation of the rock depend on the characteristics of the embankment soils, and both factors influence the design of the filter, which may or may not be the next link in this vital chain, depending on the interceptor type. If the continuous underdrain blanket connects directly to the collector, no filter is needed. If, however, the continuous blanket is not used, a filter is required. Traditionally, filters have been constructed of carefully graded soils. Today, some of the specially woven synthetic fiber mats are being used as the filter medium, and they seem to be doing a good job. These are the same type of materials just discussed as underlay protection for linings laid on crushed rock continuous drains, although the weight range is usually less.

The integrity of the underdrain, even the reservoir itself, hinges on the filter, an apparent surprise to all "do-it-yourselfers." The filter's function is to filter out embankment soils so that they will not migrate into the interstices of the trench rock or the pipe itself. Their transportation from the embankment into the trench spells trouble for the reservoir. This type of difficulty can be spotted easily at the underdrain exit pipes by observing the water clarity. Cloudy water can mean trouble, the degree of which is somewhat in proportion to the degree of cloudiness. If larger particles, heavy sand, or even small particles of gravel or rock are coming through, the reservoir could be in immediate danger. Certainly it will have a short life expectancy if the symptoms continue. Soil removal means settlement, which, if unchecked, can cause lining malfunction. When this happens, additional water from the reservoir itself is available to help accelerate the further deterioration of the embankment. The filter, despite its importance, is an item often neglected in the underdrain complex. Its omission is responsible for triggering many malfunctions.

Next in the underdrain system is the conveyor. This is merely a plain (nonperforated) pipe that carries the water from the perforated underdrain collection pipes through the embankment. The major requirements are that it have enough capacity to handle the total volume of water picked up by the underdrain and that it be strong enough to support the overburden loads imposed by the weight of the embankment on top of it.

After the underdrain has done its job in collecting the water, the pipe network must convey it to a point where it can be safely disposed of without creating additional problems. The underdrain pipes must operate at zero pressure, discharging their contents into a channel, open drain, or concrete structure that has an exit drain. The exit drain capacity must exceed the maximum total flow expected from the underdrain, so that no water will back up into the underdrain piping itself.

Some manufacturers' bulletins show underdrain lines terminating in a standpipe. This may look like an attractive arrangement to the printer or the stockholder, but unfortunately appearance is not what governs reservoir performance. Actually, this type of underdrain termination may be worse than no underdrain at all. First, it places full head pressure on the bottom of the underdrain trenches, which may not be able to cope with the pressure involved. Second, a standpipe arrangement will not indicate leakage rates. All it will do is prove that water seeks its own level.

Like all water storage facilities, the cut-and-fill reservoir requires some standard appurtenances. These include an inlet, outlet, overflow, mud valve, and sometimes level indicating equipment and emergency spilling devices.

The inlet is normally a concrete structure with a pipe of the proper size introducing water at some point in the bottom of the structure. The cross section of the structure in plan is several times larger than the cross section of the supply pipe, thus allowing water to enter the facility at low velocity. Sometimes the inlet is located on the slope adjacent to the toe, and entering water is directed parallel to the bottom of the reservoir. Inlet pipe sizing depends on the size of the reservoir and the speed with which it must be filled for consistency with normal operating schedules.

Outlet structure design is similar in principle to design of the inlet just described. In fact, the two structures are often combined with inlet and outlet water streams controlled by proper valving. It is good practice to cover the outlet with a safety screen or grill to prevent large pieces of extraneous material from being drawn into the outlet pipe. This is a particularly important safety feature if the pipe is large enough to take a swimmer or scuba diver who might be doing an underwater inspection of the structure. Outlet pipe size must meet the requirements of the reservoir owner's customers, including, if applicable, any necessary water flows for fire fighting. Obviously, the outlet should be located at the lowest elevation point in the reservoir, although in a surprisingly large number of cases this rule is violated.

The overflow takes on a variety of designs from a simple pipe extending through the embankment to a costly structure. The pipe invert, of course, is set at the maximum water level with an allowance made for water depth at maximum spillage rate. The capacity of the overflow should be at least as great as the inlet supply rate. For covered reservoirs the overflow outlet pipe must be screened to prevent the entrance of bugs and rodents. Another system is to make a water seal, similar in principle to the gooseneck pipe employed to seal the kitchen sink against odors from the sewer pipe to which it is connected.

The mud drain is a flush-mounted pipe placed at the lowest point in the reservoir and sealed either by a plug or a valve arrangement. As its name implies, the silt and fines that settle out of the water may be flushed down this pipe during reservoir cleaning operations. The mud pipe is connected directly to the sewer or some other location outside the formal distribution pipe network. It saves the expense of hauling or pumping out the mud and silt. The pipe varies in size but is usually in the range of 12 in. in diameter for the average size reservoir.

Level indicating devices are made in a variety of shapes and sizes, many being fabricated by the reservoir owner himself. They may be visual indicators, or they may signal depth automatically through the use of air pressure or electrical means. Many facilities are constantly monitored for water depth so that pumps or valves may be actuated at appropriate times.

For large reservoirs many state and federal safety agencies require emergency spilling facilities. These may take the form of an extra-large outlet structure that can be activated manually or automatically to release large quantities of water should a portion of the facility malfunction for any reason. For economic reasons the extra-large structure may be built as the outlet, and lead pipes from the structure's large emergency pipe may be tapped to supply water to the distribution mains. In this case a valve prevents water entry into the spilling device during normal operation.

A protective fence should surround the reservoir. Its specifications are usually dictated by various agencies responsible for the safety features of open storage facilities. If the reservoir is holding potable water, either raw or finished, additional safety features are required. Current practice is to build the fence on the outer embankment, so that the bottom of the fence is at least 8 ft below the berm's top elevation. The fence is usually topped with various arrangements of barbed wire to further discourage entry.

Tanks

Wooden

Wood was a very popular material for use in water tank construction in the early 1900s, and its popularity continued well into the 20th century. At that time most of the water tanks used for railway water service were constructed of wood, and their use in this area did much to spur the growth of the wooden tank industry. At present the rate of building new wooden tanks has diminished considerably because of the economic and technological advantages of other types of holding systems. Wood construction does offer some advantages for small tanks but the decreasing demand for such tanks has likewise been a factor in the decline of the wooden tank business.

Five types of wood have been used for water tanks, with cypress heading the list for the longest average life span, as shown in the following tabulation:[1]

Type of Wood	Average Life Span (years)
Cypress	40
Redwood	30
Cedar (eastern)	30
White pine	20
Douglas fir	16

Seventy years ago, wood offered several advantages in tank construction. At that time, concrete and steel were not very common, and there was a decided lack of technology in connection with their efficient use. In contrast, woodworking was a long-practiced art, and the joining and fitting of the staves was done with great skill. The staves were produced in a factory and properly coded and numbered for fast assembly in the field.

Early cylindrical wooden tanks were constructed as outlined in a very complete specification, "Gravity Tanks and Towers," put out by a Boston fire insurance company before 1919. Aside from a change in tensioning hoops from flat bands to round rods, little change in construction procedure was made through the years. The staves varied in thickness up to about 3 in., depending on the size of the tank. A groove was cut at the bottom of the staves for receiving the edges of the planks that made up the bottom of the tank.

Tensioning bands were produced from wrought iron or mild steel with tensile strengths up to 60,000 psi. Working stress was kept at not more than 12,500 psi. Charts were available from which the proper number and spacing of the hoops could be obtained. This spacing was about 21 in. between hoops at the top of the tank, decreasing to a 3 in. spacing at the bottom of a 20 ft high tank. Because staves were limited in length to 20 ft, tank heights were always within this limit. Early tanks ranged from about 12 to 30 ft in diameter. Later, however, when water companies began building the tanks, diameters were increased to accommodate the larger quantities of water required. Even so, it is unusual to find a wooden tank in this service that measures over 100 ft in diameter. Most are within the 60 ft range.

Hoop tension was achieved by tightening nuts on specially designed lugs. Tensioning could be increased at any time, merely by additional turning of the tightening nut. Many early failures resulted when the originally designed flat tensioning hoops were used. Corrosion of the hoops next to the tank went unnoticed until the tank literally fell apart. A change to rods with circular cross sections did much to eliminate this problem.

On the credit side of the ledger, wood is free from the rust problems that plagued steel construction at the time, and the wooden tank has certain inherent protection against frost and cold weather, which cause problems for the concrete tank. When the fibers of the wood in the tank are thoroughly saturated, decay is practically impossible. Also, if the wood is perfectly dry, there is little chance for decay. It is the in-between condition that can cause the wood to deteriorate. Even then, as the table on p. 26 shows, the wooden tank has a respectable life span, especially if the better aging timbers are selected.

The choice of wood for the tank depended on the area in which the tank was to be constructed. In the western part of the country, redwood or fir was utilized as the major building material. Cypress was used in the East, and pine in the South. In any case the wood in the tank swells when wet and tends to seal itself against seepage losses. This action does not do a complete job of prevention, but most deficiencies in this regard are countered by tightening the tension rings. Both actions together reduce the seepage to an amount that has no economic significance as far as the value of the water lost is concerned.

Another factor, however, is of much greater concern. When the tank is made of redwood, water seeping through the tank leaches from the wood a secretion that stains the outside of the tank. The resulting stain is very dark in color, usually almost black, and is not a pleasant sight. If the tank is painted, the deposit prevents the paint from performing its intended function. Moreover, if homes are near the tank, the discoloration problem seriously affects public relations. Adjacent homeowners continually complain about the unsightly appearance of the tank.

In some cases redwood leachings can become an even more serious problem in that they actually can cause discoloration of the water delivered to the customer. This always produces a fast reaction in the form of complaints, since no customer will tolerate any visible threat to water quality. In such cases hydraulic linings have been employed to insulate

FIGURE 2.2. Installing a one-piece, 20 mil PVC drop-in lining in a redwood tank near Oakland, California.

the tank staves and wooden columns from the water itself, a rather unique application in some respects (Figure 2.2). If no water touches the wood, no secretion is produced. Since the lining decreases the water content (and hence the swelling) of the wood, it is necessary to tighten and retighten the stave hoops after this lining application to compensate for the resulting shrinkage of the wood.

Steel

Although historians disagree as to the exact time when iron was first smelted, the consensus of opinion places the event at about 1500 B.C.[2] Despite this early beginning, things went slowly, as the technology of steel making proved to be very complex. In addition, the massive equipment and enormous power required to roll steel into plates were equally slow of attainment.

In short, the jump from iron to steel did not happen overnight. Again, history is somewhat vague as to when this took place, and its occurrence is further clouded by the complexity of the change. High-quality steel was produced in the ancient past, and in 384 B.C. Aristotle, described its manufacture.[3] The steel was made in India and the metal shipped to Damascus, where it was tempered and hammered into swords of "Damascus steel."[4]

For steel to be useful in the fabrication of all-steel water tanks, it must be in plate form. This requires that the steel ingots be rolled into thin sheets. Although evidence shows that hot rolling of steel took place in the late 16th century,[5] the process was crude and yielded steel plate of very poor dimensional tolerances. It was not until 1926 that the first continuous hot strip rolling mill began production in the United States at the Columbia Steel Company (later Armco) in Butler, Pennsylvania.[6] This mill produced strip 36 in. wide. The process was licensed, and within 10 years 27 continuous wide strip mills were producing 14 million tons annually. The sheets were produced from 0.04 to 1.25 in. in thickness, from 26 to 96 in. in width, and in lengths up to 2000 ft.[7]

As the steel industry learned how to roll plates into thin sections, it began to look into the potential business in the waterworks field. Quick inroads were made into this area because of the apparent advantages of steel-type construction over concrete and wood. No spalling, no cracking, and no problems at wall–floor junctions were some of the selling points. The steel industry also boasted that standard designs could be furnished to the engineer, and this did have an effect in speeding up the specification process. On a first-cost basis, the steel tank seemed to offer decided advantages over its concrete competitor, particularly in the smaller sizes.

It soon became apparent, however, that the steel tank did not provide the low-cost, carefree facility predicted by the seller and hoped for by the owner. Corrosion, then as today, was a very serious problem. It strikes on four major fronts: (1) where the steel is in contact with the water, (2) where the steel is in contact with the moist air above the waterline, (3) in the foundation areas wherever the steel touches the soil, and (4) in areas of contact between dissimilar metals.

The seven mechanisms known to be responsible for the corrosion of steel[8] are listed below, along with appropriate subheadings:

1. Pitting corrosion.
   A. Metal-ion cells.
   B. Differential–aeration cell.
   C. Passive–active cell.
   D. Crevice corrosion.
   E. Deposit attack.
2. Galvanic corrosion.
3. Uniform attack.
4. Cracking.
   A. Corrosion fatigue.
   B. Stress corrosion cracking.
5. Selective attack.
6. Erosion.
   A. Erosion–corrosion.
   B. Impingement attack.
   C. Cavitation erosion.
   D. Fretting corrosion.
7. High-temperature corrosion.

Although the seventh item of the list is not normally involved in the corrosion of a water tank, all of the other factors are possible. It is the first three items, though, that play the most important role in the corrosion of steel used in water holding facilities.

The corrosion of steel is an extremely complex problem, involving all combinations of the relationships listed above. In addition, a number of rate factors play a dominant role in determining how fast the action takes place. The corrosion rate of steel is influenced by the following factors.[9]

1. Acidity.
2. Oxidizing agents.
3. Protective films.
4. Temperature.
5. Velocity.
6. Heterogeneity of the metal.
7. Other factors.
   A. Bacteria.
   B. Light.
   C. Sonic energy.

The corrosion of steel is essentially electrochemical in nature. The driving force for the reaction is voltage (called emf, for "electromotive force"), and the phenomenon is likened to that occurring in a battery. There is an anode and a cathode; the latter is said to be more "noble" than the former. In the presence of good conductive water (and most drinking water is of this type), numerous small corrosion cells are set up within the tank. Each of these has an anodic and a cathodic part. The anode portion is less noble than the cathode, and hence it is eaten away by the flow of current induced. If the cathodic area is large compared to the area of the anode, the latter will be corroded away at a very rapid rate.

Most people normally think of this electrical cell corrosion effect as galvanic corrosion. In a true classification of corrosion, however, this term is reserved for the action that occurs when two dissimilar metals are immersed in the same electrolyte. Any pair of dissimilar metals under these conditions will usually differ in potential (voltage or emf.)

A corrosion cell may be set up by any one of the five means listed under "Pitting corrosion" in the outline of the seven mechanisms for corrosion. As an example, consider a small crack in the steel. Within the crack there is an area of low oxygen concentration, and just outside the crack an area of higher oxygen concentration. The latter is more noble than the former, and corrosion of the pitting variety (type 1-C) will occur within the areas of the crack. The metal involved is the same throughout, and the liquid electrolyte (water) may also be essentially the same throughout; nevertheless there is enough difference in the vicinity of the crack to set up a corrosion cell. It should be noted that the effect of the corrosion in this case will not be visible since it occurs within the confines of the crack.

Since most corrosion in water tanks is electrochemical in nature, one way to prevent corrosion would be to block the flow of current that causes the anodic portions of all cells to corrode away. This is actually done, and the process is called cathodic protection. It was first utilized by Sir Humphry Davy in 1824, when he tried to control corrosion of the copper plate on warships by utilizing the current from the galvanic corrosion of zinc.[10] In the present commercial process the main idea is to allow some less noble, sacrificial metal (the anode) to be corroded instead of the more noble steel (the cathode) of the tank. The sacrificial anodes were originally zinc, magnesium, and aluminum. Many advances in cathodic protection technology have since occurred, and permanent anodes of platinum or similar types of metal have been developed.

Cathodic protection supplements the coating system applied to protect the inside of the tank. It is the second line of defense against corrosion. The lining material helps to cut the cost of the current required. Even though the tank is coated, the current is needed to compensate for pinholes, loss of bond to the steel tank, or any other factors that contribute to the discontinuity of the coating. In fairness to both systems, it must be said that the coating system also supplements the cathodic protection.

The interest of tank manufacturers and suppliers in trying to solve the corrosion problems of steel water tanks by coating them can be fairly described as intense. Literally hundreds of materials were tried, first experimentally and then by actual field installation. All methods of application were utilized, but perhaps the most popular technique was spraying. The primary reason for this approach was cost consideration: spraying is basically a low-cost application technique. Then, too, there were plenty of potential applicators.

It was not particularly difficult to find materials that resist the action of water; and when they were applied to steel samples in the laboratory, the required resistance was not difficult to obtain. Field applications got under way, and a host of products were offered to the market. But the rosy predictions based on the laboratory work did not materialize when applications were made under the less than desirable conditions in the field. Manufacturers of the products blamed the applicators, and the applicators blamed the manufacturers. Every reservoir or tank owner is quite familiar with this pattern of events. It soon became apparent that the application of these various coating materials by spraying was not as easy as first thought. It was some time, though, before the true nature of the problem was recognized as being closely related to metal preparation.

Spraying is a much more technical process than meets the eye. What is misleading is the fact that paints have been applied to houses and other buildings of various types for years, and with generally good results.

Heavy emphasis on do-it-yourself markets imply that the painting process is a simple one. It would seem to be a very short step, then, to go from house painting to steel tank painting. Actually, there is a great dissimilarity between these two seemingly similar processes. With a thorough awareness of the technology involved, the differences in the processes become understandable.

With house paint the coating material can be designed to "breathe." It can have just enough porosity to pass moisture vapor or solvent vapor without passing any liquid water. Even though the degree of porosity of the film may be extremely low, the wood substrate can relieve some of the back pressure from trapped water because it is permeable and can absorb water. With steel, on the other hand, the substrate does not have the capability of absorbing pressures due to trapped water or solvent vapor behind the coating membrane. As a rule any vapor entrapment will show up as a blister sooner or later. The coating system for the steel tank should be designed not to pass any water in either direction.

In addition, another mechanism works against the steel tank coating. In this type of phenomenon there must be present a semipermeable membrane that will permit only the passage of water molecules through it. If such a membrane separates pure water from a water solution, water will diffuse through the membrane from the pure water side into the water solution faster than it diffuses from the solution into the pure water. The passing of water through the membrane in this manner is called osmosis.

Many thin spray coating systems applied to steel tanks act as the osmotic membrane. Water inside the tank acts as the pure solvent, and a speck of rust or other water-loving contamination trapped between the steel and the coating acts as the solution. Osmotic pressure will force water through the membrane into the area of the hydrophilic material, until the solution so formed at this point is infinitely diluted. The incoming water will raise a bubble in the coating at the point in question. If the osmotic pressure becomes high enough, the membrane will be ruptured. Whether a rupture occurs will depend on the physiochemical relationships at the point, but any blistering of the coating because of osmosis will usually result in membrane breaks.

At the ruptured point, water from the tank will be in contact with the steel tank wall. The area may be a small one, but it is surrounded by coating that may be well bonded to the metal. Here are the conditions for the creation of a passive–active type of corrosion cell (type 1-C). The bare metal is active and will be anodic, whereas the coated area around it will be cathodic because it is passive as a result of the coating. The anodic area will no doubt be quite small compared to the cathodic area, and hence corrosion can be quite rapid. To further complicate the situation, the existing coating tends to hide the area of attack. This subtle camouflage may allow corrosion to continue for some time without giving any outward evidence until a considerable amount of steel is affected. If this situation once develops, and the tank is not properly protected by cathodic means, the tank is in more trouble with the coating than would otherwise be the case. With no coating at all, the owner can at least observe the trouble spots during the inspection periods.

Although true galvanic corrosion has not been discussed in detail, it follows that the tank designer should keep dissimilar metals from contact with each other in all areas of the tank. If some contact of this type does occur, the owner should separate the metals with a divider and either coat the least noble of the two or make sure that the area of the anodic part of the couple is large in comparison to the cathodic portion.

Above the waterline of the tank are air and water vapor. This does not mean that passive conditions exist. On the contrary, the problems here are as great as they are below the waterline. The propensity of steel to corrode in air increases rapidly after the relative humidity passes about 65%. In addition, the air may contain traces of corrosive gases that speed up the process of attack. In general, the same types of corrosion cells may be set up on the lower side of the roof as below the waterline. An electrolyte is present, and that is all that is required. There is no second line of defense for the roof area since it is not protected cathodically.

In areas where the soil touches the steel, additional cells of corrosion may be set up. Normally, this occurs only beneath the bottom of the tank, and cathodic protection can help there. Laying the steel on a well-drained rock course will lessen the possibility of attack. The problem can be serious, and many steel tanks have required the installation of new bottom plates because of corrosion from the underside.

Masonry

The masonry tank for water storage was a popular form of construction at one time. In fact, the further back we go in history, the more popular it was.

To compensate for their structural weaknesses, very early tanks were built on the principle that any destructive force could be overcome if enough mass was designed into the wall. Such was sound engineering logic in ancient times when rulers had an abundant supply of low-cost slave labor. It is true that the slaves had to be housed and fed, but these costs were kept to an absolute minimum. Such a system for funding was hard to beat. It eliminated the need for cost control and value engineering since no one could improve on the cost of slave labor.

From a seepage control standpoint, too, masonry construction represented sound engineering. The building blocks were large and dense, and the resulting mortar path between them was a long one. Hence initial seepage losses were relatively low. Later, or sometimes immediately after construction, bitumen-type linings were applied to the inside of the tank. Actually, they served primarily as construction joint seals, rather than linings, because of the impervious nature of the rocks used.

Today, high labor costs and modern designs have almost eliminated masonry tank construction except in very special cases. Masonry design is difficult to reinforce, yet because of its poor structural stability such reinforcement is generally required in current construction practice. Early tanks of the present era were often built of rough stone, with even rougher texture mortar joints. This surface condition translates into an expensive cost of preparation of the wall to receive any future prime lining system. For this reason it is not economically feasible to line the small masonry tank. Tanks that leak are usually abandoned or operated at reduced water levels.

In its limited applications today, masonry construction for water holding facilities generally makes use of cement or cinder blocks. This approach eliminates form work and is claimed to be a less expensive method of construction. Walls are of limited height and often are extensions of existing facilities for the purpose of increasing their capacity. Some new facilities have been built using block construction to increase the holding capacity of cut-and-fill reservoir designs. In both cases reinforcement is used to attain the necessary structural stability.

The block design mentioned above requires a lining of some type to keep seepage losses within respectable limits, since the blocks are quite porous. During construction extra care must be used to ensure that the mortar joints are given a smooth finish, flush with the inside face of the blocks. This is particularly important if thin membrane systems are to be utilized as the lining.

With the advent of steel and modern concrete tanks, the masonry designs rapidly declined in favor. Their main claim to fame is their early origin. They also played an important role in the transitional development patterns of the concrete tank, which is discussed in the next section.

Concrete

The concrete tank as we know it today was developed through a process of evolution. At one point in the process the development actually reversed itself in the direction of the cut-and-fill reservoir. Early tanks made limited use of cement. The cement was made up into a mortar and used between stones to construct the vertical tank wall. Thus the reservoir was a built-up masonry type of structure. As a finishing operation a thin plaster mix was sometimes troweled onto the resulting surface to give added watertightness and a more pleasing appearance. This type of construction did the intended job when seepage losses were tolerated as a necessary evil and when labor costs were low. Walls were very thick by today's standards, often exceeding 12 in; 2 ft wall sections were not at all uncommon.

Early technology centered around the vertical-wall structure that is referred to as a tank. This is not surprising, since above-grade tanks are cousins to cisterns and wells. The building of the latter structures is an ancient art, and so the adaptation of built-up masonry construction to above-grade tanks was a logical step. It was also a step that could be taken piecemeal, in that a portion of the tank could be set into the ground while the remainder extended into the air. As experience was gained, higher tanks could be constructed, thus decreasing unit storage costs.

By the turn of the century structures had headed skyward. In 1907 in Anaheim, California, a 180,000 gal elevated reinforced concrete tank was constructed. The 38 ft high tank was 30 ft in diameter. It rested on top of a series of concrete columns, 16 in. square and 74 ft high. This was bold construction in its day, not only because the tank was so high but also because the walls were only 5 in. thick at the bottom, tapering to 3 in. thick at the top. Current reinforcement schedules then in publication did not give details for tanks over 15 ft in diameter or 15 ft in height.[11]

Even before 1800 some tall structures were being built out of rock and mortar combinations. A good example was the Edystone lighthouse at Cornwall, England, built in 1756 by John Smeaton, a noted British engineer.[12]

The development of portland cement was closely associated with waterworks facilities. The first "portland" was not like today's product, which is defined by the American Society of Testing Materials as:

> . . . the product obtained by finely pulverizing clinker produced by calcining to incipient fusion an intimate and properly proportioned mixture of argillaceous and calcareous materials, with no addition subsequent to calcination except water and calcined or uncalcined gypsum.

In contrast to the above definition, Aspdin, the inventor of portland cement, calcined his raw materials only until the carbonic acid was entirely removed.[13]

The real inventor of portland cement, as defined above, was I. C. Johnson, the superintendent in an English cement plant. His calcining operation served merely to remove the carbonic acid, but one day, by accident, the materials in the kiln were heated to semivitrification. Johnson decided to pulverize this material and make a paste out of it. The resulting cement was hard and looked like Portland stone, from whence it derived its name.

The use of portland cement in waterworks facilities had a very shaky beginning, because the original block fell to pieces when immersed in water. Johnson was so ashamed of his failure that he hid the product in a cellar on the outskirts of the city near a river. Out of curiosity he returned to the cellar in a few weeks. The damp atmosphere had acted upon the free lime present and had slaked it. When he tried his experiment again, the concrete did not disintegrate when placed in water.[14]

Johnson's early failure foreshadowed many that were to follow, a pattern characteristic of any new material or process. Troubles were due to improper mix and curing conditions, lack of proper reinforcement steel, design inadequacies, and a generally poor knowledge of how concrete and its reinforcement react to various environments. With the advent of more effective communications and the development of trade associations, many problems have been solved. It is still necessary, though, to conform to good design practice coupled with high-quality construction standards and to evaluate carefully all factors that will influence the final overall design.

The manufacture of portland cement was introduced into this country in 1871, with the first production at Allentown, Pennsylvania, hitting the market in 1874. However, the material had been imported from Europe since 1865, and because of the high quality of the foreign cement it was difficult for domestic production to obtain a foothold for some time.

The cement industry in America actually began much earlier, spurred on again by demands from the waterworks industry. The Erie Canal was started in 1817, and the following year saw the beginning of the cement industry here. But this product was not portland cement. Instead, it was natural cement, first discovered in the United States near Fayetteville, New York, a year after the canal construction began. Natural cement, as the name implies, is formed naturally. Certain argillaceous limestones containing various percentages of lime, silica, and alumina are quarried and burned at low temperature. The resulting product, after grinding, is the natural cement of commerce. The fact that the low-temperature kiln operation drove off only the carbonic acid distinguished this process from the one employed for the manufacture of portland cement.

Before reinforcing steel was available or its use was well understood, the designer, if he went into the air with his tank, could logically do so only by employing masonry tactics. To be on the safe side, original designs included heavy wall sections. This feature was copied later by most designers, as innovation was not a popular trend. Designers and builders stuck to things that had worked in the past.

Then designers had the idea of placing the retaining walls on sloped earth, allowing the latter to support the tank walls. In this approach the tank is no longer a tank, but becomes a reservoir. The "walls" were still brick and mortar, rock and mortar, or, later, unreinforced poured concrete. Combinations were also used, particularly ones utilizing a concrete overlay on top of brick or rock construction. Exactly when these construction principles were first utilized in modern times is unknown, but many facilities of this type were built in the years before 1920.

The concrete tank evolution process, which seemed to culminate in the cut-and-fill reservoir, did not stop there. Instead it reversed itself and went back in the vertical direction whence it came. This happened for a number of reasons. Excavation methods were very crude and costly. Trucks did not exist, nor did heavy earth moving equipment. Also, though less of a problem then than now, the cut-and-fill reservoir required considerably more ground space than did the tank. For a while, however, the latter argument was counterbalanced by the fact that building materials, construction techniques of the day, and psychological factors somewhat limited the height of tank construction.

During this period of evolution, formed and poured concrete tanks made their appearance. Again, walls were very thick. Even though reinforcement was introduced, it was crude in comparison to that used today. It went to both extremes, being either thin wire mesh or very heavy, twisted cross sections. The concrete quality varied greatly. Some of these structures are still in excellent condition, the concrete being hard and dense, whereas in others the concrete is of very poor quality, appearing to have suffered because of various combinations of poor mixing, low cement content, or chemical degradation processes. Cracks in these structures were common, this being a particularly serious difficulty with respect to cold-pour joints. Builders were not very familiar with the problem, and cement mixing, which was done by hand, could not keep up with efficient pouring schedules.

Tanks were often set into the ground, or at least a major portion of their depth was in the ground. For this reason seepage went unnoticed, or perhaps it was just that no one bothered with remedial measures. On some old tanks, though, evidence of crack repairs using concrete and asphaltic materials is visible. In a large reservoir in northwest Pennsylvania, dates inscribed by repair crews in some of these crack patching materials go back to 1915, attesting to man's battle for seepage control long before the more sophisticated lining systems became available.

There was a difference of opinion with respect to the overall design of early concrete structures. One school of thought decreed that the structures should be built with vertical walls, whereas another believed that sloping walls were better. The latter group argued strongly. Form work was eliminated, they said, and reinforcement was not needed. Although these things represented only a small dollar savings by today's standards, they still accounted for an appreciable percentage of the total job cost. An equally enticing advantage of sloped wall construction was its apparent simplicity, for both the designer and the constructor. Of course, they were soon to find that these tanks, more properly designated as cut-and-fill reservoirs, present a batch of problems of their own. These center around the subject under study herein, namely, how to keep the water inside the structure from leaking out of it. Builders of vertical-wall tanks were well aware of leakage problems in their creations, and they seemed to find some comfort in the fact that they were not alone in this respect.

To cut the cost and improve the quality of concrete tanks, reinforcement steel was added. This reduced the concrete mass by absorbing and restricting tension loads. As all designers know, concrete is very weak in tension and the function of the reinforcing steel was to compensate for this weakness. During this period, construction of concrete tanks evolved in two directions: (1) the reinforced tank, and (2) the prestressed tank. The latter type developed along two lines, which will be examined next.

Originally, the walls of the prestressed tank were flat segments, cast on the job site. Later, curved panels were fabricated. In either case the panels were then tilted up into proper position to form the tank. The geometric configuration was a polygon, each facet of which represented a tilt-up panel. Various methods were used to tie in the panels to each other. Usually, a ring wall and a prestressed top anchor beam were part of the system used. Panel joints and intersections were sealed with appropriate gaskets, waterstops, and mastic combinations. With this method of construction any shape of tank could be built. There was a tendency, however, to build in the form of a circle to take advantage of the increased structural strength in this design, particularly when taller water tanks were built.

In 1908 the first patent in the United States was issued for prestressing a circular tank by wrapping wire segments around it. Turnbuckles were utilized to tighten the wire bands in an attempt to keep the concrete walls in continuous compression, regardless of the depth of the liquid inside the tank. This concept was said to be an improvement over the cast concrete tank. Even though the latter was reinforced too, tank loading usually produced tension cracks that were the source of future troubles. Despite the patent 14 years passed before the first large-scale application of this prestressing principle occurred. Success could be achieved by this method, and often was, but careful design and construction techniques were required.

In 1942 a machine was developed and patented that can continuously and automatically wrap high-tensile wire around the concrete tank shell. The wrapping begins at the top and proceeds down the tank shell until it reaches the bottom. The wire spacing is adjusted continuously during the wrapping process to place reinforcing at a closer spacing near the bottom than at the top and thus counteract the greater tension loads found there. As with the rod reinforced tank, the process places the concrete under continuous compression. The technique does the same job that the earlier patent outlined, but it does it much faster and more effectively. The completed structure is referred to as a wire wrapped tank. It is also often called a Preload tank, after the name of the firm that obtained the 1942 patent.

In 1943, in Europe, another method of constructing prestressed tanks appeared. After some experience in waterworks construction there, the system was introduced into the United States in 1961. This method bears the name "Pritzker" after its developer. In the Pritzker method arched, prestressed panels are cast at the job site. When assembled, the convex surfaces face the contents of the tank. Added stiffness in the vertical direction is achieved by thickened sections at both edges of each panel. Thus, in addition to vertical prestressing, the panels take advantage of the arch dam construction principle. Internal loading due to the liquid within the tank induces compressing forces in the concrete, just as in the arch dam. As with the wire wrapped tank, the concrete is under compression at all times, resulting in a more efficient use of the concrete.

Regardless of how the prestressing is done, the finished tank is ready to place in service after the concrete has sufficiently cured. The tank is given no interior treatment at that time, since the concrete structural design serves also as the lining. There is almost no limitation on diameter, but small tanks are quite expensive in terms of cost per gallon of liquid held. As the diameter of the tank increases, roof support columns are added to help in holding up the roof. Seismic movements are handled by mounting the wall on a heavy-duty bearing pad, and then lacing this structure with strong cables that loop through lugs in the floor slab. This permits movement of the shell with respect to the floor, but the movement is stabilized by the cables.

As data accumulated on experience with the prestressed tank, it became evident that under certain conditions the reinforcement steel was eroded by the seepage water that passed through the concrete. Owners were slowly learning this, having first been made aware of the problem by structural difficulties in tanks holding sewage and other materials with some corrosive tendencies toward steel. Then water tanks themselves became a source of concern. Instances of attacks on steel reinforcing bars by corrosion were often visible at areas where the concrete tank had spalled on the outer walls or on the underside of the roof. Gradually, the important parameters responsible for the corrosion were established: the pH of the water, dissolved oxygen, and the type of steel.

The plain reinforced tank, which preceded the prestressing technique, seems to have lost favor among owners of water storage facilities. This type of tank appears to make less efficient use of concrete and reinforcement than do the prestressed tanks. Reinforced concrete structures for water storage require a number of joints because of placing procedures. They serve to reduce stresses caused by differential settlement or variable loading and to minimize cracking due to shrinkage or to temperature changes. To handle all of these problems effectively and economically is not an easy task, and this fact is partly responsible for the gain in popularity of the wire wrapped tank.

The effect of freezing on concrete tanks constitutes a problem of considerable magnitude. Concrete has some porosity; and when it is subjected to high water heads, water permeates throughout these pore structures. It is easy to see the effect this will have when the excess water is subjected to freezing temperatures. An expansion of the water during freezing will start a slow breakdown of the concrete. The action is known as spalling, and unless it is arrested the entire tank may end up as an unusable water holding facility. At best, expensive repair bills can result.

Another problem experienced with all types of concrete tanks involves the junction between the wall and floor members. This is a moving joint because the tank is slightly larger in diameter when full than when empty. It is this location that has caused so many problems in connection with concrete tanks. Sometimes the trouble occurs immediately and delays final acceptance of the tank, and sometimes the problem develops later. Methods used in an attempt to solve this problem must take into account the movement at this junction point.

For areas of the tank that seep water slowly, there is some possibility that this slow water loss will diminish with age if the water is very hard, or if lime is added as part of the water processing operation. The hardness of the water, or the calcium content, sometimes tends to seal up the tank because of deposition of dissolved salts in the interstices within the concrete, slowly clogging up these passages. This will not happen if leakage is very rapid. Unfortunately, it is not something that can be relied upon with any degree of confidence. If it happens, the owner should consider himself lucky. To most owners, automatic seal-up is but a dream.

In early cement tanks leakage was sometimes relatively heavy at construction joints. With newer concrete bonding techniques, monolithic design, and ready mixed concrete, however, this is becoming less of a problem.

Little success has been realized in trying to patch cracks or local areas of seepage in concrete tanks. The patches are usually effective at first, but before very long the degree of control decreases, usually in association with bonding problems at the interface. The problem is complex in that the bonding may be too good in some cases. In practice, the bond

FIGURE 2.3. A bonded lining system in a sewage treatment plant near San Francisco. This special EPDM—type lining was 100% bonded to the concrete side walls.

must be good but flexible, and it must be able to do considerable "bridging" since the crack is usually larger when the reservoir is full than when it is empty. Underwater studies show this clearly in many cases; it has been found that, when the tank is empty, no cracks are apparent in a particular area. When the tank is full of water, however, the same area may show cracks of considerable magnitude. The final problem is when and where to stop the crack repair procedure. Cracks generally do not end abruptly, but trail off to zero width.

Continuous, impermeable linings offer an advantage in repairing concrete tanks (Figure 2.3). They cover the entire area and thus leave no random places uncovered. If the proper thickness is chosen, the lining will adequately bridge the small cracks that enlarge as the reservoir fills with water.

References

1. "Railway Roadside Water Tanks for Locomotive Supply," *Railway Age Gazette,* Nov. 19, 1915, pp. 955–56.
2. Douglas Alan Fisher, *The Epic of Steel,* Harper & Row, New York, 1963, p. 11.
3. *Ibid.,* p. 21.
4. *Ibid.,* p. 22.
5. *Ibid.,* p. 49.
6. *Ibid.,* p. 145.
7. *Ibid.,* p. 146.
8. F. L. LaQue and H. R. Copson, *Corrosion Resistance of Metals and Alloys,* Reinhold Publishing Corp., New York, 1965, pp. 7–40.
9. *Ibid.,* pp. 67–80.
10. H. Davy, *Philosophical Transactions of the Royal Society of London,* Vol. 114, 1824, p. 151.
11. Association of American Portland Cement Manufacturers, *Concrete Review,* Vol. 4, No. 2, Bull. No. 23, Philadelphia, 1910.
12. Robert W. Lesley, *History of Portland Cement Industry in the United States,* International Trade Press, Chicago, 1924, p. 1.
13. Lesley, *op. cit.*
14 *Ibid.,* p. 2

3

Flexible Linings

It would probably be safe to say that at one time or another almost every conceivable material has been tried as a lining. This would be particularly true if the material formed any kind of film, was sprayable, or was available in sheet form. Even noncontinuous films would qualify, as this class of lining is still on the market today.

From the early days there was emphasis on trying to keep water in holding facilities of various types. All civilizations wrestled with the problem, beginning with the earliest ones and continuing right on until today. The ultimate solution to this perplexing problem has not yet been found.

In recent years the designer has had a greatly increased field of materials and processes from which to choose a lining. Each of the products now on the market possesses certain advantages and certain disadvantages, and its most efficient use depends on established principles of design, construction, and facility operation.

The following pages discuss the various materials now in common usage, in accordance with the classifications shown in the Introduction. Chapter 3 treats flexible linings, Chapter 4 discusses rigid linings, and Chapter 5 handles those that fall in neither classification. Combination groups are considered in Chapter 5.

The term "flexible linings" is quite broad and means different things to different people. The layman will think of the thin membrane systems, whereas the farmer will think only in terms of a plastic, usually polyethylene. Only a soils engineer or someone else with broad experience in the reservoir field will think of compacted earth as a flexible lining, but it does fall into this class, even though it is a noncontinuous type.

The common, generally accepted flexible linings are as follows:

Polyvinyl chloride and polyethylene (PVC and PE).
Chlorosulfonated polyethylene (Hypalon).
Chlorinated polyethylene (CPE).
Butyl rubber, ethylene propylene terpolymer (EPT), and ethylene propylene diene monomer (EPDM).
Polychloroprene (neoprene).
Elasticized polyolefin (3110).
Asphalt panels.
Compacted earth.

All but the last are further designated as continuous, flexible, impervious membranes (see Table I.1).

Polyvinyl chloride film was the first used flexible, continuous membrane system with large-sheet capability. It was also the material used in the greatest quantity, largely because of its ability to provide the contractor with large sheets over 40,000 sq ft in size (Figures 3.1 and 3.2). With this background it became the best known hydraulic lining membrane.

To the uninitiated, "plastics" covers both polyvinyl chloride and polyethylene. It is true that both materials are plastics, and, chemically speaking, they are rather close cousins. But here the similarity ends, as will be seen in the following discussion.

Before plunging headlong into a study of the various lining materials, it will be advantageous to examine the manufacturing and processing steps involved in their production. These operations, although differing among materials, are more or less related to each other. They were borrowed from rubber processing technology because the rubber industry utilized its already acquired film producing skills in the production of the first membrane lining materials.

Manufacturing

Resin Production

All thin membrane linings are made from materials that are the result of a chemical reaction called polymerization, wherein small molecules join together with themselves or other molecules to form long-chain polymers. The reaction occurs in a pressure vessel termed a polymerizer or reactor, which is equipped with a stirring mechanism and is temperature controlled. The small molecules fed to the reactor as liquids are called monomers. This class of chemical is usually flammable, explosive, and

FIGURE 3.1. A one-piece, 15 mil PVC lining for an Arizona mine for copper leaching operation. This piece is about 40,000 sq ft.

toxic. Furthermore, at normal temperature and pressure it is usually a gas and hence must be handled in liquefied form at low temperatures and high pressures.

Although polymerization techniques vary, in the United States suspension polymerization is the most popular method. Europe now uses it, having changed recently from the once popular emulsion method. The U.S. method cuts costs, particularly the cost of drying the finished resin.

When ethylene monomer is fed to the reactor, polyethylene resin is produced. If vinyl chloride monomer is fed in, polyvinyl chloride resin is the product. In addition to monomer, the polymerizer is also charged with other ingredients to make the reaction and product generation proceed along the desired lines. These materials may include (but are not limited to) emulsifiers, wetting agents, dispersing aids, stabilizers, and a catalyst system. The last of these controls the speed of the reaction and is often responsible for getting the polymerization reaction started.

FIGURE 3.2. A one-piece, 20 mil contoured PVC drop-in lining for a domestic water reservoir in southern California. Since a roof was installed later, the lining needed no protective earth cover.

Once it begins, the reaction is usually accompanied by the evolution of a large amount of heat, which is dissipated by a cooling system built into the reaction vessel.

A combination of monomers may be mixed together in the polymerizer, and with this technique there is literally no limit to the variety of properties that may be built into a particular resin system. Strangely enough, two polymers that are not compatible with each other in the processing stage may often be polymerized together (copolymerized) to produce a useful copolymer (two polymers). Three monomers may be polymerized together to obtain a tripolymer; four will form a terpolymer; and so on. Combinations of four or more monomers, however, are not often used, as the resulting system is too hard to evaluate because of the interactions and the considerable amount of experimental time and expense consumed in determining the best mixing ratios from among the astronomical number of combinations possible. Although some predictions may be made concerning what properties may be expected upon mixing two known monomers, many surprises occur. Sometimes the copolymer has properties better than those predicted, but other times the properties are much worse than would be expected. In short, predictions in this field do not always come true, and possible outcomes are particularly complex to analyze when three or more monomers are involved. The only way to be sure of the results is to try the combination and see what happens.

The final properties of the polymer depend largely on the choice of monomer, but the reactor conditions, the type of emulsifiers used, the catalyst system, and the other ingredients in the mixture before the start of the reaction are also important, as well as the type of processing operation that follows later in the converting plant. After the reaction in the polymerizer is completed, the liquid mixture is separated from the polymer granules and the latter are washed, dried, deflocced, screened, and packaged for shipment to the processing plant. The final product for plastic materials such as PVC and PE is white and resembles granular or powdered sugar in appearance, depending on the polymer type and manufacturing process.

Hypalon is produced by reacting ethylene with chlorine and sulfur, producing a vulcanizable rubber with some unusual properties. Chemically, the polymer is referred to as a chlorosulfonated polyethylene. It is shipped in bale form directly to the processor. Two grades are manufactured, Hypalon-45 and Hypalon-48. The latter has better resistance to oil but is more difficult to process.

Chlorinated polyethylene results from the chlorination of PE, which basically is a building unit containing two carbon atoms and no chlorine atoms. Polyvinyl chloride is made from a monomer that contains two carbon atoms and a chlorine atom. Thus it can be seen that some similarity would be expected between these two polymers. Both are electronically bondable, a property attributed to the presence of chlorine. Likewise, the CPE polymer would be expected to have fairly good solubility, like its cousin, PVC. Such is not the case, however—proof that predictions in this field often turn out to be erroneous.

Butyl rubber results from a copolymerization reaction. The basic monomer is isobutylene, which is polymerized in the presence of smaller amounts of isoprene. The latter monomer has two reactive groups* in the molecule, which cause it to polymerize in a very special way. The structure of butyl rubber is interspersed with isoprene units. After polymerization these still contain a reactive grouping, and it is at this

* Chemically speaking, the active group is referred to as a double bond. It has the property of accepting sulfur in such a way as to form a link between adjacent molecular chains during the vulcanizing process. This cross-linking process gives butyl rubber its rubberlike properties.

grouping that vulcanization occurs. No such mechanism is present in the case of PVC or PE polymers.

Early butyl rubber sheeting did not have the good ozone resistance predicted. It cracked at fold and stress points on sun exposed berms of reservoirs. Although the chemists (compounders) could minimize this problem by using certain additives, new polymers did much to help. One of the first of these was an EPT rubber† introduced by Du Pont in 1963. At about the same time U.S. Rubber brought out a polymer that it designated as EPDM. Both of these materials, although more expensive than the butyl rubber polymer, can produce, with proper compounding, a finished material lower in cost than butyl rubber compounds. This is possible because an unusual property of the polymers is utilized: both EPT and EPDM have the capability of absorbing relatively large amounts of loading materials and oils. At the same time, they suffer much less from the adulteration process than does their butyl rubber counterpart.

The Du Pont polymer was described as a terpolymer (four polymers) of ethylene, propylene, and an unidentified chemical referred to as a nonconjugated diene,* while U.S. Rubber described its material in terms of the starting point (i.e., an "ethylene, propylene, diene monomer"). Neither company specified the exact nature of the unnamed monomer, but it would appear from the above explanation that the two are close cousins.

Both EPT and EPDM are superior in outdoor aging resistance to the straight butyl rubber material, according to the manufacturers, but compounding has a great deal to do with the final outcome. Although these polymers do seem to surpass butyl in some respects, the unwritten law of compounding† is at work again, and in regard to other properties butyl is superior.

The terpolymer compounds suffered at first from excessive shrinkage in sun aging, although their ozone resistance was excellent. Careful compounding and reductions in oil additives tended to diminish the problem, however, at little or no sacrifice in ozone resistance.

Neoprene is the product of the controlled polymerization of chloro-

† Trade named Nordel.

* A conjugated diene is one that has two double bonds, or chemically reactive groupings, separated by a single bond linkage. A nonconjugated diene also has two chemically reactive double bonds, but they are not separated in this manner. The latter will polymerize in a different manner from the former, yielding a polymer with somewhat different properties.

† The unwritten law of compounding is stated thus: "A compounding change that enhances one property of a material often results in sacrifice of performance in one or more of the other properties."

prene. This monomer is closely related to isoprene, the natural rubber building block. It differs from the latter by having one chlorine atom in place of a methyl group. Neoprene's flame resistance is a result of this chlorine atom substitution.

The resin production technique for 3110 has not been publicized but is presumed to involve the copolymerization of an olefin and one of the synthetic rubber monomers. The final resin has been defined as an elasticized polyolefin, indicating that the plasticizing (softening) action is accomplished by means of a nonmigrating polymer, and hence the outdoor aging characteristics should be considerably better than those of PVC.

Although we do not normally think of asphalt as a resin or elastomer, it is a plastic by one definition. It is included in this section because it is used in a thin membrane lining system.

Asphalt is found, in varying proportions, in most crude petroleums. It is a black to dark brown cementitious material that becomes liquid when heated. It has a density somewhat less than that of water. At room temperature asphalt may vary from a solid to a semisolid material. The major constituent of asphalt is bitumen, a mixture of hydrocarbons of natural or pyrogenic origin, all of which are completely soluble in carbon disulfide.

Almost all of the asphalt produced in the United States comes from the refining of petroleum. Sometimes this process of stripping off the volatile components occurs naturally, and then asphalt may be recovered directly from the natural deposit. Such is the case on the island of Trinidad, British West Indies, where the best known relatively pure asphalt occurs in a 100 acre lake. A more extensive but shallower deposit is found in Bermudez Lake in Venezuela, while an exceptionally pure but very brittle asphalt known as Gilsonite is obtained in Colorado and Utah by mining veins of the black mineral. Another well-known example of natural asphalt, though not of commercial importance, is found in the La Brea tar pits in Los Angeles.

The asphalt used in prefabricated panel production is air blown at elevated temperatures in the refinery to give it the desired characteristics for use in waterworks facilities. It is bulk shipped to the factory, stored in heated tanks, and from there is pumped to the production equipment as needed.

Processing

From the chemical plants the resins and polymers are shipped to the processor. Although the details of the processing operations vary from plant to plant, the types of operations are basically similar. In the past the tendency was for each plant to specialize in one type of membrane production; a plant would process either PVC or PE or butyl rubber, and so on. This accounted for the fact that a manufacturer's sales efforts were focused on the one type of material he processed. In most cases the various resins and polymers tend to be incompatible in the processing steps, so that the specialization was partly in the interest of maintaining a practical setup. Lately, however, there is a tendency for manufacturers to make more than one type of membrane. In these cases the mixing and processing rooms are separated from each other by as much distance as possible to prevent accidental intermixing of the raw materials.

The processing plant, sometimes called the converting plant, takes the resin and compounds (mixes) it with other ingredients to obtain the desired properties, such as hardness, strength, age resistance, and color. The instructions that indicate how much of what to add and when to add it are called a recipe and resemble, in general detail, a kitchen recipe for a cake or pie. The person who develops the recipe is called the compounder. The product resulting from the recipe is referred to as a compound and in both the plastics and the rubber industry is based on 100 parts of resin (or rubber) by weight.

In the plastic film processing plant two basic methods are utilized. One is known as extrusion; the other, as calendering. The choice of equipment depends on which type of resin is being handled and the characteristics desired in the finished product. Most commercial resins may be processed by either method, although there are some exceptions. Polyethylene cannot be calendered, so it is always run on an extruding machine. For lining work PVC is processed on a calender, although, in the majority of cases, extruders, internal mixers, premixers, and open-faced rubber mills are also utilized. Because many steps used in PVC processing are typical for most plastic and elastomeric materials, these will be described in some detail to promote a better understanding of processing in general.

Internal mixers for PVC processing are heavy, jacketed vessels with high-powered, heavy-duty mixing blades and a hydraulically operated ram that exerts pressure on the mixing materials. The dry materials are premixed in a blender and then charged to this mixer, which is referred to in the trade as a Banbury, after the name of the man who developed it. Premixers and Banbury equipment are available in assorted capacities of from 400 to 2500 lb or more. The mixing cycle varies from about 3 to 12 min. The Banbury mixer is jacketed for heating (or cooling, in the case of rubber or other curing-type compounds). The heat, in combination with pressure, speeds up the plasticization (solution) of the PVC resin within the plasticizer.

The Banbury charge drops out through the bottom of the equipment after mixing is complete. It falls onto a set of two open-faced rolls, referred to as a rubber mill. The two rolls, horizontally mounted and temperature controlled, are 18 to 28 in. in diameter and about 6 ft long. They turn at different speeds, thus giving tearing, shearing action to the batch to assure better homogeneity. Because of the speed and temperature differential of the rolls, the batch knits together and sticks predominately to one roll, from which the operator manipulates the batch back and forth for better mixing control.

Next the hot plastic mass (about 340° F) is fed directly to the calender. Sometimes, however, the stock is first sent through an extruder, which further mixes the batch, rebuilds its heat content, and, by means of a stainless steel head screen, eliminates any foreign material from entering and damaging the costly calender rolls.

An extruder is a machine that resembles a meat grinder in principle. When used alone for film formation, raw resin, properly compounded, is fed into the machine at one end, directly from a premixer or a rubber mill. The material is forced through the barrel of the extruder until it emerges at the other end. The barrel is heated to carefully controlled temperatures. A die determines the shape of the product that issues from the front end of the equipment. When the extruder is used in conjunction with a calender, a solid tube is produced. In the case of PE processing the die is designed to produce a large-diameter, thin-walled tubing. As this hot tubing leaves the machine in a vertical direction, internal air pressure expands it to the desired marketable width. A lay-flat tubing is produced, which is next slit and packaged for shipment to the ultimate user. An extruder, along with the auxiliary equipment for producing the lining material, will cost from $10,000 to $100,000 or more, depending on size and degree of automation. Most equipment in use today will produce from 100 to over 1000 lb/hr, depending on size and product being manufactured.

A calender is a much larger and more expensive machine, and with windup mechanisms and all auxiliary equipment will cost $1 million or considerably more, again depending on the dictates of the buyer. In principle, it is quite similar to the rolling mills that produce thin metal sheet and strips in a steel mill. As a matter of fact, the technology of producing plastic sheeting borrowed heavily from the experience developed in the steel industry. There, in 1924, John Butler Tytus of Armco found that the only way to roll steel plate successfully was to do so with "crowned" rolls, that is, rolls that are slightly larger in diameter at the center than at the ends.[1] This compensates for roll deflection due to the tremendous pressures exerted in the process, and at the same time causes the material being calendered to properly track through the equipment. Because of Mr. Tytus the crowning of the calender rolls in the plastics industry was anticipated before the equipment was built, but the final steps of crowning are undertaken in the processing plant itself. The crown is partly built in at the factory that produces the rolls, and partly "ground in" at the processing plant as the rolls sit in the calender framework.

The amount of crown required is dependent on the width of the roll, the temperature required to calender the compound, the hardness of the batch stock at the processing temperature, and the thickness of the film being produced. Crown dimensions represent a compromise, because theoretically there is only one exact crown for each compound and thickness of film run. During calender operation the crown slowly wears off and must be replaced by regrinding from time to time. The roll recrowning procedure at the processing plant involves several days' time and considerable expense. Since it cannot be done for every compound or thickness change, the processor tries to run similar items on each particular calender. He cannot generally run good quality 10 mil film on a calender crowned for 35 mil film. These restrictions have been eased somewhat by the development of a crossed-axis calender. Here one roll may be pivoted in the horizontal plane around its center point to effectively open up the film thickness capability at the edge. This amounts to an adjustable crown and gives the manufacturer increased flexibility with respect to the range of products he can process. As can be imagined, he pays rather heavily for this innovation.

The calender uses heavy steel rolls that vary in width from about 50 up to 120 in. at the present time. Their diameter changes with their width and ranges from 24 to 36 in. The rolls are cored for precision temperature control, as temperature is a very important part of the process. The maximum film width that the equipment is capable of producing is usually about 6 in. less than the roll width to allow for edge trim. Although calenders may be built with three, four, or five rolls, most manufacturers feel that the four-roll machine is the best compromise. Up-to-date calender equipment will run thin hydraulic films of PVC at speeds exceeding 100 yd/min. On a 60 in. finished width, this amounts to over 16 acres of lining material in an 8 hr shift. This statistic explains the reluctance of manufacturers to run special widths or thicknesses unless large quantities are involved. Today, PVC film is run in four standard lining thicknesses: 10, 12, 15, and 20 mils. The last two are the most popular gauges.

Polyethylene processing is restricted to premixing and extruding operations, followed by a blowing step for greater effective width. This material is run in a number of standard thicknesses from 2 to 8 mils for hydraulic purposes. Ten mil films are generally too stiff to perform well in the lining field.

Hypalon is processed in a manner quite similar to that described for polyvinyl chloride resins, except that lower calender roll temperatures are used. The fact that the compound has a sharper melting point than PVC is one of the reasons why processing is more of a problem. Higher temperatures also cause a reaction that fosters the formation of pinholes. To avoid this processing difficulty and others, some manufacturers have been forced to resort to lower processing temperatures, which produce sheet with considerable surface roughness, a defect known in the trade as cold streaks. The alternative, running on a hot calender, is also bad. In addition to the problems already mentioned, premature curing can result if temperatures are too high. When this occurs, the material must be scrapped, as it cannot be removed from the equipment in usable form.

The nonreinforced membrane is produced in two standard thicknesses, 20 and 30 mil, and it can be made in almost any color desired. As might be expected, the use of colors for industrial linings has been of limited interest. Most users do not want to attract attention to the lining for fear of possible damage by vandals; then, too, colored membranes are more expensive. Black is the best aging color for most lining products and, with one manufacturing exception, the only color in which the butyl membrane can be made. It is certain that black will be the only color that enjoys much favor in Hypalon sales to the lining trade.

Hypalon compounds have a high specific gravity, nearly 1.5, the highest of the lining membrane materials. This property makes it suitable for use in brine ponds, which can float out unburied, low-gravity lining systems.

In the early days of this product manufacturer's specifications for unsupported sheet listed tolerances for pinholes. Some allowed as many as one for every 15 linear ft of material as it left the calender. This type of tolerance was also common for 10 mil PVC films in the early days, and even today thinner gauges in this range carry pinhole tolerances. As with PVC films, the difficulty with pinholes in Hypalon membranes is closely related to equipment contamination, although poor carbon black dispersion and excessive processing temperatures are also major contributors to the problem. Greater attention to equipment cleanout and the use of laminating techniques for building up film thickness have reduced the severity of the pinhole problem.

Hypalon leaves the plant in rolls up to 60 in. in width and from 300 to 500 ft in length. Like PVC film, its first trip is to the fabricating plant, although the trend in the industry is for each manufacturer to fabricate in his own plant. This assures better control of joints and of course reduces cost. It is also an advantage to the eventual user by eliminating one possibility for evasion of responsibility in the case of defects.

Chlorinated polyethylene is produced in sheet form for direct sale. It then goes to a factory fabrication operation, where it is built up into large sheets. Traditionally, this has been done in a fabricating plant not under the direct control of the processor.

The processing of butyl rubber follows some of the general steps outlined for PVC film production. The one basic difference is that butyl compounds must be "cured" (vulcanized). Such things as premixers, Banbury mixers, mills, extruders, and calenders are common to both types of materials, although extruders are generally not found in the calender train. They tend to build up heat, and with curing-type materials this is to be avoided as much as possible.

Vulcanization is a molecular cross-linking process brought about by the application of heat and pressure in the presence of a vulcanizing agent, usually sulfur or a sulfur bearing compound. In the process the elastomeric compound is converted from a sticky, thermoplastic material with poor physical properties into a strong, elastic material not affected by temperature extremes. Although not as resilient and "snappy" as natural rubber, the synthetics (butyl, EPDM, etc.) do not suffer as much as natural rubber from the effects of ozone degradation.

After the batch is mixed in the Banbury, it is dropped first onto a rubber mill to complete the mixing cycle and then onto a calender, where the material is sheeted out to the desired thickness. In contrast to plastics processing techniques, mixing, milling, and calendering are done at low temperatures to prevent premature curing. Since the rubber at this stage is tacky, the sheeting from the calender is dusted with soapstone or mica dust to prevent it from sticking to itself. A fabric liner is sometimes used for this purpose.

From the calender the sheets may be built up into larger widths and cured in one piece. As an alternative, depending on the manufacturer, the narrow widths may be cured first and large sheets put together later by a special process that vulcanizes the joints together. In this system the calender-width material may be cured in bulk or on a continuous basis, depending on the plant layout.

Curing times and temperatures vary from plant to plant. Generally, temperatures will range from 240 to 380° F at times from 10 min to over 1 hr, depending on the total bulk weight of the material being cured, its end use, and the equipment available.

The processing of EPDM and EPT is quite similar to the operations just described for butyl rubber. Likewise, the processing of neoprene follows a similar pattern.

The processing of 3110 material, on the other hand, is considerably different from the methods used for the other lining materials. It is produced in 20 ft wide seamless sheets on a blown film extruding machine, cut to proper lengths, and rolled onto tubes for shipment directly to the job.

Although asphalt panels do not seem to be very closely related to PVC or butyl membranes, the processing steps have many similarities. The first step in the panel process is the introduction of asphalt, mineral fillers, and fibers into a heavy-duty, continuous, open mixer, operated in the 250° to 300° F range. The mixer is not unlike a Banbury mixer in design except that, being open, it has no ram. The asphalt portion represents 65% by weight of the total mix. The mineral content is adjusted to give a minimum specific gravity at least a few points higher than the specific gravity of the liquid being held. Standard material has a specific gravity of between 1.1 and 1.2, but higher density grades are made when the lining is to be used in brine ponds. The high asphalt proportion in the panel is in contrast to the 3 to 8% found in asphalt concrete (paving). The fact that the mixture is a rich one has a lot to do with the properties of the product. The panel is watertight because of the high asphalt content and lack of voids. This accounts for its superiority over plain asphalt concrete when it comes to containing water and waterborne products.

The hot material from the mixer is fed to a piece of equipment similar to the roller-head die used in vinyl plastics processing. This is, in reality, an extruder in which the exiting product is shaped (premolded) by two rotating rollers instead of by a fixed, slotted die plate. As in vinyl processing, this method is more efficient than calendering when thick sections are being produced.

Two let-off rolls of the appropriate asphalt impregnated fabric, felt, or glass mat allow these materials to be fed through the equipment, meeting the two outer surfaces of the panel as it is in contact with the rollers of the die. Immediately upon exit of the panel from this laminating process, the top and bottom surfaces are simultaneously coated with hot asphalt, and the entire construction then travels submerged in a water bath until cooled sufficiently for proper handling.

Upon leaving the cooling bath, the panel strip goes first through a slitter and then through an automatic cutoff machine. Panel widths vary from 36 to 48 in., depending on the manufacturer. Panels are custom cut to fit the job layout and vary in length from 10 to 30 ft. Slit-edge material is fed back to the mixer and reprocessed. As the panel continues down the conveyor, the top surface is dusted with mica or a similar material to prevent blocking during shipment or storage. After inspection the production pieces are conveyed directly to flatbed trucks or rail cars or onto skids for stock. There is a limit on how long the material may be stockpiled because of its tendency to block (stick to itself). Panel production varies in thickness from ¼ to ⅝ in., but most of the material used for hydraulic lining is made to have a final thickness of ½ in.

The core of the panel, which is all of the construction except the two fabric layers, is held together by the mineral fibers and filler materials originally fed to the mixer. To provide added strength and to facilitate handling the panel before its installation, the panel is packaged between the two asphalt impregnated fabrics. After the installation is completed, and the panel "set" into the subgrade texture, the importance of the outer fabric layers is diminished. Without them, however, unloading the panels at the job site would be a difficult, if not impossible, undertaking.

Compounding

Compounding is one of the most important parts of the lining story. For this reason a little background on this subject will shed considerable light on the properties of linings and the ways in which they may best be utilized.

Compounding, or the art of working out the recipe, is said to be a study in compromise. No recipe can be developed that has all of the desirable properties the customer would like to have. The reason is that all physical and chemical properties are interrelated. Each time one particular property is improved, something else tends to suffer, a rule that applies to both plastics and elastomerics. Although it is beyond the scope of this book to discuss the details of the interrelationships involved, it is important to note that they do occur. Therefore to write a material specification based on taking the best value for each property from among several suppliers' bulletins is an erroneous, though often repeated, procedure. Such a specification has no practical value because it would not be possible to make a compound that would meet it.

To go back to the beginning, PVC resin lay dormant for many years after the polymer was first synthesized in Germany over 50 years ago. It was a hard, horny type of material, and no one could figure out what it could be used for. In the early 1930s, however, a discovery was made that placed this material in the limelight: a chemical was found that would convert the raw resin into a useful, flexible material. The resulting product proved to be an excellent barrier against corrosion, and its first industrial applications were in this category.

The breakthrough was the discovery that the resin, when mixed with certain liquid materials, could be altered if the mixing was followed by the application of heat and pressure. The alteration produced a very useful material, one that had great flexibility, excellent resistance to abrasion,and good strength. It was not affected by ozone and resisted an impressive number of common chemical solutions. Ozone was the culprit that had been harassing rubber articles for years, causing them to crack and degrade quickly in open exposure. Since the process was discovered in the laboratories of the B. F. Goodrich Company by Dr. Waldo Semon, there was great rejoicing among the firm's tire experts. The perfect answer to rubber degradation in tires had been found! Unfortunately, this was not the case, and although initial experiments did seek to utilize this new wonder material in tiremaking, it was soon concluded that PVC was not destined to invade this market, except for small toys. These studies showed that there is more to tiremaking than producing ozone-resistant materials!

The "oily" type of chemical that transformed PVC resin into useful articles was known in the trade as a plasticizer, or softening agent. To a large degree it is the key to the material's behavior. Other substances are also added to the plastic compound, such as pigments for color, light stabilizers to inhibit the effect of ultraviolet rays of the sun, heat stabilizers to counteract the heat effect of sun exposure and high processing temperatures, and extenders whose prime function is cost reduction.

Since the plasticizer determines in large part the characteristics of the final product, it is well to examine this portion of the mixture in more detail. Although a number of materials have been used in this capacity, perhaps the most common one is a chemical named dioctylphthalate, commonly known as DOP. Most nonchemists refer to it as "oil" or "dope." It was the chemical first used by Dr. Semon and today is still rated as an excellent all-round plasticizer.

A true plasticizer is in reality a vehicle into which the resin may be placed in solution, but it is a very unusual type of solution. Such a chemical component, mixed with the resin in the presence of heat and pressure, causes plasticization to occur. The resin and the plasticizer go into mutual solution with each other; but, unlike a normal solution process, the resin does not precipitate out as the solution cools down. The resulting mixture is a thermoplastic one, that is, it becomes stiffer as the temperature decreases and it gets soft and sticky with a loss of strength as the temperature rises. The compound may be heated and cooled through many cycles, however. Unlike rubber or other cured materials, it can be reprocessed through the calender again and again, although the properties of the compound degrade slightly with each pass through the equipment.

One great advantage of PVC (and also PE) is the ability of the material to accept an almost unlimited range of bright colors. This property sets it firmly apart from the elastomers. Except in the swimming pool lining trade, however, the color advantage has not been utilized to any great extent in the lining industry for two reasons. First, these lining systems are generally covered to prevent mechanical damage and degradation due to weather. Second, even when no burial is used, pretty colored linings tend to be attractive focal points for vandals with resultant damage or disappearance of parts of the linings. Another point technical in nature, is also involved. Black linings, because they block ultraviolet ray penetration, generally age better outdoors than colored ones and in addition are somewhat less expensive to produce.

The compounding of polyethylene is a considerably different process from that employed by the vinyl industry. One of the most important differences is that the PE material does not require a plasticizer; in fact it cannot be plasticized to yield a soft, flexible material. This fact is either a blessing or a curse, depending on which criteria are used for the evaluation.

The primary "blessing" is that no plasticizer is available to migrate or leave the compound. Hence this eliminates one of the principal reasons (though not the only one) why PVC films tend to stiffen with age: they stiffen as they lose plasticizer, either by water extraction or by heat volatilization or by both mechanisms. This then reduces their ability to "give with the blow." Their resistance to shock loading diminishes, they decrease in tear resistance, and shrinkage forces come into play. These changes are accompanied by an increase in tensile strength and hardness. As we shall see later, tensile strength is one of the least important properties of a material insofar as its effectiveness as a hydraulic lining membrane is concerned. Because PE contains no plasticizer, it escapes many of the evils associated with the presence of such a substance.

The "curse" assumes a number of forms. Polyethylene has a fixed stiffness characteristic that, within rather narrow limits, cannot be changed. As PE film increases in thickness, it becomes stiffer, to the point where 8 mil film is about the limit of its usefulness as a lining; thicker materials are too stiff. It is not possible to plasticize PE resin and still end up with a useful film. In other words, the material is resistant to almost everything. In fact, it is so resistant that liquid adhesives are not effective on it. They will not stick to the surface of the film even when powerful solvent systems are utilized. Polyethylene is nothing more than a high molecular weight wax, and it tends to exhibit many of the characteristics of wax—poor abrasion and poor crease resistance, but excellent resistance to water, water vapor, and a host of chemical solutions.

Polyethylene has a narrow melting point range in contrast to the wide range exhibited by PVC materials. This makes heat welding of PE film an extremely tricky process to control. Electronic bonding techniques, in the strict sense of practicability, are not possible with PE film. About the only way this material can be put together is by means of a sticky tape or extruded gumlike material, but even then the joint is not considered permanent. The PE manufacturers have bypassed the field fabrication problems and at present still do not market an adhesive system in conjunction with their product. This situation has forced the applicators to experiment with different materials in order to get the job done. The degree of success has varied, but most of these evils lie buried by the mandatory earth cover requirement.

On the economic front PE has a real advantage with respect to first cost. It does not require expensive calender setups, and the compounding step is much less complicated than its counterpart for PVC. These things all help to keep the cost down, but of even greater significance is the material's low density. This, coupled with a low cost per pound, translates into an unbeatably low cost per square foot, which is the basic criterion used in economic evaluation of initial cost outlays for linings.

On the other side of the ledger, PE film has extremely poor outdoor aging characteristics. Its major competitor cannot take advantage of this weakness, however, because PVC also exhibits very poor aging performance. Although the latter is not as inferior in this respect as PE, the difference is not great enough to be of practical significance. Neither material should ever be used in open exposure to the elements, except in temporary facilities. Desert exposure will limit the useful life of PVC thin films to a maximum of 2 to 3 years, and this time is considerably less in the case of PE films. Both will do somewhat better in cool, cloudy climates.

Although both materials become brittle at low temperatures, installations may not suffer excessively from this effect, so long as impact (shock) loadings are not present. High temperatures are also a problem; manufacturers' bulletins to the contrary, neither film should be used in continuous service where solution temperatures are above 125° F.

The cost advantage gained by PE's low density is offset by a technical disadvantage due to the same property. This lining material is sometimes placed in situations where the liquid being held is higher in density than the seepage control membrane. If nothing goes wrong, there are no problems. If, however, water gets beneath the lining because of a leak or groundwater intrusion, the lining may tend to float out, since it is less dense than the water in which it is immersed. This created serious problems in the early days of PE use. Current practice requires that a substantial earth cover be placed over the top of the PE lining before it is placed in service. The cover not only protects the lining from the effects of sun aging, mechanical damage, and vandalism, but also prevents it from floating should water get behind the lining for any reason.

The compounder has been almost helpless to do anything about PE's low density. The resin will not take much loading without degradation of its physical properties. Since PE cannot tolerate any loss of physical strength, there is little anyone can do to improve the acceptability of the material in brine pits.

The compounding of Hypalon follows patterns that are similar in many respects to those of PVC compounding. There are stabilizers, lubricants, colors, and loadings, although loadings cannot be as heavy as with EPDM-type materials without experiencing marked losses in physical properties. Attention must be given to the prevention of premature curing during the processing steps if the factory is producing uncured Hypalon membrane.

The compounding of CPE resin for lining applications follows a somewhat different course from that described for PVC. The principal difference is that CPE does not require a liquid plasticizer, so that the only materials that need be used are color pigments and processing aids.

The compounding of butyl rubber contrasts with that of PVC. At the synthetic rubber plant, the polymer is packaged in large blocks and shipped to the processing plants. There it is fed to Banbury mixers and continues onward through a series of processes not unlike those described for the PVC materials. Although the types of ingredients added in compounding are different from those used in plastics, they have a great many of the same types of functions within the compound. Plasticizers and softening oils are added along with loadings (clays, diatomaceous earth, carbonates, silicates, etc.) for cost reduction and reinforcement. Stabilizers and antioxidants impart stability toward processing conditions and subsequent outdoor aging, and certain lubricants are incorporated to aid processing. It should be noted that in elastomeric compounding the plasticizer concentration is normally quite low compared to the concentrations used in PVC compounding. If too much plasticizer is used, loss of plasticizer upon aging can cause some of the same undesirable effects already noted for PVC-type materials.

From this point on, butyl rubber compounding deviates from the compounding employed in the plastics industry. Carbon black, clay, carbonates, and silicates are added to the batch, mainly for purposes of reinforcement. In plastics work, on the other hand, these materials are normally added to reduce the cost of the compound or to impart a color. Carbon black of the proper particle size and shape exhibits an unusual property when added to rubberlike materials. Within rather broad limits the more carbon black added, the stronger and better are the overall properties of the compound. No parallel to this exists in the case of plastic materials, whose properties generally degrade as any type of loading is added. Of course, this peculiar property of carbon black explains why rubber is basically a black material; the carbon seems to be required to bring out the best properties of the rubber itself. Although other materials also exhibit reinforcement properties in elastomerics, carbon black seems to be the most effective for this use.

To complete the rubber compounding sequence, accelerators are added to alter the rate at which vulcanization takes place, and sulfur is used in the compound as the vulcanizing agent. Although other chemicals, such as certain sulfur compounds and some nitrogen bearing chemicals, do effect some curing cycles, plain sulfur does by far the best job.

The compounding of EPT and EPDM materials is similar to the procedures outlined above, with the exception of their greater affinity for loadings. This trait will usually produce a final compound cost below that attainable for straight butyl rubber, because this affinity for greater loading concentrations is not accompanied by the expected loss of physical properties that occurs at the same concentration of loading in an all-butyl-rubber compound.

Chemically, it is extremely difficult to tell the difference between a butyl compound and one containing EPT or EPDM. For this reason the materials are often used interchangeably by the processors, and there is a trend to eliminate designations that define the type of polymer in favor of code designations defining the properties of the finished compound.

Compounds of the butyl and EPDM class of materials have been produced in a few restricted colors on an experimental basis. So far, however, membrane users have not expressed much interest in this accomplishment, partly because of the higher cost and the washed-out appearance of these products in contrast to the bright colors of Hypalon and the plastic materials.

The compounding of 3110 is much less complicated than that of PVC or the other elastomers previously described. Little else is added to the base resin and the high molecular weight plasticizer other than pigments and, on occasion, minute amounts of processing aids.

Except for proportioning the asphalt content and the quantities of mineral filler to yield the right specific gravity, the compounding step for asphalt panels is much less complicated than that for PVC or the elastomeric compounds. Mineral and glass fibers, along with the outer fabric laminates, give necessary tensile strengths. There is no need for heat stabilizers during the processing steps, as the ingredients spend only a short time at the higher temperatures.

Since some asphalts are not compatible with each other, each manufacturer formulates from his particular asphalt sources the grades that ensure panel, adhesives, and mastics will all be compatible. The user should make certain that all panel lining materials and accessories come from the same manufacturer. Basically, for a typical manufacturer these include (1) an asphalt cement of 180 to 220°F melting point (50 to 60 "pen")* for use as a hot-applied adhesive; (2) an asphalt cement of a consistency that will be compatible with the weather conditions expected at the job site; (3) a primer, which is an asphalt cut back with naphtha, for use when panel seals to concrete surfaces are required; (4) a liquid adhesive system for panel-to-panel fabrication; and (5) asphalt mastics, heavy-bodied solutions used in conjunction with mechanical seals at structures and as adhesives. Asphalt mastics may be compounded for use underwater; when so formulated, they bear the designation "UW." The primers are sometimes formulated to enhance their adhesion to damp surfaces. Generally, the mastic materials contain mineral fibers to increase their strength.

The chemist in charge of asphalt formulations must be certain that no coal tar, phenols, or similar materials get into the mixers, as these will impart taste and odor to potable water and thus render it unfit for use. His counterparts, the compounders of plastics and elastomers, must likewise watch for undesirable materials in their recipes, such as lead compounds and certain easily extractable plasticizers.

Factory Fabrication

The PVC processing equipment produces lining films of a narrow width, usually ranging between 54 and 72 in. Except in unusual cases, these

* The penetration ("pen") value represented the standard test run on asphalts before 1974. It was the measure, expressed in tenths of millimeters, of the penetration of a standard needle vertically into the asphalt under a load of 100 gr applied for 5 sec. The test was run at 77°F. Since January 1974, asphalts have been graded by the marketing agencies on the basis of the consistency of oven aged samples (aged residues), rather than on the consistency of the asphalt as introduced into the mix. Five such grades are recognized: AR-1000, AR-2000, AR-4000, AR-8000, and AR-16,000. Although no pulications precisely relate the AR grades to the previously used penetration values, it would appear that the AR grade ranges listed above correspond roughly to the 200 to 300 and 40 to 50 "pen" grades, respectively.

widths are too narrow for use in lining large hydraulic facilities because of the extremely high cost of construction labor. At the present time the processing plants send their calender output to a fabrication plant, where it is then put together in larger sheets for shipment to the lining contractor at the job site. Under such a system the double handling of the film between factory and fabrication plant is not efficient from the standpoint of cost and delivery time, and considerable effort has been expended in attempts to get the industry to modernize this procedure. Many processors of other lining membranes do their own fabrication, but to date none of the PVC lining producers has taken this step. The problem is being restudied again in the wake of the increased demand for PVC for pollution control linings.

Because of a very special property of the material PVC film lends itself to seam fabrication by electronic bonding techniques. The basic molecule, when placed in an electric field, acts like a magnet. It will orient itself with its negative part pointing to the positive pole of the electric field and the opposite portion pointing toward the negative pole. If the polarity of the electric field is reversed, the molecule flips or turns around in an attempt to realign itself in the opposite direction. The plasticizer plays an important part in this mechanism by allowing the molecules within the structure the mobility necessary to make these back and forth movements. If the polarity of the field is reversed at a high rate, as with the high frequencies of the radio range, the flipping back and forth of the molecules to realign themselves at each cycle change creates internal friction within the compound, and heat is generated at a very rapid rate.

The electronic bonding machine contains four basic components: (1) a transformer to convert the available electrical supply to a very high voltage, (2) an oscillator tube to develop the high frequencies required, (3) a pair of electrodes, and (4) a pressure system to clamp the film being bonded to a uniform high pressure between the electrodes during the seaming cycle. Normally, the top electrode bar is ¼ to ½ in. in width and varies from 24 to 48 in. in length. The bottom electrode is a large, thick metal plate, which also serves as a working area for aligning the film between each sealing cycle.

Machines for this type of bonding on hydraulic films normally operate on 220 V circuits and put out from 6 to 10 kW of power. They run at about 50% efficiency, the losses being mostly in the form of heat. The cost of a new machine will run from about $5000 to $20,000 or more, not including materials handling equipment, which is generally custom built. Commercial machines vary somewhat, but most operate in a radio-frequency range of about 27 MHz.* When the machine is actually bonding, power is on for 1 to 3 sec, the time required to bond the PVC film thicknesses used in hydraulic lining work. Most machines are constructed with built-in shielding against stray radio waves, but additional shielding is usually required if the machines are to operate near an airport. This guards against the possibility of their interfering with aircraft radar installations.

The bonding operation involves the placement of the two pieces of film together in proper alignment to produce a lap joint. The pressure cylinder, air or hydraulically operated, then brings the top electrode bar down onto the joint area against the horizontal backup plate, which serves as the bottom electrode. The current is then turned on, and almost immediately the film softens and melts together to form an excellent seal. Automatic controls turn off the power at a preset interval. The film is heated uniformly throughout, although it is somewhat hotter at the center than at the outer surfaces, the latter areas being cooled by conduction of heat through the electrodes. Both electrodes are heavy pieces of metal to ensure uniform contact and dimensional stability and do not, themselves, increase much in temperature during the run. Depending on the total thickness of the film being sealed, the current is generally on for no more than 3 sec, so the electrical portion of the process is rather fast. After the seal is made, pressure is released, the material is moved ahead for the required distance, allowing for a slight bar overlap, and the cycle is repeated.

By bonding seams in this manner and using the proper materials handling setup, PVC linings may be made economically in large pieces for shipment directly to the job site. Fabricating plants sometimes place restrictions on the maximum width for which they will fill orders, but it is not uncommon for sheet widths to reach 200 ft or more. There are generally no restrictions on the lengths of the pieces. Single 1 acre pieces (about 40,000 sq ft) have been made many times, and single pieces 125 ft wide and 600 ft long were once used to line sewage ponds in Alaska. If 15 mil thickness is assumed, such a lining will weigh 7,500 lb. Although this is not the top limit in size, it does illustrate what can be done by the process. Factory labor rates are much lower than those paid construction workers, and at the same time the electronic joints are made under the ideal conditions of the factory, rather than in the field, where such an advantage is not present.

A further advantage of the process is that it produces homogeneous

* 27 MHz = 27 megacycles per second = 27 million cycles per second (Hz).

joints. No adhesive system need be evaluated, and the joints will not let go as long as the material itself is not physically attacked by the solution being held. Electronically bonded joints have a good quality record, although, as with any process, attention to quality control within the factory is required. It is possible to produce a "false joint," one that looks sealed but actually is not and can be separated with only minor effort. However, these joints can be readily detected and remedied with normal inspection procedures. Compounding is important to the success of electronic bonding processes. The processing plant must always be informed if fabrication is to be by electronic means, so that compounding may be done accordingly. Electronically bondable films should not contain conducting carbon black, metallic dispersions, or metallic printing inks, and certain inorganic loadings are restricted as to both type and quantity.

As long as the compound does not change appreciably, the machine setting requires little or no adjustment. Changes in the settings are required, however, if the thicknesses to be sealed are changed or if a different compound is being sealed. Each time there is a plastic material roll change, the operator must be particularly watchful, as this could reflect a different compound or batch from the factory.

This method of seaming PVC plastic film is not applicable to sealing material on the job site except under very special conditions. The machine itself is heavy and sensitive to dust and vibration. Since the electrical characteristics of the process demand that the generating equipment be located in close proximity to the electrodes, the latter cannot be utilized as a portable entity. If the equipment is taken into the field, it must be protected against all aspects of the environment. Machines have been set up in a large trailer and taken to the job in this manner, but in most cases a permanent factory is utilized.

In the early days of vinyl film, attempts were made to "iron" the material together by pressing it between two hot metal platens or rollers, or by using an ordinary electric hand iron. Polyvinyl chloride is a poor conductor of heat, however; and before the films would stick to themselves, their outer surfaces became so soft and sticky as to be unmanageable. Although it is possible, using extreme care and a nonsticking buffer film, to make the process work, it is extremely slow and difficult to control.

More recently, machines have been developed that utilize heat welding principles. One of these impinges a hot air blast between the two pieces of film being sealed. This is immediately followed by a pair of small pressure rollers through which the heated seam passes. Another technique places a hot knife (an electrically heated resistance element in the shape of a flat blade) between the two portions to be sealed, again followed by a squeeze roll device for pressure. Although both processes are somewhat tricky to operate, particularly on thin films, they seem to be more easily adapted to sealing PVC with fabric reinforcement. Again, it is important that the processor know whether heat welding is to be done, as a compounding adjustment may be made to facilitate the overall process. In any event thin, nonreinforced films are sticky and weak when hot and so present difficult handling problems. Start-ups and stops are particularly bothersome in both of these processes.

The methods described above should not be confused with a heat welding process applied to heavier sections (¼ in. or thicker), which may be fabricated in the field by means of a technique not unlike that employed for the conventional welding of metals. In this case, of course, the welding rod is a special PVC base material. This process, although requiring skill, is much easier to control than heat welding on thin films for somewhat the same reason that makes welding extremely thin metal plates difficult.

In contrast to the situation with PE, one of the properties of the PVC system is that good adhesives can be developed to adhere the material to itself. Part of the effectiveness here is due to the film's being attacked by the solvents contained in the adhesive. This, together with the cohesion of the adhesive particles, makes the seam. As with anything else that involves PVC, the compound is a part of the process and must be reckoned with in developing a good adhesive system. The adhesive resins must not be attacked by the plasticizer system, and there must be no tendency of the plasticers to be more soluble in the adhesive resins than in the PVC resin; otherwise they will migrate and decrease the strength of the bond. Since PVC adhesives may show some degree of selectivity, it is well to evaluate adhesive effectiveness on actual production run film to be used in the eventual lining job.

There is no factory fabrication process for unreinforced PE films, as no adhesive will stick them. Heat welded films for the lining trade are difficult to produce, and at the present time this unreinforced lining material is sold only in standard sized, seamless pieces. These sizes range from 20 × 40 ft to 40 × 100 ft. However, longer lengths can be made on special order.

The original Hypalon films lent themselves to a wide variety of fabrication techniques. They were put together by means of suitable adhesives, solvent welding, heat welding, and electronic bonding. These were the same methods by which early PVC films were fabricated.

The type of factory seaming done depends on four factors: (1) type of Hypalon, (2) compound, (3) processing techniques, and (4) age of the membrane. No flat statement concerning the best method of fabrication can be made until all of these factors have been thoroughly evaluated. The various manufacturers' products tend to be very selective in regard to factory and field seaming techniques.

The bulk of the Hypalon produced today for the lining trade is uncured and is effectively put together in the factory by means of a Hypalon solution applied to the fresh solvent washed seam area. The resulting joint is stronger than the base material and cannot be pulled apart after it has once set. Full strength is developed in about 24 hrs, although considerable strength is produced in less than 1 hr.

Lesser amounts of Hypalon lining membranes have been put together with adhesives,* hot bar (induction heated), and electronic bonding. These processes have experienced difficulties, probably because of the curing nature of the Hypalon, which renders it much less susceptible to adhesive attack than in the case of the Hypalon solution method.

There is no standard size for factory fabricated pieces, and most producers will make whatever size the customer needs. As a guide, pieces approaching 20,000 sq ft of 30 mil materials are produced quite regularly; 12,000 sq ft seems to be a more efficient size when dealing with 45 mil thickness. The former pieces would weigh 4400 lb, and the latter would run to 4000 lb, not counting the skid.

There is no factory fabrication of CPE lining materials at the point of manufacture. The calendered film is sent to custom fabricators, who make the narrow sheets into larger pieces for shipment to the installation contractor. Adhesives have generally been used for this work, along with electronic bonding techniques. One unusual property of CPE materials is that special formulations may be electronically bonded to PVC films.

When butyl rubber is fabricated into large sheets before the material arrives at the job site, this is done in the processing plant. Large sheet buildup is closely associated with the production operation, and this fact, combined with the heavy weight of the sheets, dictates that the work be done there rather than in a specialty fabricating shop, as is the case with PVC materials.

The initial sheet production comes from the calender in narrow widths, varying from 36 to about 60 in. Because it is sticky, a liner is used to separate the plies on the take-up roll. From this roll the sheet is pulled out and cut into strips of appropriate lengths. Their edges are then stuck together, utilizing a rubber solvent and the natural tack of the butyl.

*Adhesives are made using a variety of synthetic resins, tackifiers, and additives dissolved in a combination of solvents. These are in contrast to the Hypalon solutions, which contain only Hypalon and one solvent with minute amounts of special additives.

Since these seams are very weak until they are vulcanized, they are handled with extreme care. The large piece, again protected from itself by means of a liner or soapstone, is next vulcanized in a large curing room. The curing cycle may run as long as an hour, or even more, depending on the size of the fabricated sheet.

Another fabricating process starts with the cured sheet in the original calender widths. Large pieces may be built from these narrow widths by producing seams just as they would be made in the field, utilizing the manufacturer's recommended field seaming tapes and/or adhesives. One disadvantage of this method is that more care must be exercised in the handling and packaging of the large sheets for shipment to the field. As with field seaming, the freshly made joints are very weak for some time after fabrication, and this fact must be borne in mind during initial handling in the factory. The seam should not be subjected to undue stress in the shipping container, lest it pop open in some areas.

A third method of fabricating in the shop involves the use of a vulcanizing operation. Here the cured sheets are joined to each other by means of a vulcanizing press. The equipment consists of a large flatbed table with a hydraulically operated ram to give pressure to the joint during the vulcanizing operation. The process amounts to an old-fashioned inner tube patching operation but on a much larger scale. The press is equipped for heating to give the necessary curing temperature. Although the technique gives good results, it is slower and more costly than either of the other methods.

The factory fabricated pieces are put up on large rolls that may weigh over 1000 lb. The standard width for nonreinforced butyl sheeting varies with the thickness of the sheet. For 1/32 in. material 45 ft widths and 100 ft lengths are common, but for 1/16 in. material the width is kept at 20 ft or less, while the length can be 100 ft or more. Odd-shapes may be fabricated, thus often avoiding scrap generation in the field. The newer entries into the field have not announced their intention to produce large fabricated pieces in the factory.

The factory fabrication of neoprene lining material is almost identical with the butyl and EPDM processes just described. Neoprene sheets have stronger factory joints than do butyl or EPDM sheets because neoprene has better adhesive qualities than most other systems. In fact, neoprene, because of its excellent tack, is the base material for many industrial adhesive systems.

There is no factory fabrication for the 3110 lining membrane at the point of manufacture. All fabrication is done in the field, or the pieces are sent to another factory, where they are joined into larger sheets. Factory seaming is done by means of a hand-held heat welding machine, which will be discussed in more detail in the section entitled "3110, Elasticized Polyolefin."

Asphalt panels are not fabricated into large sheets in the factory because of their heavy weight and restrictions on the shipping of large pieces that cannot be rolled or folded to accommodate common carrier shipping regulations.

Reinforcement

Soon after the various lining membranes were introduced, they also became available with fabric reinforcement. This was not surprising, since the reinforcement of rubber goods as a means for enhancing certain physical characteristics of the finished product was a well-known practice.

Fabric reinforcement usually* increases the tensile strength of the lining. It also facilitates handling and seaming the membrane both in the factory and during field installation. However, these advantages, although obvious, are of secondary importance. Actually, the three most important reasons for the use of fabric in lining membranes are to provide:

1. Stability against puncture.
2. Stability against shrinkage.
3. Increased tear resistance.

Reinforcement of early linings utilized heavy, closely woven cotton fabrics. The technology of coating them had already been developed in connection with the manufacture of vinyl upholstery for the furniture and automotive trades. Initially, the fabrics were given a prime coating to effect adhesion to the PVC film until it was discovered that preheating the fabric would allow it to be combined with the film directly on the calender. Later, the process was improved by means of a high-speed laminator that preheated both film and fabric simultaneously and then applied controlled pressure. In the case of butyl-type membranes, the fabric prime coat continues to be the best method for combining, as it allows lower processing temperature before the curing cycle. The problem here is that nylon tends to embrittle during the high temperatures

*Fabric reinforcement does not always increase the tensile strength of a lining. In some cases a fabric could increase the tear resistance of a construction; but if it had less equivalent tensile strength than the membrane itself, it would not increase the tensile strength of the reinforced item.

of the cure; therefore, when it is used in the reinforcement fabric, the curing cycle must be carefully controlled to conserve the nylon's strength.

All types of fibers have been tried in the reinforcing fabrics, cotton, jute, nylon, polyester, and glass being the ones that commanded the most attention. It was soon found that cotton, jute, and glass fabrics presented difficulties because they were water sensitive and, in addition, could not resist the attack of soil microbes.

Cotton does a reasonably good job against sun exposure alone, but when in contact with soil organisms it is quickly attacked, even though it is coated for lining use. Depending on the type and integrity of the coating itself, the fabric deterioration rate is relatively fast. The construction usually loses most of its strength from reinforcement in less than 5 years. Copper and mercurial mildewcides do a good job of arresting the degradation process but are seldom used in the lining field because of their high cost. They are also toxic to man, a fact that rules them out for potable water lining use. In the case of jute and glass other considerations contributed to their lack of general acceptance.

Although nylon has some drawbacks, as mentioned above, it remains the most popular reinforcing fabric for lining membranes. Its principal attraction is its ready availability and its resistance to water solutions and soil organisms. Its disadvantages are its poor resistance to acid solutions and sunlight, and its lack of adhesion toward any of the membrane polymers. In regard to acid solutions, wicking can put acid into contact with the fibers. Depending on the type of acid, its concentration, and the temperature, the nylon can be destroyed in a matter of a few months. Although sunlight is also detrimental to nylon, the speed of deterioration depends on the weight of the fabric, the thickness of the coating, and its color. White and pastel coatings less than 10 mils in thickness allow too much ultraviolet radiation to pass through to the nylon, causing degradation of its physical properties. Loss of enough strength to render the reinforcement useless depends on the original tensile strength of the fabric and the magnitude of the stresses to which it is subjected. Under the high stress experienced by inflatable reservoir covers, the service life of a 5 to 6 oz/sq yd fabric reinforced construction is limited to about 10 years. For lining uses where stresses are lower, the life of a comparable item will be longer, although there are no data as to how much longer. Dacron (polyester) fabrics in similar use will last longer, but again there are not enough of them in service to allow meaningful predictions.

Since polyester fabrics exhibit good acid and sunlight resistance, they have challenged nylon's dominant position. Because polyester yarn is not as strong as nylon, however, constructions of equal strength are more costly. Like nylon, polyester yarn exhibits no adhesive qualities toward the lining membrane materials.

In the lining field PVC and PE films are traditionally sold on a nonreinforced basis, although on a few occasions the former has been reinforced with heavy nylon fabrics (5 to 6.5 oz/sq yd) for use as special-purpose one-piece drop-in linings for small vertical-wall tanks. Polyethylene film has been reinforced with lightweight, open-weave nylon fabrics for use in general construction work. Although some of the reinforced PE membrane found its way into the lining trade, it generally does not perform well in this use and manufacturers do not recommend it for this application. In this case both the membrane and the fabric are poor sun agers.

In early Hypalon and CPE reinforced membranes, adhesion was the major problem. The solution was a simple one after manufacturers learned that tear resistance and adhesion, not tensile strength, were the important parameters in reinforced lining technology. The adhesion problem was overcome by adhering the membrane to itself through the interstices of the fabric weave, a technique called strike-through. Originally troubles were encountered whenever the weave was too tight or the yarns were too heavy (or both) to permit development of adhesion by this means.

The argument over reinforcement versus nonreinforcement of thin flexible lining membranes has raged for several years. Some manufacturers produce one or the other, whereas some make both types. The position taken in this argument by each producer obviously depends on what he can do from a manufacturing standpoint. Good arguments can be advanced for both sides, but even the unbiased experts fail to agree on the best course to follow.

Du Pont's 3110 lining material is not sold on a reinforced basis, although experimental quantities have been made with reinforcement. In contrast, all asphalt panels are reinforced with glass or mineral fibers.

When concrete is used as a lining in a reservoir, it is usually reinforced. Here the reinforcement is steel, and its function is to prevent stress and curing cracks from developing in the lining. On slope areas the reinforcing bars also help the concrete to carry the tensile loadings more effectively.

Gunite linings, when reinforced, utilize wire mesh. Without this reinforcement Gunite may present problems on steep slopes. Although these linings are sometimes placed in this type of service on an unreinforced basis, moderation here will probably be repaid by greater serviceability and longer life.

Physical Testing

Each of the membrane manufacturers has a physical testing laboratory. These units are kept busy running tests on raw materials, on process control, and on final product quality. They also develop the data utilized in sales and marketing brochures and specifications.

The most common physical test run on reinforced lining membranes is tensile strength. Ironically, this is the least important of the physical properties associated with lining materials. Two types of tensile tests are run, the grab tensile and the strip tensile. Most specifications utilize the former because higher results are obtained, although the grab tensile test was actually developed for technical reasons.

When a 1 in. wide sample of coated fabric is placed between the 1 in. wide jaws of a (strip tensile) testing machine, a physical phenomenon takes place as force is applied to the sample. Because of the natural crimp of the warp (lengthwise) yarns, those on the edge of the sample tend to "pop out" to a relaxed condition, thus transferring the load to fewer yarns than are actually present in the 1 in. wide piece. The result is that the tensile strength is a correct measure, not of 1 in. of the fabric, but of something less than 1 in. Unfortunately, the number of yarns that will pop out is unpredictable. The type of fabric, yarn size, coating, and sample preparation are important variables in regard to what happens. Because of this technical problem, the grab tensile test was developed.

In the grab tensile test the sample width is 4 in., while the testing jaws are still 1 in. wide. "Pop-out" is eliminated, and the ultimate strength at break is a more natural one in that yarns outside the jaw edges also contribute to the fabric strength. It is obvious that grab values are higher than strip ones, but there is absolutely no way to predict how much higher. It is imperative that the specification values given by the manufacturer for tensile strength indicate which test was used. If no mention is made, it will probably be safe to assume that the grab tensile method was utilized. In both cases the rate of separation of the jaws is 12 in./min (approximately).

Because the unreinforced lining systems are so weak in tear resistance and because this parameter is so important with respect to lining performance, special problems occur in this area of testing. The biggest problem is that most manufacturers have avoided reporting tear values at all. Those that do report them often do not define the type of tear test used. This is an unfortunate oversight, because without test identification tear values are worthless. There are several basic tear tests, which give data with a considerable spread in values. Little or no relationship exists between the values produced by the various methods.

TABLE 3.1. Tear Test Comparator

Name of Test	ASTM Test No.	CCCt191b Test No.	Testing Head Speed (in/min)	Type of Material Tested[1]	Remarks
Trapezoid	D-751	5136	12	R	
Tongue	D-2261	5134	12	R	American National Standard Test No. L-14.207-1971
Elmendorf	D-1922	5132	[2]	U[3]	
Graves	D-624[4]		20	U	
	D-1004[5]		2	U	

[1]R = reinforced and U = unreinforced material.
[2]This is a shock test. A pendulum falls, and its energy tears the film and registers the tear resistance on a graduated scale.
[3]Test is occasionally used for very lightweight coated fabrics.
[4]For vulcanized rubber molded specimens.
[5]For plastic film and sheeting.

Table 3.1 lists some of the important aspects of the more common tear tests currently being run for evaluating membrane lining materials.

The trapezoid and tongue tear tests are both run on pendulum-type machines with the movable jaw traveling at 12 in./min. Both are run on reinforced materials, and the results are reported in pounds (or kilograms), as with tensile values. As a rule the trapezoid test will give appreciably higher results than the tongue test, but there are exceptions.

Nonreinforced film linings are run on the Elmendorf machine to determine their tear values. Results were originally reported as grams per mil (1 mil = 0.001 in.) until it was determined that the correlation between different thicknesses of material tested was poor. Now the results are reported in grams, and the thickness is stated.

The Graves method of testing is applied to thick premolded specimens but has been used also to measure the tear resistance of thinner membrane lining materials on some occasions. The results are reported as pounds per inch of thickness. Thus a sample 0.080 in. thick that tears at 32 lb will have a Graves tear of 400 lb/in. It should be remembered that this is the only common tear test for which results are reported as a direct function of the sample thickness. The Graves method is not used on reinforced materials, since they have no true thickness.

Elongation (at break) values are obtained at the same time that tensile strengths are run. They vary over a wide range, depending on whether the membrane is reinforced or nonreinforced. In the former case expected values fall in the range of 10 to 30%. Unreinforced membranes range anywhere from 100 to about 600%. Lining membrane performance on the job does not depend on elongation values, but is much more closely related to tear, especially tear in the vicinity of a nick or other imperfection in the film.

Low-temperature testing is another confusing area, in that two basic tests are run. They give values that vary by 20 to 50°F. Obviously, the tendency is to report on the cold-bend "mandrel" test, ASTM D-2136, as it gives lower values. Basically, the sample is allowed to come to equilibrium in the cold test box and then is bent around a ⅛ in. mandrel. The slower the bending rate, the lower will be the low-temperature value. The other, more meaningful test is based on impact of the sample in a loosely folded condition. Results are much more consistent, but the figures are much poorer (i.e., higher). The tests do not correlate.

Some manufacturers of the butyl family of membranes have consistently given the permeability ratings of their products, that is, their resistance to the passage of water vapor under controlled conditions. Since this test has nothing to do with the suitability of a membrane as a lining, it has gradually fallen from favor.

Three tests that are very important to the integrity of a lining system are seldom if ever run: (1) resistance to delamination between plies, (2) seam strengths in both "peel" and "shear," and (3) shrinkage. In the case of seam strengths material should be evaluated in fatigue tests in both hot and humid environments and also immersed.

Most membrane suppliers run a series of standard immersion tests in a representative variety of acid, alkali, and salt solutions at various concentrations and temperatures. A number of common oils (vegetable, animal, and mineral) are checked, together with a few solvents. These tests are indicators of performance, but the owner should also run them in his pit under actual operating conditions. Unfortunately, pit contents are usually somewhat unknown, so discretion must be exercised in these cases.

The ultimate success of the evaluation rests on the general integrity of the manufacturer and the lining contractor, and the willingness of the owner and engineer to utilize the services and experience of these specialists. The right combination here is worth more than the most impressive array of specification values printed on pretty paper.

Types

Polyvinyl Chloride and Polyethylene

It is appropriate to begin the discussion of lining systems with the plastic membrane materials. Of modern day impervious lining systems, they have accounted for the highest proportion of the business. Because of their relatively low cost, they enjoy a good popularity rating with the man who is responsible for holding down capital expenditures.

There are many different types of plastic sheet materials, and all of them have been tried at various times for the lining of reservoirs, canals, and related hydraulic facilities. The initial attraction here is that most plastics exhibit a high degree of water resistance coupled with a reasonable price tag. In the modern-day search for linings, these materials would seem to be an excellent place to start, although, as we shall see later, there is much more to a lining system than its water resistance and initial cost. In fact, with respect to the latter it is well to emphasize here that there is no such thing as a low-cost lining. If much area is involved, anything is expensive. In addition, the land, supporting structures, and appurtenances entail costs.

The search for plastic linings was stimulated by the profit motive. Manufacturers were attracted by the heavy publicity describing the market potential for these types of materials. Then, too, many of them had well-developed production capabilities, as they were already marketing similar products for other industries. Although every known type of plastic was evaluated for use as a lining, most were eventually discarded for one reason or another. Two have survived, and it is with these that we will begin our discussion.

The materials referred to are polyethylene (PE) and polyvinyl chloride (PVC). Although they differ considerably in their properties, many users do not distinguish between these two plastics. They are similar in appearance, and both are thermoplastic, which no doubt accounts for the fact that they are often placed on equal footing in lining specifications. Like any lining system, each has very specific advantages and disadvantages, and there are few situations for which both can be recommended.

In the preceding sections on resin production, processing, compounding, reinforcement, factory fabrication, and testing the two materials were discussed in considerable detail. It will be helpful, however, to summarize the important points as follows:

A look at this comparison chart will tell the user a great deal about both of these lining materials. It is obvious that they should not be used on critical facilities. Temporary installations, farm ponds, shallow lakes, and related facilities can best utilize these materials. With proper caution they may be given some tough use requirements; but the designer should give careful consideration to details of design and construction, and each step of the lining installation should be closely supervised.

Advantages

PVC	PE
Higher strength	Lowest first cost
Greater abrasion resistance	Lowest shipping weight
Good adhesive system	
Largest size pieces available	
Greater flexibility	
No thickness limitation	
Good crease resistance	

Disadvantages

PVC	PE
Poor sun aging	Poorest sun aging
Higher cost	No adhesive system
Higher shipping weight	Poor abrasion resistance
	Limited piece size
	Burial required to prevent "float out"
	Very stiff, except in thicknesses under 8 mils
	Poor crease resistance

Although the following installation comments will be helpful, an experienced lining contractor should be used.

Polyvinyl chloride linings are accordian pleated in both directions (double festooned) and leave the fabrication plant in wooden crates or heavy cardboard boxes or as specially wrapped packages. In all cases they should be placed on skids by the fabricator to facilitate easy handling at the job site. The top of the package should be marked "TOP," and an arrow should show the direction of the initial unloading. If the lining is a square, rectangle, equilateral triangle, or parallelogram, this is all that is needed provided that all the pieces are the same size and shape. If they are not, the various crates should be plainly marked as to what they contain. If irregular shapes are involved to fit a specific layout, clear, concise directions and sketches should be stenciled on each crate. The lining design should show a number or letter on each piece, and the crates should correspond to these markings.

The foregoing precautions are important, because a large piece of lining in the field is not identifiable until it is completely spread out. If at that time it is found to be reversed, the job of turning it over or around represents considerable lost motion and expense. Often the lining is marked at a certain point to identify this point with respect to the site surroundings. This is a particularly valuable practice in the case of an irregularly shaped lake lining.

Polyethylene linings are shipped to the job site directly from the processing plant. The factory fabrication step is eliminated for technical reasons already discussed. These linings are usually double folded and rolled on cores. They are seldom cut to shape except at the job site, so all shipped rolls are the same size. The widths of the rolls vary from 4 to 40 ft. Since this lining system is a stock item, the length is constant, usually 100 ft. Longer pieces may be run at the factory on special order.

The size of the PVC installation crew varies with the job but usually runs from 6 to 15 men and a foreman, depending on the nature of the installation. A greater number of people makes the spreading out of the large PVC pieces and their positioning much easier, although theoretically the entire job of installation could be accomplished with only two men, if there was no concern as to the total number of days involved. The crew size for a PE installation usually is smaller than that for a PVC job of the same size. The PE pieces are smaller and thinner and therefore weigh much less than the PVC panels.

After the first two pieces of film have been spread out, the crew immediately begins the field seaming process. The crew is allocated so that, by the time the first seam is finished, the second piece is spread out and available for field fabrication. It is imperative that all edges be straight before seaming commences; otherwise, excess material must be cut out later to prevent leakage at the junction points. During the sealing operation maximum effort must be directed toward development of a technique that does not stretch the film excessively at the sealing edges.

Seaming is begun at the center of each piece and is run in both directions to the end of each section. This assures that two seaming crews may operate simultaneously on the same seam; at the same time better quality seams result from starting in the middle of the piece. For PVC linings, adhesive is applied only to dry, clean surfaces and under zero or low wind velocity conditions. Field seams are generally overlapped from 6 to 12 in. to minimize dirt intrusion from beneath. The actual seam width made is no more than 2 in. (Figure 3.3). The adhesive has been applied by brush, spray, or extrusion through a nozzle. Each method demands a technique of its own to achieve optimum results. The spray technique is tricky and requires a special adhesive formula to prevent "cobwebbing." It is seldom used commercially.

The ability to effectively put PVC together in the field with adhesives comes primarily from experience. Most adhesive systems used for this purpose are fast, meaning that they lose their solvent and attain the

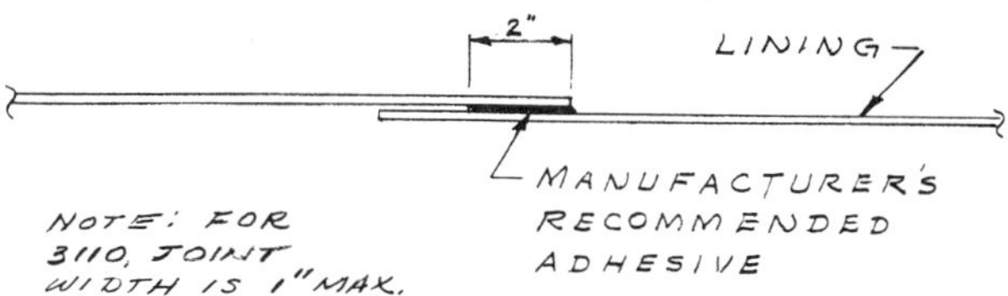

EPDM & BUTYL RUBBER:

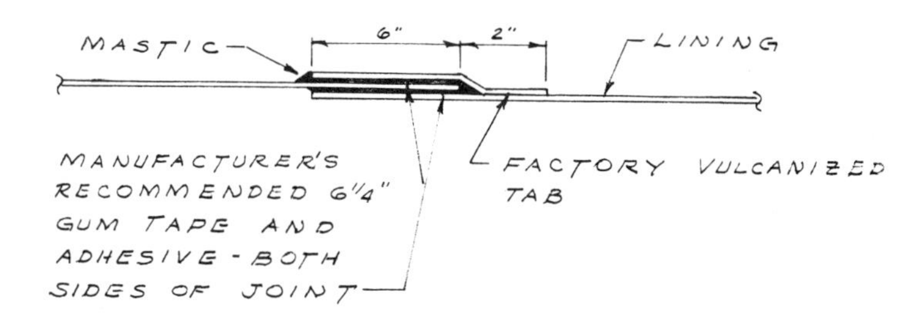

NEOPRENE:

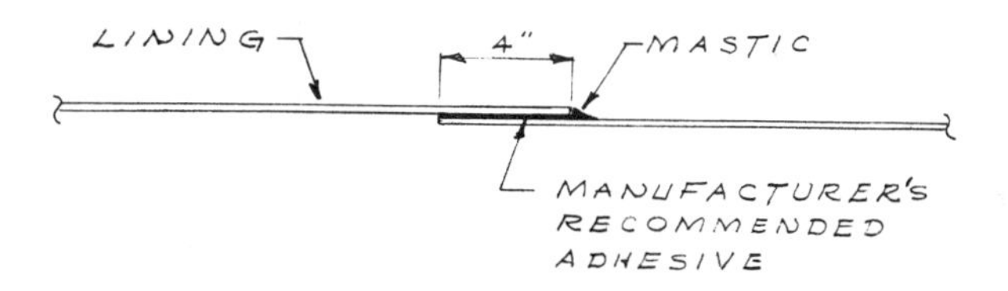

FIGURE 3.3. Typical field joint details (elastomeric and plastic membranes).

proper tack condition quickly. The bond must be made when the adhesive is at the right tack for optimum strength. As the adhesive hits the film, and the latter is attacked by the solvent, the thinner films (8 to 12 mils) tend to curl and make field seaming somewhat difficult. As the film thickness increases, this becomes less of a problem and field seaming is easier and less expensive. Films above 15 mils in thickness are much easier to handle in this respect, with 20 mil material being about optimum as far as adhesive bonding is concerned. The same adhesives applied in field seaming are also used to repair any mechanical damage that the film sustains during installation.

No really good adhesives have been developed to adhere the vinyl linings to concrete. The proven adhesive systems for this use are asphalt based and, since asphalt and PVC are not basically compatible, should not be employed in this case. Normally, the plasticizers are more soluble in asphalt than they are in the vinyl compound, so the latter is stiffened considerably because of such contact. Although this difficulty may be lessened by proper compounding, other properties will suffer as a result. The problem of sealing to concrete involves mechanical anchor systems, caulking compounds, gaskets, and field experience as to how each of these can be most effectively used. The same situation exists with respect to PE films, since no adhesive will stick permanently to them in underwater environments.

The effectiveness of the adhesive system is related in part to the age of the PVC film. It is generally found that the older and more weathered the vinyl film, the more difficult it is to utilize adhesives for repairs. The same is true with respect to the electronic bonding repair technique. In fact, the ability to repair a PVC lining by this method decreases rapidly with age. After the membrane has been sun aged for only 12 to 18 months, it can no longer be electronically bonded.

The reasons for the age-related repair problems are that PVC aging is accompanied by loss of plasticizer and chemical changes as well, both of which are irreversible. To a large extent, adhesive bonding and electronic bonding depend on the plasticizer–resin combination. This degradation of the film's repairability is an important point and one that detracts from the other good features of the material. Obviously, maintenance is required on any system, and hence the repair techniques should be effective regardless of the age of the lining.

Aged film may be repaired in some cases by treating the patch area with a slight amount of a strong solvent, such as MEK (methyl ethyl ketone),* followed by two or three adhesive prime coats.†After the primed area has dried, it may then be treated with the base adhesive, and the adhesive coated patch of the new material may be applied when both surfaces have reached the proper tack.

For PE linings the seaming procedure is somewhat different. Seams once made may be torn apart and remade, an impossibility with the PVC materials. Because the sticky tape or gumlike material at the joints is quite thick, perfect alignment is not as critical as with the vinyl film linings. There is some room for error, although it is still sound practice to try for good alignment at the joints.

*Acetone is a close chemical cousin of MEK, but it is not very effective as it evaporates too fast.

†The primer is made by diluting the adhesive about 1 : 1, using MEK solvent.

When working with PE materials, greater care must be exercised to prevent damage to the lining. Polyethylene does not have the strength or abrasion resistance of vinyl. The subgrade must be as smooth as possible and absolutely free of any sharp objects.

Wind is the principal enemy of any installation crew placing thin membrane lining systems. The secret is not to let the wind get under the film, and any crew will testify to the wisdom of this advice. There are several ways to cope with the wind, but the best and safest method is not to work under windy conditions, particularly when wind velocities exceed 20 mph. If work must proceed despite wind, the use of sand bags or (more effective) a continuous line of earth at the unsealed edges of sheets will help. The latter method is the minimum protection suggested for all exposed edges at the end of each day's work. Effective temporary protection will be obtained by turning over all exposed edges three or four times into a compound fold; an edge thus treated is much more stable to the forces of the wind than when otherwise exposed. As a good illustration of this principle, the first pullout of an accordian pleated piece is rarely disturbed by high wind if left exposed in this condition.

The edge folding technique just discussed is a natural process to some extent. When the wind blows, the edge is generally peeled back over itself one or more times. With each sequence of folding the film becomes more stable to the wind. As a result, there is usually a limit to the amount of film that can be damaged by the wind. The best policy is to play it safe at all times. Of course, when the wind approaches 30 mph, gusts may hit 40 mph or more, and the job is in real jeopardy if not shut down. Not only does the wind stir up dust, which is a problem in making quality seams, but also there is definite danger to personnel. The sail effect produced by wind acting upon a large piece of plastic is tremendous. If a workman hangs onto the edge of the film too long, he can literally be catapulted into a hospital. The film will do less damage and have less damage done to it if everyone lets go early in this game, even though it may take some time to retrieve a windblown piece.

Another technique that has been used to combat wind problems both during and after installation is to anchor a portion of the slope pieces at their edges running down the slope. If the wind does get beneath one of these sections, only the material in that section is affected. This system is effective for the completed job, whereas sandbagging cannot be counted on to be permanent or completely effective. Of course, if the wind reaches a high enough velocity, Mother Nature will win in spite of anything that can be done, with the exception of complete burial of the lining beneath a foot of soil. If soil burial is to be done, this step should be taken as soon as possible after each piece is seamed.

The point at which to begin the lining work depends on the layout, the lining quantities on the slope and bottom, and the problems presented by wind and rain. Usually, the lining is begun at the top of the slope, because there is an anchor point there. The anchor system most commonly used is a 12 to 16 in. deep trench (Figures 10.4 and 10.5), cut by a trencher or by tilting the blade on a bulldozer or road patrol (grader). The film is folded into the anchor trench so that at least 8 or 10 in. lies out on the bottom of the trench. The use of boards or pipe in the trench is unnecessary and serves only to increase the job cost. The film cannot be pulled out if 1 ft of compacted earth is backfilled into the trench.

Hypalon

Hypalon is one of the newest members of the plastics family to find use in the lining field. Although the polymer was developed in the early 1940s by the du Pont Company, it did not attain any degree of publicity as a lining until 1969, almost 7 years after its first application for this use.

The material defies accurate classification, being both a plastic and an elastomer, depending on when the classification is made. Here Hypalon is placed in the plastics family because its original properties and the handling techniques used with it resemble those of the thermoplastics.

As Hypalon leaves the processing plant, it is for all intents and purposes a plastic material, with thermoplastic properties. After exposure to light, heat, or certain liquid chemicals, however, it slowly cures into an elastomerlike material, less sensitive to heat than when it was first produced. Since the curing cycle is incomplete, the best way to describe Hypalon is as a material that is not a plastic or an elastomer, but something in between. The natural curing cycle, in the early stages at least, seems to act in a manner somewhat like the case-hardening of metal, that is, the curing or hardening of the polymer proceeds from the outside inward. Present evidence indicates that the cure rate is retarded as curing proceeds. Some manufacturers indicate that the cure is probably not more than 20 to 30% complete, regardless of the age of the compound.

The resulting aged membrane will then have two sets of properties, one that corresponds to the outer, cured shell, and one that applies to the inner portion that has not cured. The properties of the complete membrane are the sum effect of both portions of the sheet.

Hypalon is said to possess superior aging capabilities in outdoor exposure, exceeding the butyls, Neoprene, 3110, EPDM, PVC, or PE in this respect. It is resistant to about the same chemicals as butyl and, similiarly, is attacked by many of the ones that attack butyl, although it possesses better oil resistance. In general, it should be kept away from oxidizing acids (a good idea for any of the lining systems discussed in this book), aromatic hydrocarbons, chlorinated solvents, turpentine, lacquer solvents (ketones), acetates, and asphalt under certain conditions. Crude oils have been used against Hypalon, but they certainly should not contain any aromatic components. Even so, the suppliers do not recommend Hypalon in this service.

Hypalon's tendency to cure upon aging presents some field problems for the lining contractor. He must arrange to have the rolls stored in a warehouse out of the sunlight. If this is not possible, the rolls may be covered with white plastic to shield against premature curing by ultraviolet rays from the sun. As the cure progresses, the sensitivity of the surface to solvents and adhesives of the solvent-attacking type decreases. The manufacturer's recommendations concerning alternative adhesive formulations must be followed, should this become a problem.

Initial polymer offerings were said to be electronically sealable, but later it was found that the resins with the most desirable properties as a lining did not respond to the electronic bar treatment. Work continues in this direction, however, so future compounds may possess this desirable attribute.

When first introduced as a lining membrane, Hypalon got off to a disastrous start, failing three important criteria for successful lining systems. First, pinholes in the material from the processing stage could not be adequately controlled; second, shrinkage—sometimes on the order of 30 to 40%—created many problems; and third, factory and field joints possessed poor integrity.

The only sure way that the industry has found to control pinholes is by lamination. For example, a 30 mil sheet may be produced by laminating two 15 mil sheets together. The laminating process represented a major improvement. All of the first unsupported membranes carried pinhole allowances in their specifications, and no lining user would want to buy material with a built-in problem like pinholes.

For effective control of shrinkage, a reinforcing fabric can be sandwiched between the two plies during the laminating process. Although this technique never completely eliminates shrinkage, it does prevent most of it. Depending on the nature of the fabric used, shrinkage may be reduced to values below 2%. In most lining work this may be a satisfactory level.

When the membrane is laid against bare earth, the tendency to shrink is greatly diminished because of the introduction of friction. The membrane is actually stabilized by its direct contact with the earth. Contributing to the reduced shrinkage is the adhesive system used; it must be quick to develop a tough, tenacious bond. Under these conditions it actually serves to stabilize the material in that joints will not fail under shrinkage tensions. In early Hypalon membranes joint failures occurred with frequent regularity.

Black Hypalon has a greater tendency to shrink than colored membranes, white being the best in the low-shrinkage category. Light blue and light green also are better than black, as would be suspected.

According to manufacturers' original bulletins, Hypalon film is recommended for service at temperatures up to 150° F when first installed. As curing progresses, this value may be increased to as high as 185° F. Again, as with PVC temperature ratings supplied in manufacturers' bulletins, it would be well to reduce this value by 20° F until more documented experience is available.

There are several schools of thought as to how to field-fabricate Hypalon membranes. Each manufacturer has developed his own particular system, much as in the case of the butyl membranes. Some variations have also been introduced by those who have made installations.

Electronic and induction heated bars are not feasible for field use at this time. Moreover, heat welding techniques are at a disadvantage because it is hard to develop and control the required 250° to 300° F temperatures in the field, not to mention the logistics of handling the heavy equipment. One way out of this dilemma is to paint the mating joint areas with trichloroethylene solvent. This will then decrease the heat required for joint fusion down to about 135° F for about 5 secs. Such values can be obtained rather easily in the field with lamps or heat guns, and the necessary pressure can be supplied by hand rolling the film joint on a rigid backup board.

The introduction of Hypalon linings was accompanied by a zipper-type method of field fabrication. This was not the first time that this type of technique had been used. Although it was pushed heavily by the resin suppliers in the early days of Hypalon, installations today do not use it because it does not seem to offer the technical and economical advantages first envisioned. This is due in part to alignment problems, complicated valley seal layouts, and subgrade surface irregularities, all of which complicate the installation of a fixed seam system.

Hypalon's erratic behavior as a result of semicured materials being taken into the field has produced considerable confusion with respect to the guarantee. Originally, the 15 year guarantee was widely publicized as one of the merchandising features of the material; but as time went on, this aspect became clouded with uncertainty. Some manufacturers sold the straight lining with no guarantee at all, while others preferred to handle the warranty problem on an individual job basis. Even then, the warranty was differently worded from the forms previously used for butyl rubber. Later, Hypalon membranes were advertised in conjunction with a carefully worded 20 year warranty, and this seems to be the present trend in the industry.

A photograph of a 45 mil Hypalon lining being installed in a large eastern potable water reservoir is shown in Figure 3.4. Figure 3.5 shows the completed installation.

Chlorinated Polyethylene

Chlorinated polyethylene, commonly referred to as CPE, was introduced commercially in January 1965 for the lining of a brine pit in Oklahoma. It was Dow Chemical Company's answer to the vinyl and butyl-type lining materials. The feeling was that CPE would fit between these two systems costwise, and at the same time offer the designer a lining with some selected properties that were better than those of either of the two competitive materials.

The first installation was made of unreinforced 20 mil material, but since that time experience has indictated that a 30 mil minimum thick-

FIGURE 3.4. Photograph at start of installation of a 45 mil, five-ply Hypalon lining of 600,000 sq ft potable water reservoir in northern New York.

FIGURE 3.5. The 125 million gallon reservoir of Figure 3.4 lined and in service.

ness is preferable for lining service. Reinforcement was added; and until it was later removed for technical reasons, reinforced CPE was a principal competitor of reinforced Hypalon linings. With the introduction of CPE the design engineer working in the lining field had three materials that competed for this market: CPE, Hypalon, and EPDM (butyl family). The unsupported CPE membrane is more expensive than nonreinforced PVC and thus did not compete directly with this material.

Chlorinated polyethylene is strictly a plastic lining material, that is, it is thermoplastic, softening with heat and stiffening with low temperature. Unlike current Hypalon formulations, CPE may be electronically bonded, either to itself or to PVC. The latter capability has been used to advantage in combination linings with PVC on the bottom and CPE on the slopes.

The CPE basic compound has good aging characteristics and therefore was initially sold with a 20 year guarantee against degradation in open exposure. Later the guarantee was modified and eventually removed because of technical considerations, although research and development efforts to resolve the difficulties are continuing so that warranties may be reinstated.

Initially, CPE was put together with a solvent mixture containing toluene and tetrahydrofurane. Since straight polyethylene is not affected by these chemicals, it is the introduction of the chlorine molecule into the structure that renders CPE susceptible to attack by these solvents. Actually, CPE, like PE, has considerable resistance to toluene, so that the addition of tetrahydrofurane, a more powerful solvent, is required to effect a bond.

There is a difficulty with trying to bond materials with solvent alone. This technique was studied for some years in the case of PVC. Despite PVC's greater vulnerability to attack by the action of such solvents as MEK (methyl ethyl ketone, another powerful solvent), this technique never got into high-volume production. Even the use of solvent with the addition of direct heat could not produce high-integrity seams, and the process is now extinct.

As will be recalled, PVC is a very poor outdoor aging material. The resin itself is a hard material with no ductility, and hence it has to be plasticized to render it into a usable, soft film. Polyethylene, too, is a hard resin that canot be utilized in films over 8 mils in thickness because it is too stiff and cannot be plasticized. Yet CPE, without the need for plasticizers, can be made into a flexible film with excellent sun aging properties. The secret, of course, involves some complicated chemistry which will not be discussed here.

There have also been some problems with this new material in two areas: (1) seam integrity, and (2) delamination between plies of the film and the reinforcing nylon scrim fabrics. Since the seam problems have involved adhesives, CPE has apparently inherited some of the problems noted in PE. It will be recalled that PE has no adhesive system based on film attack.

"Delamination" is another way to say "inertness." Polyethylene has already been discussed in this light; it is so inert that nothing wants to stick to it. In one sense inertness is desirable in a lining system. Nothing will attack a lining that is perfectly inert. On the other hand, neither will adhesive systems. The other possibility is electronic bonding or heat welding. With PE, the former is not technically possible and the latter is so difficult as to make its use unacceptable. The reason why heat welding is difficult is that PE, in contrast to PVC, has a very sharp melting point. Even with PVC, heat welding of thin films is difficult and is seldom done because of the great amount of skill required. The situation is almost the same as welding thin pieces of metal—it is just too difficult to control the heat flow.

In most initial CPE film applications adhesives have been utilized. Although no reason has been given, it may have something to do with the compounding of the material or the type of resin being used. The term, "chlorinated polyethylene" does not indicate how much chlorination is used, and we already know that the amount is important. In short, the less the chlorination, the closer the polymer will approximate the properties of PE. Hence there is a decided skill involved in determining just how to produce by this means a film that will have all of the desirable properties of PVC-type materials without all of their disadvantages.

Chlorinated polyethylene shows good resistance to burning because of its chlorine content and its lack of plasticizer (which burns well). It performs about the same as Hypalon insofar as chemicals are concerned; strong solvents, including the aromatics, affect it just as they do Hypalon. Although the manufacturer's bulletins list service at 158° F as satisfactory, this seems to be a gray area. If such high temperatures are contemplated, it will be well to do some careful checking before proceeding with any large-scale installations.

The manufacturer stresses the importance of reinforced CPE linings on slopes steeper than 2 : 1, and 30 mil unsupported material is the thinnest recommended even on 3 : 1 slopes.

Butyl Rubber, Ethylene Propylene Terpolymer, and Ethylene Proplyene Diene Monomer

After the introduction of polyvinyl chloride and polyethylene plastic materials, there was no significant change in the flexible membrane lining field for about 15 years, until butyl rubber made its appearance. This was the first elastomer to gain prominence as a lining. Butyl was not really a new material, having been developed during World War II and used extensively since then in the production of inner tubes for automobile tires. Its principal advantage in that application was its resistance to the passage of air. As tubeless tires came onto the scene, the basic raw butyl supplier (The Enjay Chemical Company) began searching for new markets that would adsorb production lost to the new automotive trend. Part of its attention focused on the lining field. A comprehensive market survey was made, and from it tremendous forecasts were developed as to the potential usage for butyl rubber sheeting as a lining in canals and reservoirs.

From the survey data Enjay succeeded in interesting manufacturers in going into the production of butyl rubber sheeting. Slowly at first, and then with gradually increasing momentum, rubber processing firms tooled up for this lucrative enterprise. But the business did not develop according to plan.

This was not the first time in the history of linings that optimistic forecasts were made, only to be followed by a disappointing reality. The vinyl industry had already experienced this lesson, and even earlier, the asphalt panel manufacturers had been disappointed in the way that market developed. This situation resulted from the wide disparity between potential lining business and that which was actually budgeted. For one thing the forecasters placed heavy emphasis on agricultural lining markets, and these simply did not materialize. Other reasons, some associated with the psychology of linings, are discussed in Chapter 6.

The first reported large commercial use of butyl rubber lining materials occurred in 1960 despite the fact that initial polymerization took place in 1937. Part of the delay was due to the war, which absorbed all available rubber materials for the defense effort. The initial lining was described as 3/32 in. in thickness and was furnished to the job site in sheets 20 ft wide by 100 ft long. These sheets were assembled in the field, using a combination of butyl gum tape and adhesive to effect 6 in. wide seams. Individual sheets weighed over 1300 lb and were quite expensive, but the most startling thing about this new material was the guarantee. The manufacturer that introduced the material as a lining, Carlisle Tire and Rubber Division of the Carlisle Corporation in Carlisle, Pennsylvania, guaranteed the material on a non pro rata basis for 15 years against degradation in open exposure. To be sure, this fact attracted considerable interest throughout the country, although the high price dampened the enthusiasm of buyers for some time. The first job was for the Humble Oil Company at Hull-Daisetta, Texas; it was still in excellent condition more than 15 years later.

The announcement of the Hull-Daisetta installation was followed a short time later by another news release, this time from Houston, Texas. There ⅛ in. thick butyl rubber was used to encase the outside foundation of the Humble Oil Building against water intrusion. At the building site groundwater was several feet above the bottom of the foundations. Such a situation had always caused design engineers great concern because keeping groundwater out of basements under these conditions has never been an easy job. The installation was very effective, however, and at last report it too was doing an excellent job.

From these two beginnings, the new lining (Figures 3.6 and 3.7) made great strides, replacing much concrete, asphalt concrete, and asphaltic-type materials from the prime lining markets. This occurred in spite of the fact that the early butyl linings were more expensive than either of

FIGURE 3.6. Two LPG pits in Texas, lined with 1/16 in. reinforced butyl rubber and protected with two coats of a special oil-resistant, reflective aluminum coating system.

the asphaltic lining systems. They offered little competition to the plastic linings, however, because there was such a wide price differential between the two materials.

Since Carlisle's entrance into the butyl rubber lining field in 1960, several other manufacturers have come into the picture; U. S. Rubber in 1961, followed by Goodyear (1963), Hodgman Rubber* (1966), and Reeves Brothers (1970). Other manufacturers have also discussed plans for formal entry into the field, but so far no one has made the initial step.

Unlike rubber linings, which had enjoyed considerable usage in the

*Hodgman was purchased by Plymouth Rubber Company in 1969.

FIGURE 3.7. Two 1/16 in. reinforced EPDM lined pits for waste chemical treatment at a mining operation in Arizona.

chemical industry, the butyl rubber lining was said to resist the action of ozone. This fact, more than any other, made possible the 15 year guarantee. Ozone had always been a serious problem for natural rubber materials and accounted in large part for the degradation of natural rubber tires in the days before synthetic rubber materials made their appearance. There is no doubt that the liberal time guarantee did much to launch the successful endeavors to market butyl linings.

Because of its cured nature, butyl sheeting had a decided advantage over PVC and PE plastic linings, which stiffen as the temperature drops and become progressively softer and weaker as the temperature increases. Normally, these plastics are not recommended for use above 140°F, whereas thermoset (cured) butyl rubber may be used at temperatures ranging from −50 to 200° F or above. In this range butyl suffers no appreciable degradation of its flexibility or strength characteristics.

Butyl rubber is classed as an elastomeric material. Although it is rather dead in comparison to natural rubber, it is more alive than the plastics. Plastics, like elastomers, have a memory effect and tend to return to their original dimensions when external forces are removed. For plastics the return rate is quite slow, however, in comparison to natural rubber materials, which "snap back" rather quickly when external forces are removed. Butyl lies between these two extremes.

The 3/32 and ⅛ in. butyl linings did not gain markets quickly because of the cost problem. These materials, on an installed basis, often cost more than concrete. Of course, the butyl lining is much more impervious than concrete, but the race is not always to the swiftest! Concrete technology is well taught in most engineering schools, and it has a long record of experience behind it. New engineers out of college tend to favor that with which they are most familiar. In addition, they do not have strong gambling instincts, so they will go with proven materials. To combat these problems, the butyl lining thickness was gradually reduced to the 1/32 to 1/16 inch range. Here butyl competed with concrete on better than even terms.

One of the properties of butyl rubber that was emphasized by the early suppliers was its excellent impermeability, which results from its density. The term applies to the tightly packed molecular arrangement of the material, not its weight per unit volume. This density effectively impedes the movement of moisture and moisture vapor through the film. Early publicity extolled the importance of low permeance in the waterproofing field. Through confusion and misunderstanding, users began to associate low permeance with good performance in a reservoir lining. In the field of packaging, to which waterproofing is closely related, the importance of low moisture vapor transmission* rates is obvious. In packaging, the driving force for the movement of water vapor through a membrane is dependent on a differential in humidity (or partial pressure of water vapor) between the two sides of the membrane. In the case of a reservoir lining, this differential is not a significant driving force; rather, it is the head (or pressure) of water that wants to force the contents out to an area of lower energy potential. Many more important factors than MVT determine the effectiveness of a lining material.

Butyl's excellent impermeability has one distinct drawback: it is so impermeable that even the solvent systems of the adhesives do not attack it! The first entry into the field chose to solve the problem by use of an

*Commonly called MVT and reported as grams/100 sq in./24 hr/mil. The test is run under controlled temperature and humidity conditions.

intermediary material, a 6 in. unvulcanized butyl gum tape. A single-component adhesive was applied to the rubber sheeting, and when it had reached the proper tack, the tape was then laid down in this adhesive strip. The tape was supplied with a PE backing to keep it from sticking to itself in the roll.

Adhesive was next applied to the mating piece of lining; when it had reached the proper degree of tackiness, the gum tape backing material was removed and the two pieces were joined together. A considerable amount of skill is required to make a perfect joint. Two cardinal rules apply to the procedure: (1) the sheet edges to be joined should be perfectly aligned before making the seam, and (2) the edges being joined should not be stretched during this operation. The joint is made simultaneously, that is, all 100 ft of the seam is readied and the entire closure is effected at essentially the same time. When properly done, all 100 ft of the joint will fall into place at the same time; the joint literally "makes itself." As one butyl lining veteran once said, "You don't tell the butyl where it should be; it tells you!"

Not only does the quality of the field seam depend on the skill and experience of the crew, but also the design of the facility is an important contributor. Three principal factors exert influence on seam quality:

1. The side slope.
2. The surface texture of the substrate.
3. The substrate stability.

The side slope ratio plays a peculiar role in seam quality. If the slope is flat, the seam is physically easier to make, and an easily made seam is usually a better seam. As the slope increases in steepness, the seam becomes more difficult to put together. This is due to the problem of keeping the lining on the slope and also to the tendency of the worker to stretch the seam because of his weight. A seam stretched during fabrication will almost always result in poor-quality work at that point.

If the slope increases in steepness to the point where a scaffold may be utilized effectively, the seam quality may improve over that attainable on, say, a 1:1 slope (45°).

The surface texture of the substrate is extremely important. It is not possible to do top-quality seam work on a surface that is uneven, with deep footprints or eroded gullies, or one that is covered with rocks or other debris. The best practice is to rid the surface of these irregularities and then roll it smooth.

Soil stability is important too. If a man's weight will cause his foot to sink into the soil surface, this will produce stressing at points too close to the seam, at a time when the seam should be in a relaxed position.

To build as much safety factor as possible into the seam, the Carlisle Corporation went to a 6 in. joint. This was quite a departure from the maximum 2 in. seam widths being employed for the thin plastic membrane linings. The same company also introduced the 6 in. factory tongue-and-groove joint (see Figure 3.3). As the name implies, a groove is formed at the sheet edges to be seamed, by using an 8 in. flap and vulcanizing a 2 in. edge of this material onto the base sheet so that the free edge of the flap is flush with the edge of the sheet itself. By such a procedure a double-sealed joint is produced in the field, which forces leakage water to seek an exit through a 12 in. path instead of a 6 in. one.

The uncured gum tape for the joint is 6¼ in. wide. Seaming instructions indicate that field joints are to be made so that the tape is exposed at all edges. This prohibits an overhang condition, which can cause an edge to work loose because of wind or moving water. Sometimes the edge of the plain 6 in. lap joint is overlaid with gum tape or sealed with mastic to prevent the possibility of an edge peeling back for any reason. The top lay of gum tape is usually 2 or 3 in. in width. If more reinforcement is required, some of the cured butyl, in a narrower width than the tape, may be sealed onto the top surface of this overlay tape to make what is, in effect, a field fabricated tongue-and-groove joint. Usually, gum tape in exposed areas is covered by cured material, as the gum tape does not age well in the sun.

As the gum tape ages in the joint, it is said by the manufacturer to become vulcanized as part of the original sheet, although it has an indefinite shelf life when not in contact with cured butyl. The gum tape system is still in use by Carlisle today.

One defect that constitutes a continual problem with any butyl-type field seaming technique is "fish mouths." These result when one piece of butyl lining is stressed at the joint during seaming more than the mating piece. This defect may also be the result of too liberal amounts of adhesive, unevenness of application, or a combination of these effects. In any event there is formed at the seam a hump in one of the butyl pieces, so that it does not lie flat against the mating piece. This effect sometimes occurs immediately after the seam is made, or it can occur as a "pop-up" several hours afterwards. If the latter occurs, the hump will never lie flat on a permanent basis. The joint is defective, and the hump must be cut out and a repair patch applied over the cut piece. Since this type of defect increases the cost of the lining, it is best to prevent its occurrence by using the procedures outlined earlier.

As newer entries appeared on the horizon, so too did newer field seaming techniques. United States Rubber introduced the two-component adhesive system that originally used no gum tape. Later gum tape became a part of the company's system. Both Goodyear and Hodgman came out with a one-component adhesive requiring no gum tape and then, along with U. S. Rubber, cut the width of the joint to less than 4 in. Originally U.S. Rubber featured joints from 2 to 4 in. in width, but later increased the minimum width to 3 in. and the maximum to 6 in. A problem encountered in the field dictates seam width to some degree. It is difficult in an earthen reservoir to run long seams and still maintain 2 in. of width at all points, because of the overall roughness of the terrain. To assure a minimum 2 in. lap throughout, some portions of the joint would have to be made wider than this value.

Butyl rubber has presented difficulties in connection with seals to concrete structures, but it is certainly not alone in this regard. At the present time it is recommended that joints of this nature be backed up with a mechanical anchor system of some type, and that mastic materials be applied at the termination points of the lining.

One manufacturer has produced a special high-tack gum tape that may be placed directly onto the clean, dry concrete without the need of primers. The butyl lining will also stick to this tape directly, with no adhesive application required. Instructions indicate, however, that a mechanical anchor system should also be made a part of the joint, so the problem still awaits the ideal solution.

A typical piece layout for butyl rubber or EPDM linings is shown in Figure 3.8. The plans for the layout were originally developed for prefabricated asphalt panel lining systems in the mid-1950s. Later, the newer membrane products copied these layout designs, which are still in use today. Highly ozone-resistant flexible linings sometimes eliminate valley seams and allow folds to develop at these locations; Hypalon, CPE, 3110, and PVC systems often utilize this technique to reduce costs.

Neoprene

Neoprene, introduced by du Pont in 1932, was developed as a synthetic material having good resistance to oil, gasoline, heat, light, and ozone.

The use of neoprene as a lining in hydraulic facilities has been slow to develop because of its high cost and the lack of an urgent need for its properties. As a result of the cost problem it is used only in places where conditions dictate the need for its special properties. Early installations utilizing neoprene were brine holding ponds associated with oil drilling

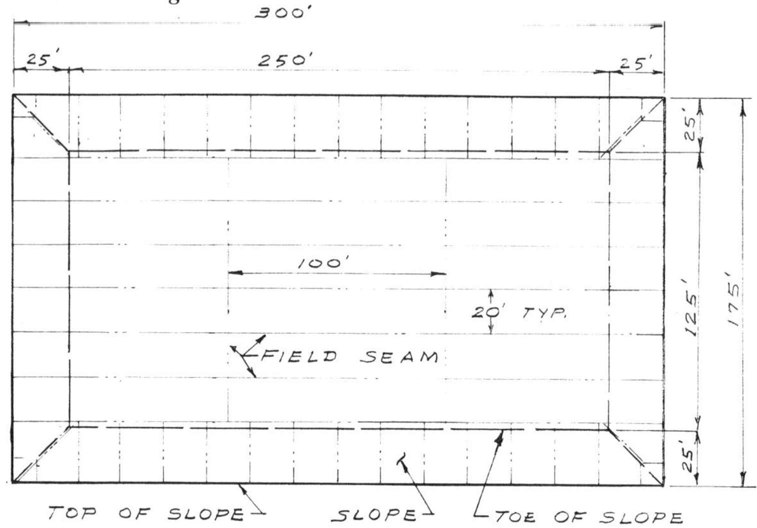

FIGURE 3.8. General layout–plan for butyl, EPDM, and neoprene linings.

operations, LPG* storage facilities, and water flooding projects (Figure 3.9).†

As pollution control activities intensified, neoprene was pressed into service as a lining for crude oil and oil residue pits. At first it appeared that this application would be ideal for the material, but since these early installations interest in this lining for crude oil has lessened. The material has poor sun aging, guaranteed adhesives could not be developed, and the resistance of neoprene in sheet form to crude oil was not as good

*LPG stands for low pressure gas. Brine is pumped into underground storage areas to force the gas out. When the storage facilities are full of gas, the brine is held in cut-and-fill reservoirs, which generally need to be lined.

†These projects store brine in lined pits. It is pumped from there into low-grade oil strata to force out oil for subsequent recovery.

FIGURE 3.9. A 1/16 in. reinforced neoprene lining job under construction in an oil refinery. The pit will hold brine contaminated with trace amounts of oil.

as originally thought. This came as somewhat of a surprise, as neoprene has had a long and successful history as a gasket material in oil pipelines.

As with other oil-resistant membrane systems, the truly difficult problem is to find an excellent oil-resistant adhesive. Resin processors and producers are quick to proclaim the virtues of an oil-resistant sheeting but are not quite so enthusiastic in their endorsement of an adhesive that will resist the action of oil over the long pull. This fact, along with the cost problem already mentioned, has helped to slow neoprene's penetration into the lining field. A lack of confidence by the manufacturer is reflected in the limited guarantee of only 5 years, compared to the 15 years offered by the butyl supplier.

Sun aging constitutes another question mark in the case of the chloroprene polymer. A considerable amount of outdoor exposure experience is based on thick sections above ground. When thinner sheeting materials are laid directly on the ground in desert locations, the aging process is much more marked.

A good example of this type of behavior may be found in the vinyl (PVC) extrusions used in aluminum prefabricated window assemblies. Some of these installations have been in service for over 15 years and are still in excellent condition. The same PVC compound in a thin film on the desert floor, however, would not last for 3 years. In the window frame application a thick section is used. It is also protected on three sides by a metal that blocks out the sun's rays and effectively conducts away any heat they generate.

Neoprene in oil pit lining installations has shown visible effects of the immersion, manifested as a swelling accompanied by softening and puckering, all occurring in less than 1 year of service. The installations are too new to determine whether the effect is passive or continuing.

In using neoprene as a pit lining membrane, the manufacturer cautions that difficulties will probably be encountered if aromatics* or chlorinated solvents* are present. Neoprene is severely attacked by these chemicals.

The lining is usually shipped to the job site in 20 ft wide pieces, 100 ft long. Neoprene rubber is a good adhesive base, and the lining is joined to itself without the need for a gum tape. Joint widths are recommended at 4 in. as in Figure 3.3.

As can be seen from the data of Figure 6.2, neoprene is considerably more expensive than any of the butyl rubber family of linings.

3110, Elasticized Polyolefin

During the years of lining development, the one request most often heard from engineers and consultants was for a lining that could be put together in the field by means of some type of heat welding procedure. In the mid-1970s E. I. du Pont Company introduced a lining system that seemed to answer this need. They named the product, "3110," and at this writing some 50 installations have been made (Figure 3.10).

The lining is produced in 0.020 in. thickness and one size of sheet, 20× 200 ft. It is the first lining membrane system that can boast the inclusion of a hand-held heat welding machine as part of the system. The machine is leased to the lining contractor on a weekly basis. It has a top seaming

*The term "aromatics" refers to such things as benzene, toluene, nitrobenzene, naphthalene, dioctylphthalate, and other benzene ring structures. Chlorinated solvents include carbon tetrachloride, chloroform, and trichloroethylene (a cleaning solvent).

Note that dioctylphthalate (DOP), the primary plasticizer for PVC resins, is an aromatic material. Close proximity of PVC lining to neoprene materials, especially in the presence of heat and pressure, could cause trouble because of incompatibility.

FIGURE 3.10. A 3110 lining installation in a potable water reservoir in New Mexico. Note the background equipment for controlling the hand-held automatic seam welding equipment.

speed of 20 ft/min, although most projects dictate lesser seaming speeds for best efficiency. Since the seaming machine is electrically driven, a power source, either 120 V/ac or a 1000 W generator, should be available at the job site.

There are some unique features of the seaming equipment. Perhaps the most remarkable is its ability to produce an excellent joint even though both edges to be seamed are smeared with wet clay just before seaming. Seams may also be made in a light drizzle, and machine adjustments permit field seaming to proceed during a wide range of ambient air temperatures.

The material has some degree of oil resistance but is not generally recommended for use in this type of environment. Its chemical resistance appears to be better than that of any other lining system in general use today, these linings having held such substances as hydrofluoric acid, carbon tetrachloride, and trichloroethylene in varying low concentrations. Despite these credentials based on case histories, it would be well to run careful evaluations and immersion tests for each particular application.

The 3110 lining is said to have excellent resistance to sun aging, and no protective sun cover is recommended. Nevertheless, the material is sold without a guarantee. This is a departure from tradition in connection with lining systems introduced within the past 15 years. Such a policy is feasible if the in-place lining and performance costs are low enough, a fact that can be verified only after the system has been on the market for a few years.

Seals to structures are made utilizing two-part epoxy-type resin systems, usually in conjunction with the common mechanical anchoring devices.

Repair procedures are somewhat complex and involve the use of flameless hot-air guns, high-temperature-resistant transparent Kapton film (from the du Pont Company), Perclene/Triclene solvents (from Dow Chemical Company), seam rollers, and related materials. The existing material to be repaired is first heat welded to some cardboard placed beneath the lining. The repair patch is laid over the damaged area, overlaid with the Kapton film, and heat is applied from the flameless heat gun. Film softening is accompanied by application of pressure, using the seam roller.

The manufacturer recommends that 3110 material be placed on slopes of 2 : 1 or flatter and that solution temperature be no more than 160° F. The installation work can be done at any temperature, although experience has shown that an ambient temperature of 60 to 85° F is preferred.

The du Pont organization restricts the sales of its 3110 material to qualified and experienced lining contractors, following a developing trend in this direction on the part of most reputable manufacturers.

Prefabricated Asphalt Panels

Entering the decade of the 1950s, engineers were forced to choose from a limited list of materials for the control of seepage from water holding and conveyance systems. By then, wood was beginning to drop out of the picture. Concrete, Gunite, and steel seemed to be the acceptable materials, along with the compacted clay lining. The first three, in reality, were structural materials but were being pressed into duty as linings because the designers had no other choice.

Asphalt had been tried many times as a lining, originally in the form of asphalt concrete and later as a built-up membrane similar to that employed on roof decks. When such materials for roofs were laid on the ground, the felts were attacked by soil microorganisms and soon these linings lacked the necessary structural support. In addition, the sun aging properties of the thin asphalt membrane systems were not good.

Most engineers were not happy either with 6 in. concrete as a hydraulic lining material. Seepage control at construction joints continued to be a problem. In addition, cracks due to improper cure caused trouble, along with cracks caused by temperature expansion and contraction forces. Reinforcement steel was used in efforts to counteract the effects of bending forces caused by normal subgrade settlement and intermittent filling and emptying cycles. In addition to all these problems, concrete linings were still expensive.

A great amount of Gunite lining work was being done, particularly on projects sponsored by the U.S. Bureau of Reclamation.* It was much more economical than its poured counterpart but still suffered from crack problems.

The first product introduced as a continuous, flexible, impervious membrane lining, *specifically* for the control of seepage, was the ½ in. prefabricated asphalt panel. The first reported usage was in 1951 for the lining of an irrigation ditch near the town of Cotulla, Texas. The following year the material was first applied in a reservoir, while in 1953 it was first used as a pollution control lining in a refinery near Dollarhide, Texas. The linings for these early reported projects were manufactured by the Gulf Seal Corporation in Houston, Texas. Shortly after the initial introduction of the panel lining, three other manufacturers joined the field; Envoy Petroleum Company, Long Beach, California;† Philip Carey Manufacturing Company, Cincinnati, Ohio; and W. R. Meadows Company, Elgin, Illinois. The linings produced by all of these manufacturers were essentially based on the same type of product.

Despite the obvious need, acceptance of the asphalt product was surprisingly slow. Its ½ in. cross section tended to shock many engineers who had grown accustomed to viewing a 6 in. section of concrete. Even when compared to 3 in. asphalt concrete, the new lining seemed very thin. Using something only ½ in. thick did not seem to be a very logical approach, and few engineers wanted to take this sort of gamble during the early development days of the panel. They also wondered about the

*Out of its Denver headquarters, lining work involving government money was recommended and controlled as to specifications.

†Envoy Petroleum was recently purchased by Asphalt Products Oil Company of Long Beach, California. Panels are now sold under the name of Envoy-APOC and APOC.

sun aging properties of asphalt, remembering the history of built-up asphalt membranes in earlier hydraulic structures. True, the panels cost less than the concrete lining, but they were more expensive than asphalt concrete. The fact that they were impervious did not seem to be too important at the time. Although growth was slow, it finally began to accelerate by the mid-1950s. In 1958 and 1959 the upward growth trend became evident with the installation of several large jobs for big-name water facility owners.

Like all materials, panels have a number of advantages and disadvantages, and careful analysis of these factors will guide the user as to how this type of lining may be best utilized. The panel is a flexible, continuous, impervious membrane system. It is rugged in nature and resists the action of vandals to a very high degree. An ordinary pocket knife or razor blade is not normally capable of cutting through the material. This has proved a worthwhile attribute, particularly in recent times.

The U.S. Bureau of Reclamation took a great interest in this lining through its Lower Cost Canal Lining Committee. Although they used it several times in various locations, it did not enjoy really widespread application in USBR backed projects because of its high cost. Its heavy weight meant high freight costs too, an additional drawback for budget-conscious, poor water districts.

When panels first came out, no particular attention was given to subgrade texture, so the lining was laid on almost any type of subgrade that appeared. Crushed rock, stubble, and other surface imperfections were thought to cause no trouble. It was quickly learned that, although the panels could resist a great deal of abuse in the form of rough-textured subgrade, there were definite limits. Crushed rock did not prove to be an acceptable subgrade texture, as portions of the panel under a head could be pushed into the spaces between the rocks. If these spaces were large enough, they could cause rupture of the panel due to extrusion. As soon as these facts became known, more attention was given to the subgrade texture, to the point where rocks, stakes, broken glass, and like protuberances were removed before the lining was laid. Stake holes were filled too, and "no rapid changes in subgrade elevations" became the general guideline as to how good the fine grading must be. Under proper conditions the panel can be bent rather sharply, but the generally accepted design technology has been based on a minimum radius of 12 in. at all transitions between surfaces.

The panels are rather heavy, weighing about 3 lb/sq ft. They exhibit flow properties during their early life, particularly on steep slopes (1 : 1) in hot weather, but this tendency gradually decreases with age. Once the water has been introduced into the facility, it pushes and molds the panels into the irregularities of the subgrade. This acts as a stabilizing force against any further movement. The flow tendency of new panels would seem to preclude their use as a lining on vertical surfaces, and, such a view is partially correct. Although successful panel installations have been made on vertical surfaces, the cost of this type of installation is usually high. A transitional type of lining is sometimes used for vertical walls in excess of 10 feet in height.

The tendency of the lining to exhibit slow movement in response to deformation forces is a decided advantage when it comes to stray rocks or objects in the subgrade. The asphalt lining literally surrounds the object, which becomes embedded in it. This engulfing process does not damage the asphalt panel even when the rock or foreign object is ½ in. in thickness; no cutting action occurs even though the rock may have sharp edges. The asphalt does not exhibit serious "necking down" at the point of contact with the object. This property is more important than might be realized. In large jobs it is often impossible to remove every rock. Even on excellent subbases, a stray rock may remain undetected.

One unusual feature of the panel lining is its conservative strength and elongation values. The ½ in. membrane has a tensile strength of barely 500 psi and an elongation of 35%. These values are far below those for any of the newer thin membrane materials. As will be seen later, the success of any lining system is dependent on many parameters. Strangely enough, tensile strength is not one of the important ones.

When panels were first marketed, they were sold to anyone who wished to buy them. This was not to prove a very effective policy, since trouble-free installation of the panels required more experience and skills than was originally assumed. Such has been the case with lining systems in general, although at first each manufacturer tends to regard application as the easy part of the process.

As each new lining was introduced, manufacturers touted its "ease of installation with inexperienced crews." Later, these claims were modified, and several top manufacturers now stress that "installations shall be by approved, experienced applicators." Often specific experience clauses are included in specifications, and lining bidders must document their past experience in the field.

Original panel layout for lining jobs was developed by utilizing simulated models. There are several effective ways in which the lining may be placed, and most of these variations were tried in the early installations. Some of the basic layout techniques developed in this work were applied to all of the later lining systems. Examples are the practice of running all long joints parallel to (or down) the slope instead of across it, special valley layouts, reinforcement "scabs," and the goring of panels lying on curved surfaces.

Two general layouts evolved from this work. In one, panels were laid out with all edges and ends butted tightly together. Cover strips (8 in. minimum width) were then laid across all panel intersections. To prevent water "short circuits" at all strip junctions, an overlay patch or "scab" piece was installed on top of the cover strip at appropriate places.

The other layout lapped the panel ends and butted their edges (Figure 3.11). Cover strips were laid flush into the panel ends, each cover strip overlapping the preceding one by at least 6 in. In this method the cutting of extra "scab" pieces was eliminated (see Figure 3.12). Cover strips varied in width between 8 and 9 in. and in thickness from ¼ to ⅜ in.

FIGURE 3.11. An asphalt panel lining in a California desert setting. This fire protection reservoir features a one-ply ½ in. lining with heat welded cover strips on the slopes, transitioning to a two-ply ¼ in. construction on the bottom.

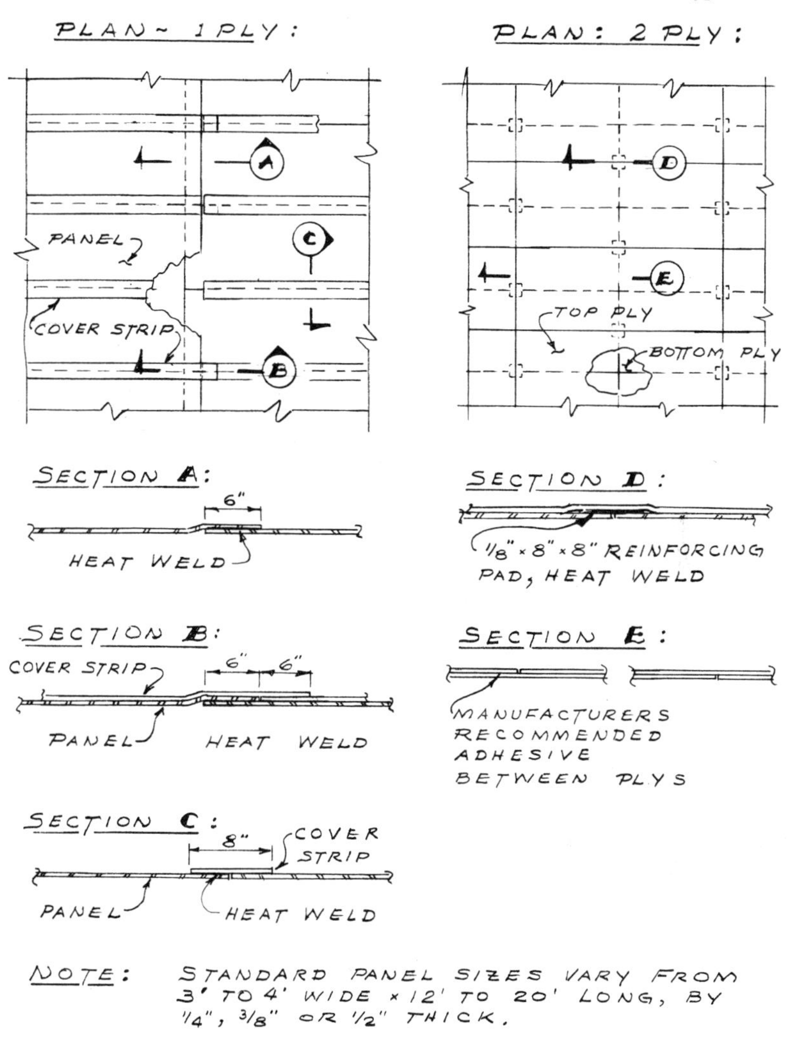

FIGURE 3.12. Typical field joint details for asphalt panels.

In irregularly shaped lakes a simple "Dutch lap" type of joint was sometimes used. This eliminated the need for a cover strip and was generally a faster and more efficient layout than the other two methods described.

The panels and strips were put together originally with a mopped application of hot-applied asphalt adhesive. A gas-fired kettle was used to melt the adhesive, and it was transported to the required areas by hand in insulated buckets.

In most layouts there are one or more appurtenances. Panels must be fitted carefully to these structures and roof column support bases. Many methods of cutting panels for this work have been evaluated, but the old technique of hand cutting with heavy-duty linoleum knives is still the fastest and most efficient. It must be remembered that rocks (mineral fillers) are embedded in the panel. These, plus the gumming and clogging tendencies of the panel make sawing very ineffective. Portable power shears with a ½ in. bite are not yet available from industry.

Asphalt itself does an excellent job of waterproofing when immersed in water. It experiences no visible effects from immersion. Water solutions of most chemicals cause no trouble, nor do any of the chemicals used in water treatment. Unfortunately, asphalt materials do not age well in desert areas of the country. As long as water covers the panels most of the time, they do an excellent job. Such is not the case, however, on exposed berm areas. Here the asphalt slowly begins to lose its more volatile fractions, becoming less ductile in the process. The loss of these components is accompanied by some shrinkage.

In attempts to compensate for problems in exposed areas, the original installations were often sprayed with a top coat of hot asphalt, a procedure that continued for several years. The practice expanded, and for some time even the covered reservoirs were given this treatment. The extra coating helped to protect the panel from the effects of the sun aging process. Although the coating itself was affected by a form of surface cracking, this was not a serious problem as far as lining performance was concerned, since the coating did not play a part in the actual sealing of the structure.

Part of the aging process of the panels involved the impregnated felt outer laminates. As they aged on berm areas, they showed some tendency to shrink. This shrinkage effect was noticeable at all exposed edges. In some cases the felt shrinkage forces were strong enough to cause edge curl. Panel edges were not normally exposed, but the edges of strips above the waterline were subjected to this aging phenomenon. The hot asphalt coating tended to eliminate the problem. When it was not used, rejuvenation of the affected areas by means of special asphalt paint or primer alleviated most of the difficulties. Treatments of this type were required every 3 to 6 years if the installation was to give good service.

Later, other top coating formulations were developed, using leafing-type aluminum pigments to plate out as an efficient shield against the damaging ultraviolet rays of the sun. These inexpensive barrier coats did much to improve the outdoor aging characteristics of the lining above the waterline, extending its life there by many years. At first, they were applied over the top of the hot asphalt coating. Because this coating had some tendency to cold-flow slowly down steep slopes, in addition to its surface cracking problems, recoating was necessary every 3 to 5 years.

Eventually, hot asphalt top coating lost favor and was abandoned for a rather unusual reason. When the original panel installations were 4 or 5 years old, bubbles began to appear on the surfaces of some of the panels. A study showed that this condition appeared only in some covered reservoirs and not in those that were unroofed. Panel core material was unaffected, the bubbles being formed only between the panel and the hot asphalt top coating. It was further found that, in addition to lack of light, water depth, circulation, and chemical content were also important factors. Good circulation and treated water seemed to diminish the problem, whereas darkness and stagnant water conditions aggravated it.

An interesting and important clue to the bubble problem was observed. In affected reservoirs, bubbles were also found in the paint coatings on immersed steel pipes and structures to a degree in direct proportion with their severity and frequency on the panels. The variable was light. In deeper water, bubbles on panels and paint coatings were more severe. In shallow water, bathed in light from vents in the roof, no bubbles were found on either panels or steel coated structures.

Biochemical testing linked the panel problem to an airborne mold called *Alternaria,* commonly found on tomatoes and potatoes. In addition, a natural phenomenon known as osmosis was at work. The latter was known to be of importance in failure mechanisms associated with steel coating systems and is discussed in more detail in Chapter 7. Fortunately, neither the *Alternaria* nor the felt decomposition products caused any foreign taste or odor in the water itself.

In summary, water passed through the thin asphalt top coating by osmosis. This internal water pressure behind the lining caused some of the uplift forces that formed the initial bubbles. The penetration of small quantities of water through the thin top coating membrane by osmosis also fed the *Alternaria* with the moisture it required to function. In this moist environment, it attacked the cotton of the felts, releasing as by-products water, carbon dioxide, and other gases. These gases also supplied some of the energy necessary to blow the bubbles. Eventually, the bubble walls cracked, flooding the cavity with water. Subsequent foot traffic broke down the bubble walls further, scattering small pieces of the top asphalt coating and exposing the core. Once the felts were consumed, no further degradation of the core took place.

The formation of the bubbles did not affect the integrity of the lining. It did cause an unsightly appearance, however, and there was always the possibility that small pieces of asphalt would get carried into the outlet lines. This point seemed to cause water companies the most concern, since, as mentioned above, bubble conditions were never associated with water quality problems, such as bad taste or odor.

After the bubble problem proved that the hot asphalt coating added nothing to the integrity of the lining, the coating was removed from the specifications. The industry also set to work to remove the cellulose felts from the construction. This was first accomplished in 1965, and thereafter glass fabrics were used as the outer laminate layers on all lining constructions that went into closed reservoirs. One year later, Envoy Petroleum eliminated felt construction on all linings, switching 100% of its production to the all glass reinforced panel. Bubble troubles disappeared, and other manufacturers followed suit in converting to glass construction or at least making it available in cases involving covered storage facilities.

Meanwhile, water facility owners began complaining about the use of cover strips on the panel linings. These made cleaning operations difficult. Silt and debris that collected on the reservoir bottom was hard to sweep to the mud drain because it had to be elevated over cover strips every 3 or 4 ft.

As a result of this criticism, a two-ply ¼ in. construction was developed. In this field fabrication method the top ply serves as the cover strip for the bottom ply. Sprayed hot asphalt adhesive proved difficult to use between panels, as larger areas had to be sprayed and the top ply flipped into position before the adhesive cooled. Partly for this reason, and partly for safety and speed of application, heat welding of panels was developed, and it soon replaced the use of the hot asphalt adhesive. The process was used on jobs involving cover strips too. Although the two-ply construction increased the lining installation cost somewhat, owners liked the smooth finish that resulted, and by 1968 most large reservoir jobs utilized this type of panel lining construction, at least over the bottom areas of the structure.

With increases in adhesive technology a system was developed that joined the panel plies by means of a cold-applied liquid adhesive system. It proved to be faster than the heat welding process and was adopted for general use shortly after the general acceptance of the two-ply system. The adhesive is slow to act and hence does not develop good strength fast enough for use on open, exposed slopes steeper than 2½ : 1. When these conditions prevail, heat welded construction is used on the slopes because the development of good adhesion is almost instantaneous with this process.

Panels may be put together using cold-applied, bodied mastic, and this method is often used on small jobs. Here, again, full development of seam strengths is slow, usually requiring at least 8 or 10 days under ideal conditions. Water should not be put into the facility for 10 days after completion of the lining.

One of the advantages of the asphalt panel is its ability to form tough, long-lasting seals to concrete structures. This worthwhile attribute was developed with less effort than might be expected, primarily because the seal is more of a natural process than one requiring intense technical development (see Figure 10.7).

When asphalt primer is painted on a concrete surface, it exhibits an extremely high degree of penetration into the pores of the concrete. The solvent carrier used for the adhesive has high wettability on the concrete, even though it is not a waterborne system. On the contrary, the surface must be very dry before it is primed.

After the primed surface is dry, asphalt-type mastics will adhere tenaciously to it, and this is the prime factor in the integrity of the seal at that point. Naturally, the mastic has the same tenacious properties with respect to the panel lining itself. The high adherence of mastic to both lining and primed concrete is really the major key to good performance in this area. There is one more important point, however, and this one holds for all linings; the seal must be made in the plane of the lining. For this, a 6 in. minimum width shelf should be available.

The prime–mastic concrete seal is slow to attain its full strength, as some of the solvents of the mastic must dissipate before this can be realized. Several days are required for this process, and the curing continues for some time thereafter. When it is complete, however, the bond of the lining to the concrete is practically indestructible.

Seals to concrete on sloped or vertical surfaces are always mechanically pinned to the concrete to stabilize the lining during the lengthy curing period. On horizontal surfaces, no pinning is necessary. The mechanical anchor consists of pins driven by percussion through 1½ in. diameter galvanized washers, 6 in. on centers, all encased in mastic. Seals should always be made in the plane of the lining with no "coving" permitted unless the lining is firmly backed. The anchor into concrete at the top of the slope consists of pins driven by percussion through continuous metal bars. The pin spacing is usually set at about 16 in. In the very beginning, redwood header strips were used. They did not perform well, however, and were soon replaced by ¼ × 2 in. mild steel bars, galvanized after cutting. Later, these were reduced to 3/16 in. in thickness so that they would better conform to surface irregularities in the concrete. More recently, ¼ × 2 in. structural aluminum bars have gained support and currently are used on most jobs.

The lining operation follows a schedule determined by design and weather considerations. Normally, lining begins at the top of the slope, and progresses downward and across the bottom in a manner that places the lining material at the lowest elevation last in the sequence. In this manner the areas most susceptible to erosion by rain are covered first. This is the method used on reservoirs that have adequate drain structures. Where these do not exist, lining is usually done at lower elevations first, in sufficient quantities to contain all drainage from the heaviest rainstorms expected. Expensive construction delays and pumping costs are thereby avoided. Final slope preparation by the earthwork contractor must pace the lining crew on these areas so that erosion damage from rain is held to a minimum. The same general procedure has been adopted for the other lining systems.

At the corners a valley line is created, and the lining on adjacent slopes terminates at this line. Here a valley seal is made, a technique that is also used for EPDM, butyl, and neoprene systems.

Compacted Earth

Historically, the lining system with the longest record of successful performance is compacted earth. It enjoys this distinction partly because its origins go back several thousand years, when it was employed in the construction of dams and irrigation works of ancient civilizations. Its good record is also the result of some specific properties that, in combination with proper design and application techniques, makes this lining one of the best overall systems ever developed. Moreover, since installation is done by earth moving equipment from raw materials on the site, the economics are very good.

The compacted soil blanket is classified as a flexible membrane, although it is not a continuous one. Such a lining system permits the continual passage of a controlled amount of water through it. Properly placed, it serves as a uniform resistance in the path of the water. Being flexible, it enjoys an excellent reputation with respect to normal subgrade settlement and is of particular advantage in areas where seismic activity is prevalent.

Despite very early beginnings the scientific approach to the use of soil as a material of construction did not begin until the 1930s. Since then much has been learned on the subject, but there is still a long way to go. Although soils engineering is a well-defined discipline, the soil itself is not so well defined. Only three things are present in the design; soil, water, and air; nevertheless, the subject is a complex one, and the literature does not provide any single do-it-yourself manual on how to design a water holding structure of this type. That is the reason why it is so important to place this responsibility in the hands of an experienced soils engineering firm.

Soils vary. This simple sentence is short, but it tells a long story. All natural materials vary from one place to another, and even in the same location a few feet may produce rather striking differences. This is particularly true in the case of soils, the variance of which cause the engineer to change his usual approach. Normally, he will design a structure using assumed values for strength. With earth as the construction medium, he must design his structure to suit the characteristics of the available materials. Moreover, after construction begins, the design may have to be modified to compensate for unforeseen changes in the soils. The contractor must also be on guard against the many pitfalls associated with earthwork. The methods he used on one job may be unsatisfactory for the next one. Sometimes, compacted soil linings utilize a companion lining system in series, designed to control wave erosion problems or to serve as a further impedance against seepage losses. Again, experience is a great ally, for both the designer and the contractor.

In the cut-and-fill reservoir, the soil must possess several attributes:

1. Proper strength.
   A. To support its own weight.
   B. To support the weight of the water impounded.
   C. To support the concentrated loads of any appurtenances.
2. Proper stability.
   A. Against deformation or settlement that might damage the soil or structures.
   B. Against excessive swelling or shrinkage of the soil.
   C. Against the above conditions, regardless of whether the soil is wet or dry.
   D. Against conditions of complete saturation.
3. Proper imperviousness of the lining member.
4. Proper permeability characteristics of the drainage members.

The strength of the soil depends on three factors: water content, void ratio, and nature of the soil. In general, strength increases as water content and void ratio decrease.

A number of problems are encountered in the use of compacted earth linings. Settlement from consolidation and elastic deformation can cause difficulties, particularly with concrete structures. In such areas good compactive effort is more difficult to achieve. As soil density rises, elasticity increases and compressibility decreases.

Shrinkage, too, can be a very serious problem in compacted earth reservoirs or dams, regardless of whether or not the facility has an auxiliary lining. It can serve as an important triggering mechanism for eventual failure. Low shrinkage depends on high density and minimal water loss from the lining membrane.

For this type of lining proper design must ensure that swelling is kept to a minimum. In general, clay-type soils are prone to lose strength when wet. The action also depends on the density and mineralogy of the soil; swelling tendencies increase with higher density. Swelling is best controlled by proper soil selection, by prevention of water intrusion into the soil, and by making the lining so dense that adequate strength exists even in the weakened condition.

Permeability characteristics are important in trench-type underdrains and in all hydraulic facilities that are protected with continuous drainage blankets. In both cases these are the second line of defense, and they must remain in good operating condition at all times if they are to make their proper contribution to the optimum overall performance of the holding system.

Soil linings must be controlled by proper choice of ingredients, moisture content, and compaction. These parameters bear a relationship to those involved in the technology of PVC plastics already discussed. Development of the right soil mix is equivalent to the compounding step, water is analogous to the plasticizer, and compaction serves as the processing (continuous molding by the calender) step. The water content of soils has a function almost parallel to that of the plasticizer in plastics. Without water, optimum properties cannot be developed in the soil. As water is added to dry soil during processing (compaction), the properties improve, but too much water causes a decrease in strength. The situation is exactly the same with the PVC plasticizer. There is a ratio that will develop optimum properties of the compound for a particular use; all other concentrations yield less than the best characteristics.

To control the main ingredients in compacted earth construction, it is first necessary to arrange them in some logical order so that a meaningful specification may be written. Soils have been classified in several different ways, depending on who is doing the classifying.

The Bureau of Public Roads Classification System was developed in 1929. It has been modified many times and is used for evaluating soils for highway subgrade and embankment construction. The Civil Aeronautics System classifies soils according to their suitability for airfield subgrades. In many states special highway department systems are adapted to the native soils commonly encountered in the particular state.

In 1953 the Unified Soil Classification System was introduced.[2] This method is based on modifications of the Airfield Classification System developed by A. Casagrande[3] in 1948, as a means for identifying and grouping soils for military construction. The system is also effective in classifying soils for use in the construction of embankments, earth dams, and compacted earth reservoir linings.

The lack of a simple method for classifying soils is not a technical oversight. The situation exists because the various engineering disciplines involved with soil design use different criteria in their analytical approaches to the problem. As a simple example, a highway engineer on a fill project is looking for soil properties considerably different from those that would be of concern to the designer of an airport runway base or a reservoir holding 150 ft of water depth. Any classification system is further complicated by the great number of soil types and gradations and the astronomical number of combinations that they can produce.

A soils engineer designing a cut-and-fill reservoir is mainly concerned with two things, the strength of the soil and the control of seepage forces. Although he uses the information afforded by the Unified Soil Classification System, he is vitally interested in shear strength, density, and permeability, which are defined as follows.[4]

1. *Shear strength* of soil is the resistance it offers to deformation. It is often defined in terms of unconfined compressive strength, in the case of cohesive soils, but several types of shear tests have been developed. The use of strength results depends on other soil properties: moisture content, mineralogy, history, testing techniques, and evaluation methods.
2. *Density,* for cohesionless soils, may be found by comparing the actual void ratio with the range in void ratio from loose to dense for that soil. This measurement is as important for cohesionless soils as strength is for cohesive ones.
3. *Permeability coefficient* is a constant that defines the ease with which water will pass through the soil. It is dependent on the soil type, water temperature, and flow direction. It is usually expressed as centimeters per second (cm/sec) and hence has velocity dimensions. The coefficient varies from 0.1 for coarse soils to 0.0000001 for the so-called impervious clays. The test is run at 68° F. At 104° F, its value is 50% higher, while at 32°F it is only about half as much as at room temperature.

TABLE 3.2. Unified Soil Classification (After U.S. Waterways Experiment Station and ASTM D 2487-66T)

Field Identification Procedures (Excluding particles larger than 3 in. and basing fractions on estimated weights)				Group Symbols [a]	Typical Names	Information Required for Describing Soils	Laboratory Classification Criteria		
Coarse-grained soils More than half of material is *larger* than No. 200 sieve size[b] (The No. 200 sieve size is about the smallest particle visible to naked eye)	Gravels More than half of coarse fraction is larger than No. 7 sieve size (For visual classification, the ¼ in. size may be used as equivalent to the No. 7 sieve size)	Clean gravels (little or no fines)	Wide range in grain size and substantial amounts of all intermediate particle sizes	*GW*	Well graded gravels, gravel-sand mixtures, little or no fines	Give typical name; indicate approximate percentages of sand and gravel; maximum size; angularity, surface condition, and hardness of the coarse grains; local or geologic name and other pertinent descriptive information; and symbols in parentheses	Use grain size curve in identifying the fractions as given under field identification	Determine percentages of gravel and sand from grain size curve Depending on percentage of fines (fraction smaller than No. 200 sieve size) coarse grained soils are classified as follows: Less than 5% *GW, GP, SW, SP* More than 12% *GM, GC, SM, SC* 5% to 12% *Borderline* cases requiring use of dual symbols	$C_U = \frac{D_{60}}{D_{10}}$ Greater than 4 $C_C = \frac{(D_{30})^2}{D_{10} \times D_{60}}$ Between 1 and 3
			Predominantly one size or a range of sizes with some intermediate sizes missing	*GP*	Poorly graded gravels, gravel-sand mixtures, little or no fines				Not meeting all gradation requirements for *GW*
		Gravels with fines (appreciable amount of fines)	Nonplastic fines (for identification procedures see *ML* below)	*GM*	Silty gravels, poorly graded gravel-sand-silt mixtures				Atterberg limits below "A" line, or *PI* less than 4 — Above "A" line with *PI* between 4 and 7 are *borderline* cases requiring use of dual symbols
			Plastic fines (for identification procedures, see *CL* below)	*GC*	Clayey gravels, poorly graded gravel-sand-clay mixtures	For undisturbed soils add information on stratification, degree of compactness, cementation, moisture conditions and drainage characteristics			Atterberg limits above "A" line, with *PI* greater than 7
	Sands More than half of coarse fraction is smaller than No. 7 sieve size	Clean sands (little or no fines)	Wide range in grain sizes and substantial amounts of all intermediate particle sizes	*SW*	Well graded sands, gravelly sands, little or no fines	Example: *Silty sand*, gravelly; about 20% hard, angular gravel particles ½-in. maximum size; rounded and subangular sand grains coarse to fine, about 15% nonplastic fines with low dry strength; well compacted and moist in place; alluvial sand; (*SM*)			$C_U = \frac{D_{60}}{D_{10}}$ Greater than 6 $C_C = \frac{(D_{30})^2}{D_{10} \times D_{60}}$ Between 1 and 3
			Predominantly one size or a range of sizes with some intermediate sizes missing	*SP*	Poorly graded sands, gravelly sands, little or no fines				Not meeting all gradation requirements for *SW*
		Sands with fines (appreciable amount of fines)	Nonplastic fines (for identification procedures, see *ML* below)	*SM*	Silty sands, poorly graded sand-silt mixtures				Atterberg limits below "A" line or *PI* less than 5 — Above "A" line with *PI* between 4 and 7 are *borderline* cases requiring use of dual symbols
			Plastic fines (for identification procedures, see *CL* below)	*SC*	Clayey sands, poorly graded sand-clay mixtures				Atterberg limits below "A" line with *PI* greater than 7

	Identification Procedures on Fraction Smaller than No. 40 Sieve Size						
		Dry Strength (crushing characteristics)	Dilatancy (reaction to shaking)	Toughness (consistency near plastic limit)	Group Symbols [a]	Typical Names	Information Required for Describing Soils
Fine-grained soils More than half of material is *smaller* than No. 200 sieve size (The No. 200 sieve size is about the smallest particle visible to naked eye)	Silts and clays liquid limit less than 50	None to slight	Quick to slow	None	*ML*	Inorganic silts and very fine sands, rock flour, silty or clayey fine sands with slight plasticity	Give typical name; indicate degree and character of plasticity, amount and maximum size of coarse grains; colour in wet condition, odour if any, local or geologic name, and other pertinent descriptive information, and symbol in parentheses
		Medium to high	None to very slow	Medium	*CL*	Inorganic clays of low to medium plasticity, gravelly clays, sandy clays, silty clays, lean clays	
		Slight to medium	Slow	Slight	*OL*	Organic silts and organic silt-clays of low plasticity	For undisturbed soils add information on structure, stratification, consistency in undisturbed and remoulded states, moisture and drainage conditions
	Silts and clays liquid limit greater than 50	Slight to medium	Slow to none	Slight to medium	*MH*	Inorganic silts, micaceous or diatomaceous fine sandy or silty soils, elastic silts	
		High to very high	None	High	*CH*	Inorganic clays of high plasticity, fat clays	Example: *Clayey silt*, brown; slightly plastic; small percentage of fine sand; numerous vertical root holes; firm and dry in place; loess; (*ML*)
		Medium to high	None to very slow	Slight to medium	*OH*	Organic clays of medium to high plasticity	
Highly Organic Soils		Readily identified by colour, odour, spongy feel and frequently by fibrous texture			*Pt*	Peat and other highly organic soils	

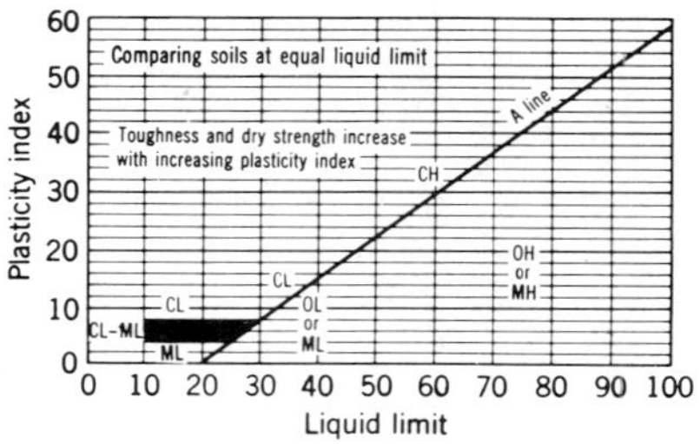

Laboratory classification criteria for fine-grained soils:

Plasticity index (0–60) vs. Liquid limit (0–100)

Comparing soils at equal liquid limit

Toughness and dry strength increase with increasing plasticity index

A line

CH

CL

OH or MH

OL or ML

CL-ML

ML

Plasticity chart for laboratory classification of fine grained soils

From Wagner, 1957.

[a] *Boundary classifications*. Soils possessing characteristics of two groups are designated by combinations of group symbols. For example *GW–GC*, well graded gravel-sand mixture with clay binder.

[b] All sieve sizes on this chart are U.S. standard.

*For soils having 5 to 12% passing the No. 200 sieve, use a dual symbol such as GW-GC.

Although these parameters are the most important, other factors, such as soil structure, composition, texture, moisture content, and the strength of the underlying soils on which the reservoir embankments will rest, are also of interest to the soils engineer.

The employment of a competent soils engineer is a mandatory first step in the design of a compacted soil lining. This will eliminate the need for the owner or design engineer to concern himself with the complexities of the various soil classification systems. Nevertheless, it may be helpful, in trying to understand the problem at hand, to examine one of the systems in condensed chart form. For this reason the Unified Soil Classification System is shown in Table 3.2.

The table introduces the reader to some unfamiliar terms, such as D_{10}, PI, and plasticity chart. To define the first, it is necessary to refer to the grain size chart (Figure 3.13). Here are plotted the grain diameters versus the percentage of the sample finer than the grain size indicated. For additional clarity the chart includes U.S. standard sieve sizes and, at the top, one convenient arrangement that breaks soil down into recognized groups. The latter grouping is useful to the engineer and helpful to the layman but is strictly arbitary in scope. The grouping is somewhat misleading with respect to clays, because some soils finer than 0.002 mm (millimeter) contain no clay, whereas some clay minerals are larger than 0.002 mm.

The *effective size* of soil grains is labeled D_{10}, or the diameter corres-

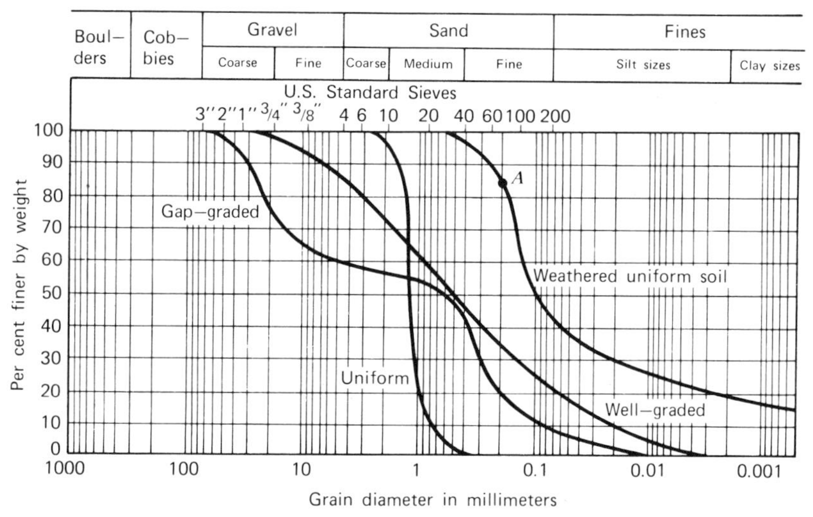

FIGURE 3.13. Grain size chart and ASTM–ASCE grain size scale.

ponding to 10% on the grain size curve; D_{60} refers to the diameter that intersects the 60% line; and so on. The ratio $D_{60} : D_{10}$ is the uniformity coefficient, C_u. When C_u is less than 5, the soil is said to be *uniform;* when greater than 10, *well graded.*

The Plasticity Index, or PI, represents the range in water content through which the soil is in the plastic state. Defined in another way, it is the difference between the liquid and plastic limits. The liquid limit (LL) is defined as the water content of the soil at which a standard trapezoidal groove (cut in moist soil and held in a special cup) is closed after 25 taps of the cup on a hard rubber mat. The plastic limit (PL) is the water content of the soil as it begins to crumble when rolled by hand into a ⅛ in. diameter thread. Figure 3.14, the plasticity chart, shows the relationship between the plasticity index and liquid limit for fine-grain soils.

Not only is the soils engineer important to the success of a compacted earth lining, he is also instrumental in ensuring the success of all other lining systems used in cut-and-fill reservoir design. Although lining and subgrade stability are dependent on each other to a certain extent, subgrade stability has a greater effect on lining than vice versa.

In many cases the soils engineer will work closely with a trained

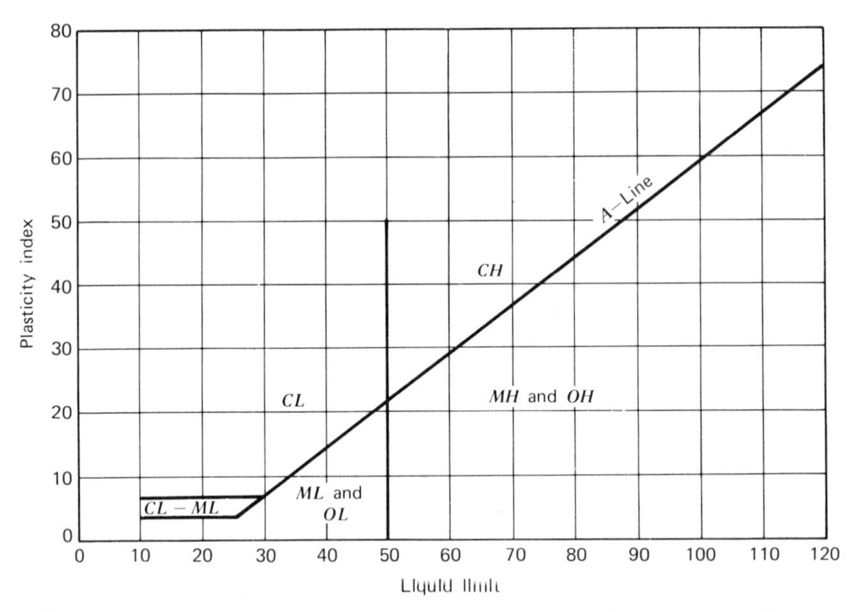

FIGURE 3.14. Plasticity chart for the classification of fine-grained soils. Tests made on fraction finer than No. 40 sieve. (After A. Casagrande and the U.S. Waterways Experiment Station.)

geologist or other professional of similar background. This combination will ensure that proper geological information is available concerning the nature of deposits underlying the reservoir site.

The geologist will be looking for historical evidence of faulting or flooding. Defects in the underlying strata, including cracks, fissures, and caves, are of particular interest, along with possibilities of seismic activity in the site area. The search for foundation stability involves a knowledge of boring depth requirements, often overlooked by the uninitiated. In general, the rule of thumb states that dam portions should have exploratory excavations at least twice as deep as the embankment is high. The same rule applies for the portions of cut-and-fill reservoirs that lie entirely in fill.

To find use as a lining material, soil must be compacted or densified. In early times this was done by trampling the soil under foot or by tamping it with heavy logs. Today this step is accomplished by mechanical means. Several methods are available, and the one selected will depend on the soil analysis and the final specifications.

Compaction amounts to a reduction of void areas within the soil. The action occurs as a result of particle reorientation, fracture, or distortion, or by various combinations of these things. The soil resists compactive effort because of cohesive forces, friction between particles, and capillary attraction. All three of these blocking actions are reduced by the addition of water. Too high a moisture content, however, will cause saturation, which will prevent further reduction in void ratio so that additional compactive effort is ineffective.

The person most responsible for early work on the principles of soil compaction was R. R. Proctor[5] The "Proctor tests" are the simplest and most wisely used in connection with his "optimum moisture/maximum density" concept.

Two test methods are available, the *standard Proctor* and the *modified Proctor,* both of which utilize a 4 in. diameter, 1/30 cu ft cylinder as the standard sample size. In the former test the compactive effort consists of delivering 25 blows of a 5.5 lb hammer, falling 12 in. on each of three equal layers of soil. The total work done on the sample amounts to 12,400 ft-lb/cu ft, which is equivalent to light rollers or thorough hand tamping.

The modified Proctor utilizes 25 blows of a 10 lb hammer, falling 18 in. on each of five equal layers of soil. Here the total work done on the sample is 56,200 ft-lb/cu ft, equivalent to that obtainable with the heaviest rollers under favorable working conditions. In moist soils the modified test will usually give a maximum density from 3 to 6 lb / cu ft higher than the standard method.

Confusion regarding soil compaction tests may best be clarified by examination of Figure 3.15. The dry density of a compacted sample of a hypothetical soil is plotted against the corresponding water content of the sample to yield the lower of the two curves shown. This is the curve referred to in discussions involving maximum density, percent compaction, optimum moisutre, and maximum moisture.

It can be seen that, with the laboratory compactive effort on the sample in question, a maximum density of 119 lb/cu ft is attainable at 12.5% water content. To achieve 95% compaction (or a dry density of about 113 lb/cu ft) of this soil, the water content must lie between 7 and 18%. If more compactive effort is applied to the sample, the curve will shift slowly upward and to the left.

The upper line is the zero air voids curve (theoretical), based on a sample containing absolutely no air. This is the limiting value above which no density–moisture curve may rise.

Dry density values obtained in the field might exceed 100% of the maximum density if more compactive effort was applied there than was equivalent to the amount applied to the laboratory sample. Naturally, this would require more time and energy and would be an unnecessary extra expense.

Control of soil density (compaction) during construction operations is essential to the proper performance of the cut-and-fill reservoir. This is done by careful monitoring of moisture, compactive effort, and soil composition. Each step is checked by the inspector, and compaction and moisture tests are run in the field on a regular schedule.

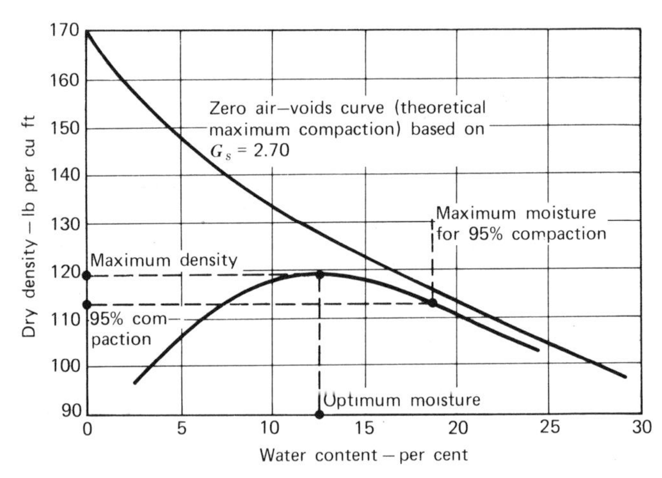

FIGURE 3.15. Moisture–density curve of a cohesive soil for one method of compaction and maximum moisture for a specified degree of compaction.

In a nutshell, compacted earth facilities have two potential degradation problems: too much water and not enough water. As a rule, moisture control during the construction period is a minor problem. Afterwards, however, the moisture content of the soil is not subject to this type of control; rather, it depends on the environment. Since the soil structure is holding water, its moisture content will rise, commonly to saturation, in the parts remaining under water. In all cases it is imperative that the structure be designed with this fact in mind.

The soil becomes saturated, not only below the water line, but also above it. This can occur either because of percolation of surface water or vapor condensation. Even if neither of these methods is active, the soil can still become saturated above the waterline as a result of capillary action. Just as water within a small tube vertically immersed in water will rise above the level outside the tube, so will water rise within the interstices of a soil formation. The height at which capillary action will force the water up above the level of the water in the reservoir depends on dissolved impurities, temperature, mineralogy, and the nature and extent of the soil voids. Theoretically, the distance may vary from a few inches in open soils to over 100 ft in dense soils. Intercepting strata, cracks, or other imperfections of soil structure are usually limiting factors, but even so, the rise may be appreciable. Thus the soil may be saturated both above and below the normal water level. The condition of saturation is emphasized in the following quotation: "It is necessary to design earth structures on the basis that soil moisture could increase to the point of saturation.[6]"

This statement is worth special emphasis since the phenomenon described is a most important factor in the success of cut-and-fill reservoir installations. The importance of good design was stated in another way by Harry R. Cedergren at the Second Phoenix Seepage Symposium in March 1968. In his talk he emphasized:

> Many of the failures of earth dams have been caused by piping along outlet pipes, beside or under spillway walls and slabs, or along other contacts between natural or compacted earth and rigid structural members. If the designers and builders of dams give sufficient attention to fundamental principles of seepage analysis and control, and the work is carried out under carefully prepared plans and specifications *seepage problems* can be virtually eliminated.[7]

Of course, these words come as no surprise to the designers of earth fill dams. It is strange, though, how the principles mentioned above are sometimes relegated to the background in the case of cut-and-fill reservoirs, particularly those of relatively shallow depth. The reason is not that designers are necessarily becoming more lackadaisical, but that the nature of those doing the designing is changing because the reasons for lining are changing. Because of government pressures for linings, particularly in pollution control areas, the design of shallow-depth facilities is slowly transitioning into a do-it-yourself affair. Although state safety regulations generally control design and construction practices for dams, there are many exceptions with respect to cut-and-fill reservoir facilities. Again, it is emphasized that the design of earthen linings, whether they be employed in dams or in small reservoirs, is no undertaking for the untrained. This section of the book is intended not to promote do-it-yourself activities, but rather to show the complex nature of the problem and the reasons why it is so important to have the job done correctly by those with proper know-how and experience.

What about the installation of a continuous, flexible, impervious membrane system on top of a compacted soil lining? Will this addition effect a relaxation of the soil design considerations? The answer, of course, is no, but many have not fully understood the significance of this negative reply.

As we have already seen, capillary attraction is one way in which a soil may become saturated, even though it is not inundated by water within the facility. There are also other ways in which this can happen, and these will be discussed shortly.

In many cases persons designing water holding facilities may have been misled by the term "impermeable membrane," which has generally been associated with lining materials such as butyl rubber, PVC, polyethylene, and Hypalon. This is not to imply that the term is a misnomer. These membranes themselves are, for practical purposes, impermeable. But the final installation might not be impermeable in the strict sense of the word. Even though the lining might be a 100% effective membrane at the time of construction, the situation with respect to its integrity could be altered with age. Normal subgrade settlement problems at structures or other types of mechanical damage to the membrane could change the situation. Outdoor aging of the membrane could also play an important role, the long-term effects of which are not yet known.

Sometimes forgotten is the fact that soil in compacted earth reservoir facilities may become saturated because of circumstances that have nothing to do with their use as water holding devices. This can be caused by an unusually heavy rain or a prolonged rainy spell. Sometimes a flood condition may exist and be responsible for bank saturation from outside the facility. A broken or cracked portion of a water transmission line can also be responsible for this type of condition. It is imperative that the soil structure be stable, *wet or dry,* regardless of whether or not an auxiliary lining is used. The latter is definitely not a substitute for good soil design.

Critical facilities with compacted earth linings should generally be protected with some type of underdrain backup system similar to that described in Chapter 2. In recent years the procedure of designing underdrain systems has often been slighted with disastrous results. This practice has become rather widespread because of a lack of understanding of the basic principles involved by owners and other inexperienced persons who attempt to bypass the services of competent, trained, and experienced soils engineers. The underdrain is a very important safety backup system that should be considered as independent of the lining.

The continuous blanket, common to modern, lined cut-and-fill reservoirs, is used with many variations in facilities protected with compacted earth linings. Often called underdrain carpets, they may or may not underlie the lining and may or may not be parallel to it. In any event they serve as a useful line of defense to arrest and safely lead away all stray water (from any source) that finds its way into the embankments.

A compacted earth lining needs another protective device against the effects of wave erosion. Wind-whipped waves generally degrade the slopes, causing them to slowly sluff away. To prevent this, some form of control measure must be applied. It can take the form of a continual maintenance program, but usually a preventive defense is more effective.

One of the most popular methods of wave erosion control involves the use of a specially designed pervious asphalt concrete blanket. This does a good job of protecting against erosion, and at the same time its open design allows hydrostatic back pressure to be harmlessly vented without danger of buckling the asphalt blanket.

Another method, the use of crushed rock or riprap, has two disadvantages. One is that the supply of suitable rock for the purpose never seems to be located in the spot where it is needed. The high cost of hauling will usually eliminate this method from consideration. The second disadvantage is that wave erosion, although reduced in scope, may still proceed behind voids in the rock blanket.

Either method may benefit from a reduction of slope steepness, which, though never effective in eliminating erosion, does lessen the severity of the problem.

Another important problem with compacted earth facilities arises from the fact that water exerts an equal force in all directions. Upward movement in a concrete foundation can occur when the upward pressure exerted by the water beneath the concrete exceeds the downward force due to its mass and external loading. This is the well-known *reverse hydrostatic* condition, whose forces have a devastating effect on linings of all types. The result of this action is lining rupture.

The same condition can occur within the soil structure. There the result is called *heave.* The soil then expands with an accompanying decrease in void ratio, often trapping a blister of water within the soil mass. The roof in this miniature water-filled cavern then collapses to the floor. By this process the blister rises to the surface. As the bubble reaches the surface, the soil appears to be cooking, and the condition is termed *boil.* Heave can occur in any soil, but boiling is limited to cohesionless (sandy type) materials. Boiling reduces the strength of the soil to zero.

Piping, the prelude to disaster in a compacted earth facility, follows the boiling step if the affected soil is washed away by water movement. The small pit so formed causes increased flow into it and a shortening of the seepage path. This, coupled with the increased hydraulic gradient, causes the boil condition to be magnified, and the pit deepens. It then begins to proceed in the upstream direction, toward the source of the impounded water, moving at ever-increasing speed.

When the "pipe" reaches the source of water, a sudden breakthrough occurs. A rush of water through the hole quickly enlarges it, as the water rushes through with its characteristic twisting action. Fine sands and silts (cohesionless soils) are most susceptible to piping, but none is immune.

Most of the preceding discussion has dealt with the problem of too much water within the earth lining, and its effects as it leaves the structure. There are also problems concerned with a lack of earth moisture. This may reduce the desirable properties of the membrane by causing cracking and slaking. The former condition is an obvious detriment to any lining system, but the slaking problem can also be harmful. Such a condition can result when a clay material is dried well beyond the shrinkage limit. When the dried area is suddenly immersed in water, it may disintegrate by flaking off soil particles and converting itself into a soft, wet mass. In addition to the slaking process itself, the membrane is more easily subjected to further degradation by wave action in the affected areas.

After becoming aware of all the potential problems with compacted earth linings, the reader may logically ask why this lining was rated "one of the best" in the opening paragraph of the section. The answer has to do with proper design and control. Although the potential problems described do exist, most can be handled, and the others may be effectively compromised. Proof that this is so lies all about us, in the form of many large earthen lined facilities with excellent performance records.

References

1. Douglas Alan Fisher, *The Epic of Steel,* Harper & Row, New York, 1963, p. 145.
2. U. S. Waterways Experiment Station, "Unified Soil Classification System," Technical Memorandum No. 3–357. Vicksburg, Miss., 1953.

3. A. Casagrande, "Classification and Identification of Soils," *Transactions of the American Society of Civil Engineers,* 1948, p. 901.
4. George B. Sowers and George F. Sowers, *Introductory Soil Mechanics and Foundations,* 2nd Edition, MacMillan Co., New York, 1965, p. 85 ff.
5. R. R. Proctor, "Fundamental Principles of Soil Compaction," *Engineering News Record,* Aug. 31, Sept. 7, Sept. 21, Sept. 28, 1933.
6. Sowers and Sowers, *op. cit.,* p. 279.
7. Harry R. Cedergren, "Control of Seepage in Earth Dams," *Proceedings of the Second Seepage Symposium,* Phoenix, Ariz., Mar. 25–27, 1968, p. 106.

4

Rigid Linings

Rigid linings are the forerunners of the continuous, flexible membranes. The various types have all been around for quite some time. Depending on how the word "lining" is defined some have been known since antiquity. This group includes the most economical (asphalt) and the most expensive (steel) of the lining systems treated in the book. As will be shown, however, the price tag alone does not make the lining. Judgment must be made on the basis of performance, and by this criterion, some rigid systems have not enjoyed good reputations as hydraulic linings.

This is not to say that rigid linings are no good as a class. They have a very definite place in the technology. What they have to offer as structural members complements the flexible systems, which do not rate well in this respect. The rigid systems serve well as protective devices over other, more fragile lining systems, and it is in this application that they have done their most efficient work. For strict control of water loss, though, the rigid members, with the exception of steel, do not rate a perfect grade. Unfortunately, the use of steel linings in cut-and-fill reservoirs has been limited because of cost and technical reasons.

Despite the above facts rigid membranes have been and continue to be used in appreciable quantities. Tradition accounts for some of this, but properly designed facilities can be built and many are in existence to prove this point. If the use of rigid linings is contemplated, however, experienced engineering help is definitely required.

Concrete

The use of concrete as a structural material in the building of water holding facilities was discussed in Chapter 2. In contrast, older cut-and-fill structures utilizing concrete, often in conjunction with stone or brick, depended not on steel reinforcement but on the earth for structural stability. In those times the engineer had a very limited selection of items that he could use as a lining member. Concrete was the best known material of construction at that time and was often used in this role. It also exhibited a hard working surface which was ideal when it came time to clean out the reservoir.

In tank construction utilizing vertical walls, the concrete serves as the lining in addition to its role as the structural member. When concrete is used as the lining of a cut-and-fill reservoir, however, it does not play this dual role. Its relationship to the facility is complicated, because slab reinforcement serves to replace it in the structural class in most cases. In the concrete lined earthen structure, however, we are primarily concerned here with its use as a lining.

As shown previously, users of concrete oscillated back and forth between applying the material as a structural member and utilizing it as a hydraulic lining. Partly for this reason, but also because of advertising claims, a lack of alternative materials in the early days, and widespread usage, concrete came to be known as a prime lining material. As engineers used it more frequently in hydraulic facilities, and as its technology was taught more and more in schools, it grew in popularity. Several engineering firms became known for their consistent use of concrete linings in the design of hydraulic facilities.

Despite the background and experience amassed with concrete in hydraulic structures, there is considerable disagreement as to its utility as a lining in cut-and-fill reservoirs. Opinions differ widely, depending on the background of the person being questioned. Firms that do considerable design work and are oriented toward concrete in this regard are adamant concerning its benefits as a lining in hydraulic structures. Others argue that concrete should not be considered as a lining for reservoirs if water loss is going to present any kind of problem.

As we have seen already, concrete is technically not an impervious material in the strict sense of the word. In the laboratory it is possible to make a test sample that possesses rather good resistance to the passage of water, but in the field this does not seem to be an easily attainable goal. Construction joints continue to present leakage problems, and field applied concrete membranes have widely varying degrees of permability. Good-quality concrete also has the tendency, in time, to degrade with respect to its waterproof qualities.[1]

At present a considerable amount of business activity is associated with attempts to eliminate seepage from concrete structures. Some of this effort is centered on new construction, and some involves the relining of existing facilities. This interest in lining is not generally prompted by economics involving the lost water as such. Rather, the urgency stems from knowledge that the leaking water may cause damage to the structure, its reinforcement steel, or its foundations and that contingent liabilities may also be present.

The following design factors are of importance to the successful utilization of unreinforced concrete as a lining material:

1. Subgrade.
2. Underdrain.
3. Side slope.
4. Concrete mix.
5. Placement and thickness.
6. Joints and waterstops.
7. Curing.
8. Initial filling.

The importance of subgrade conditions cannot be overemphasized in connection with unreinforced concrete lining. In the lining classification system, this material is one of the rigid, semiimpervious types. Although many of the flexible lining systems can tolerate some variance with respect to substrate stability, plain concrete is not one of them. This vulnerability can be tempered to some extent by the use of reinforcing steel, and linings of that type are discussed later in this section.

Concrete linings without reinforcement are particularly vulnerable to actions of frost, swelling, and shrinkage within the soils on which they rest. A careful examination of the substrate by a competent soils engineer is therefore essential before use of the site for a reservoir. A geologist should also examine the site from the standpoint of geologic factors that might render the reservoir inoperative because of subsidence, seismic activity, or the like.

Undesirable surface soil conditions may be remedied by removing the portion of the subgrade in question and replacing it with select material of the desired properties. Nonexpansive materials are used, placed in lifts not exceeding 6 in. and compacted to at least 90% of standard maximum density. In deep, critical facilities a value of 98% is often used. From 6 to 24 in. of the top soil is normally involved in this operation, although the amount can be more in the case of expansive soils, organic deposits, and so on.

If rock is encountered, this does not affect the concrete lining, although it may be altered or backfilled in local depressions to reduce the volume of concrete required. Organic deposits are also removed, and in locations where weed growth is prevalent or probable, a soil sterilant may be applied before lining.

Underdrains are used to protect the reservoir lining against damage from hydrostatic uplift, to aid in leak detection, and to prevent extraction of the earth in the embankments by water movement behind the lining. Although used to a limited degree in the past, they are now considered important by most engineers designing critical water holding systems. These include facilities of large capacity located so as to cause loss of life or widespread property damage should a breach of the structure result. Underdrains should be of the same general design as outlined in the section on cut-and-fill reservoirs in Chapter 2. Many states have set up dam safety standards monitored by an appropriate governmental group. This group usually demands proper underdrain protection for facilities above a certain size and has the power to withold an operating permit until the structure meets its requirements.

The permissible side slopes depend to some extent on the type of underlying material present. Often its angle of repose will determine the slope to which the embankments must be held. In any event the maximum slope for cast concrete linings is usually listed as 1 : 1, but values of 1½ : 1 or 2 : 1 are more commonly used in current reservoir construction. The flatter slopes offer several advantages: concrete is easier to cast, slopes are more stable wet or dry, and people or animals that fall into the reservoir can get out more easily.

The use of a concrete lining does not reduce the importance of a stable slope on which to place it. This principle has been overlooked on occasion by placing the lining on steep slopes composed of sand. Here the thought was that, once the concrete cured and water was in the facility, the lining would stabilize the sand, preventing it from sliding down the slope. This is not necessarily the case, however; and, depending on the circumstances, the slow migration of sand toward the toe area will continue after the reservoir is in service. The resulting condition can cause lining failure, particularly if seepage through the lining is dissipated at a high rate into the soil beneath. When this situation exists, no balancing back pressure is present to offset the pressure exerted by the impounded water. The concrete cannot follow the subgrade surface and hence must "bridge" the gap, creating serious bending forces in the slab. If the head of water is high enough at that point, the concrete will rupture.

Slope stability is of paramount importance for the success of any lining installation. The flexible membranes have some "forgiveness" built into them because of their stretch characteristics, but with unreinforced concrete there is a very limited amount. When such conditions exist, soil stabilization techniques often provide an effective and economical remedy.

In concrete work the recipe is called the mix and is analogous to the compound referred to in plastics work. Although many things are involved in the development of the right mix, probably the most important single factor affecting the quality of concrete is the amount of water used. The more water added to the mix, the less impermeable is the resulting concrete. There is a minimum amount of water required to react properly with the cement, and the closer the water content is held to this value, the more impermeable and durable the concrete will be. It is theoretically possible to produce a concrete lining that is completely watertight. Unfortunately, however, the water content required to achieve this end is so low as to make field handling almost impossible. For this reason this theoretical ideal is not generally reached in the field.

The amount of water to add to the mixture generally depends on the strength of concrete desired. In concrete hydraulic structures, however, the water–cement ratio is governed more by the expected exposure conditions than by strength considerations. Hence both criteria are important in designing the mix. In most cases the exposure conditions dictate a stronger and better concrete than the structural design requirements. Nevertheless, experienced engineering firms designing hydraulic structures insist on a minimum of six sacks of cement per cubic yard.

The water content (as pounds of water per pound of cement) of mixtures for hydraulic facilities ranges from 0.58 to 0.62 for mild exposure conditions down to a range of 0.49 to 0.53 for more severe conditions of freezing and thawing. If brines are present, the ratio drops to 0.44.

Air entrainment agents are recommended for all of the foregoing mixtures, and it is imperative that they be present if severe conditions or exposure to brine is anticipated. The most common air entraining agents used in concrete for hydraulic structures are Vinsol resin and agents derived from petroleum residues. The addition of air entraining agents will produce concrete less permeable and hence more durable because of the entrapment of minute, disconnected, randomly distributed bubbles. It is estimated that 300 to 500 billion bubbles of entrained air may be present in a cubic yard of concrete so treated. The entrained air serves as reservoirs to accommodate the expansion that results from freezing weather. Problems caused by freezing are relieved as the pressure generated by the expansion of water is "vented" into the air spaces without damage to the concrete. These advantages seem to more than offset the 14 to 16% strength reduction caused by the bubbles. Naturally, however, this reduction must be compensated for in the slab design.

The quality of the water used in mixing concrete is a very important consideration. It should contain no oil and should be free of acids and alkalis. Turbidity should be less than 2000 ppm, and no decaying vegetable or animal matter should be present. In short, the water should be good enough to drink. Even then, excess sulfate content should be avoided if possible. When sulfate is present in 1% concentration, a reduction in concrete strength of more than 10% may result.

Aggregates occupy from 60 to 80% of the volume of the concrete. They should be clean, hard, nonporous, and free of chemicals or coatings that would affect the bonding power of the cement paste. Washed and graded sand, gravel, crushed stone, or commercial slag is generally suitable.

Many grades of portland cement are available, and it is often possible to select the one that will give the desired properties of strength, durability, watertightness, and wear resistance. The development of optimum properties may usually be enhanced by the use of admixtures. These materials may be added for air entrainment, as mentioned earlier, for changing the rate of cure, for the evolution of heat, or for the control of alkali reactivity, to name a few purposes.

The concrete should be thoroughly mixed until it is uniform. The rule of thumb is 1 min of mixing for a batch that does not exceed 1 cu yd. For larger batches mixing time should be increased ¼ min for each additional ½ yd.

The thickness of the concrete in hydraulic structures will vary from 3 to 7 in. As water depths increase, or the structure, for other reasons, falls into the critical classification, thickness should approach the higher value. Less thickness may be tolerated when the subgrade material is granular and free draining than when it is a tight, claylike material.

There are a number of reservoirs lined with unreinforced concrete. Their performance varies widely; some have done well, and others have been beset by leakage troubles of one sort or another. In short, no general statement about their performance can be made because of the complexities of the variables and their interaction with one another.

One initial and obvious advantage of using nonreinforced concrete is that it costs less to lay than reinforced constructions. Also, according to the Portland Cement Association,[2] when waterstops are not required, nonreinforced concrete linings with correct panel size to control intermediate cracking will be hydraulically tighter than reinforced linings with larger panel sizes. This is so because the closer the joint spacing, the smaller will be the opening at each joint. This seems to accord well with statements by the late Russell Kenmir, speaking before the American Waterworks Association conference in Cleveland in 1968:

Although many satisfactory hydraulic structures have been built with the joints spaced at about 40 feet or more, the author's experience indicates that a spacing of 20–30 feet provides greater insurance against cracks developing between joints. Where unreinforced concrete linings are used, the concrete should be placed in strips 10–12 feet wide and with transverse dummy or weakened plane joints spaced about 10–12 feet apart.[3]

Of course, in the design of hydraulic structures, cracks between joints are of great concern, and much time and money have been spent on trying to find a foolproof solution to this problem.

Construction joints in concrete reservoirs are usually contraction joints. If at all possible, expansion joints in concrete reservoir linings should be avoided because movement in a concrete structure of this type can create a future maintenance problem.

Floor joints in shallow reservoirs have often been sealed with joint sealant alone. There is a tendency for such joints to develop difficulties later on, and the final design here will depend on just how critical the structure is from the standpoint of safety, or how much potential damage might result from a breaching of the lining.

Good practice usually dictates the use of both waterstops and joint sealant in floor joints, particularly in deep reservoirs. The cutoff point between deep and shallow facilities is a gradual one but lies somewhere in the 20 to 30 ft range. Below 20 ft is considered shallow, whereas above 30 ft of depth is regarded as entering the deep phase. Polysulfide and polyurethane joint sealants have been widely used in reservoir work, with the latter appearing to perform better, although movement in structures has caused leakage difficulties in joints caulked with either material. Here again there seems to be no substitute for good overall design of the entire system, from the subgrade preparation right on through the lining. None of the component parts can be used as a remedy for improper design techniques.

Vibration of the concrete placement is extremely important if a good-quality lining is to result. Specifications must be followed carefully in all phases of this work to ensure proper compaction. The common "wormholes" are almost always the result of poor compactive techniques rather than, as many people believe, air entrainment.[4]

Concrete lining should not be placed in extremely hot weather or at a temperature near or below freezing. If freezing temperatures prevail and concrete must be cast, heating is essential to ensure that cement and surrounding forms do not drop below 40° F.

Adequate curing must go hand in hand with proper temperature of casting and must also compensate for too rapid a loss of moisture. This is accomplished by application of curing blankets to the new concrete immediately after stripping the forms. When no forms are used, the same precautions concerning excessively rapid cure are in order. In lieu of the blankets, a 14 day water cure is desirable for hydraulic structures, or the concrete can be coated with any one of a number of curing compounds or impervious blankets designed to retard the rate of water loss.

During the initial filling operation the water level should not rise faster than 1 vertical ft/day to give the concrete time to adjust to the pressure. Even more important, perhaps, is the fact that this procedure allows the concrete time to absorb water. The swelling thereby induced tends to reduce the widths of any small cracks that may have been formed.

Many of the precautionary measures discussed for nonreinforced concrete when used in hydraulic facilities also apply when reinforcement is utilized. Some additional points are worthy of special mention.

Reinforcement in concrete reservoir linings is required when panel sizes are too large to reasonably control intermediate cracking. The main purpose of the reinforcing is to hold the edges of cracks together if and when they form. When this is done, larger panels can be used.

The spacing of construction joints seems most efficient if held in the 20 to 30 ft range. They serve, in reality, as contraction joints. As with unreinforced design, true expansion joints in reservoir linings are to be avoided if at all possible. Movement of this type in a lining design is sure to spell trouble. Joints can be kept to a minimum when the reservoir is roofed, but when the lining is exposed the situation is different. Below the waterline the concrete is kept cool by the water, while above it the sun assures a much higher temperature. A construction joint under this condition can hardly avoid being a contraction and expansion joint. In such cases waterstops are probably the best bet, and sometimes this type of design utilizes joint sealers too. The vulnerable part of such a partnership is the waterstop itself. If it fails, the caulking material is usually in trouble, because it cannot maintain the reservoir integrity for long periods of time under conditions of heat, cold, joint movement, and alternate wetting and drying, to say nothing of problems induced by freeze–thaw cycles. The latter may also work within any minute spaces that develop between the sealer and the concrete surface.

Gunite

The advent of airblown concrete, commonly called shotcrete or Gunite, placed concrete in a more competitive position. Here the mix of cement, sand, and water is sprayed onto the surface to be lined without the necessity for using forms. Since the equipment is more mobile, inaccessible sites may be treated. The ground surface need not be as level as is required for concrete itself because the application follows ground contours without wasting material. A hard, rough working surface results, but it can be finished to almost any texture desired. A particular advantage of the shotcrete method is that the cement mixture can be sprayed on vertical surfaces without forming. This development cut the cost and helped to make the method more popular.

Although Gunite opened the door for the deeper penetration of concrete into the lining field because of price, it brought with it most of the disadvantages of the concrete lining. Cracking during the curing process was still a problem, and of course differential settlement caused great difficulties with the lining. The introduction of wire mesh to give the Gunite more strength helped considerably.

Gunite is hard to patch, generally for the same reasons that make the operation difficult on concrete. Many techniques have been tried, but the really successful ones depend on special technology. Of course, the facility may be relined with a continuous membrane system, and this is probably the most effective long-term method of seepage control. When a crack treatment method alone is used on Gunite or concrete, initial results are often encouraging, but leakage reappears later as the patch malfunctions or cracks show up at other locations.

There is also the problem of what cracks to repair. Not all of them go completely through the lining. The fact that their end points are not well defined, trailing off to infinitely small imperfections, makes it difficult to know when to terminate the repair procedure. To be sure of a good job, all cracks must be repaired. Often cracks are not visible when the reservoir is empty, but may open up and pass water under the weight of the total reservoir head. Conversely, cracks visible in empty facilities do not always leak under a full head.

One of the great advantages of the Gunite lining is that there is almost no demand as to the texture required in the subgrade. This is true also to a limited extent with concrete (an uneven subgrade wastes concrete), but none of the impermeable lining systems can match this point. In cases where extensive and costly subgrade preparation would be required, the shotcrete lining may fill the need, even though it does pass a certain amount of water because of cracks and porosity. One good example of this is its use as a lining in very rocky areas; doubtless because of the subgrade stability, it does well in this type of application. In fact, this is about the only way such a site can be lined with any degree of economy. Not all rock is stable, however; so again the geologist may be called upon to make the final decision.

Gunite linings require good subgrade and underbank stability, and of course this is necessary in both the dry and the saturated condition. This requirement is not a deterrent to the use of Gunite, however, since all lining systems demand the same condition.

Gunite-type linings in reservoirs require underdrain systems for the same reasons that the latter are installed in conjunction with other linings. They are not as frequently used in these applications, however, probably because of past design habits.

Shotcrete first gained momentum as a lining in canals, where underdrains are not often used. In some respects canals are more favorable for the omission of drain systems than are cut-and-fill reservoirs. Lower water depths are the rule here, and this is a big help. Another favorable factor is that canals in the West are traversing permeable soils in many areas, and these tend to serve as built-in underdrain systems. Then, too, water levels tend to fluctuate less in canals than in reservoirs.

When water is trapped behind a Gunite lining because of fast drawdown, heavy rains, or increased groundwater activity, serious problems can result. If this water cannot percolate back through the lining fast enough, the reverse hydrostatic condition will cause cracking and buckling of the lining. Many installations reflect this difficulty, creating maintenance problems that are hard to solve.

A number of special relief valves have been tried in efforts to automatically take care of back pressures behind the lining. None has been very effective, however, and their use will not compensate for lack of a good underdrain system. These valves are prone to malfunction due to clogging, and where to place them is always a vexing problem.

As already mentioned, shotcrete may be placed on any slope steepness desired, including vertical surfaces. When side slopes are steeper than 1½ : 1, it is more economical to utilize wire mesh reinforcement than to risk problems associated with concrete under tension. The reinforcement is generally not less than 4×4 mesh utilizing No. 14 wire.

Sharp changes in the direction of the lining should be eliminated, as stresses at these points can concentrate cracks there. Usually a 12 in. minimum radius at intersections between two surfaces is good insurance that cracking there will be minimal.

The recipe or mix for shotcrete is less complicated than that for concrete, as only three ingredients are involved: sand, cement, and water. The ratio of sand to cement is held to rather narrow tolerances. For best results it varies from 4 : 1 (4 sand : 1 cement) to 4½ : 1. Mixtures richer in cement than the 4 : 1 ratio are seldom used.

Water addition is at the nozzle and is controlled by the operator. The amount added will run from 6 to 7½ gal per sack of cement. This corresponds to a ratio of 0.46 to 0.58 lb water/lb cement, about the same range as for poured concrete in more severe climates.

Fine aggregates should be well graded, with particle size limited to ⅜ in. for the large guns down to ¼ in. for the smaller sizes. They should be hard and clean with not more than 5% by weight of silt or clay. Machine mixing of the ingredients is the normal procedure; thereafter, they are passed through a sieve to remove lumps that might interfere with the gun operation.

Shotcrete is literally "shot" from a gun onto a predampened subgrade. The guns operate from an air compressor with 60 to 300 cu ft/min capacity. In the mixing chamber of the gun the pressure should range from 30 to 90 psi. Water is introduced at the gun, and it is necessary that its pressure exceed the air pressure by at least 15 psi. In no case should the water pressure be less than 60 psi.

The thickness of Gunite linings will vary from 1 to 4 in. for most applications, with the bulk of the installations falling in the 1½ in. to 3 in. range. On side slopes steeper than 1½ : 1, wire mesh reinforcement is required to prevent fresh mixture from sliding down the slope or wall. With good application practice almost any slab thickness may be built up, but ordinarily vertical and sloped surfaces limit thickness to about 1 or 1½ in. per pass. At thicker placement rates the fresh mass may slump. Better control of thickness is achieved if the lining is placed in ½ in. layers.

Sand and cement are mixed dry in the proper ratio and then fed to the lines, which carry the product to the nozzle. It is at this point that water enters the mixing chamber, and the entire wet mass is ejected onto the surface to be lined. The operator stands about 3 to 5 ft from the work, holds the gun perpendicular to it, and uses a rhythmic motion similiar to that employed in spray painting.

The use of expansion joints in shotcrete linings should be avoided if at all possible, for the same reason that they should be avoided in concrete linings. The feasibility of eliminating these joints will depend partly on the climatic conditions prevailing at the time of installation. If the lining is placed during the hottest part of the year, expansion joints will probably not be required. On the other hand, placement in winter months will mandate their use. Some engineers place them regardless of the weather conditions.

Contraction joints will normally be required for Gunite-type linings. The usual practice is to space these on 10 ft centers. The standard control grooves are cut about ¼ in. in width and one-third the depth of the lining. Many joint sealing compounds are available, but their use has not been associated with widespread satisfaction, particularly on a long-term basis. Materials in this use that are rigid enough to withstand the water pressure lack the flexibility to cope with movement. Many of these have excellent adhesion to the concrete surface. In most cases the adhesion is stronger than the cohesion within the concrete itself. When this condition exists, contractive forces can cause a failure of the concrete itself. On the other hand, if the flexibility or softness of the sealant is at the other extreme, impounded water pressure may force extrusion of the compound into the crack. Its dissipation in this manner will lead to water leakage at these joints.

The ideal sealant would represent a compromise between the two extremes mentioned above, but unfortunately no outstanding product of this type has been developed. For this reason a composite type of system has been used. Here the crack width is increased, and a rigid member capable of supporting the total head is placed first, set into a thin mastic–adhesive layer. On top of this, a softer sealing material is laid, the total construction being shown in Figure 10.11. The method increases the length of the path that seepage water must take to escape the facility. It is effective for this reason, not because the water pressure tends to "make the seal improve," as many people believe. As the drawing shows, this type of crack repair has a much greater chance of success if the reservoir is completely lined. In this case any hairline cracks that appear in the concrete just adjacent to the original crack will not transmit leakage water.

Another system has been used; and as long as no other cracks develop later, it is quite effective. There are many variations of the system, the most popular one being shown in Figure 10.12. A metal with sufficient strength to bridge the gap is placed over the crack. To keep it properly aligned, it is riveted on one side of the crack with flat-head percussion driven studs. This permits the plate to slip with respect to the crack as the reservoir "moves" under varying heads.

In all cases of crack repair the structural integrity of the facility should be prechecked before remedial work of this type is done. The engineer should check especially for void areas below the slab, since the lining, by restricting balancing water pressures, can be responsible for caving in the slab.

Proper curing is an essential part of the quality control program necessary for good Gunite work. The finished surface of the newly placed lining must be kept moist for 5 days, either by coating with a membrane sealing compound, by continuous spraying with water, or by covering with constantly dampened burlap, straw, or earth.

The first filling of the shotcrete lined facility should follow the precautions already outlined for concrete linings. This amounts to a restriction on the rate of filling amounting to 1 vertical ft/day.

Asphalt Concrete

To assess the use of asphalt concrete as a lining for hydraulic facilities it is necessary to carefully analyze the total design. Not only must one look at the asphalt concrete, but, more importantly, the type and nature of the substrate must also be carefully evaluated. The type of facility, its intended use, and the risks associated with leakage must also be taken into account.

Historically, asphalt based material has been used as a seepage barrier for many thousands of years. First as a caulk, then as a pure membrane, it helped the early civilizations to waterproof first their canals and aqueducts and later their baths and sewage conduits. Some of these facilities are still in use, attesting to the longevity of asphalt. Despite these fine credentials, asphalt concrete must be used properly in today's hydraulic structures if noteworthy results are to be obtained.

Asphalt concrete is an extremely versatile material. It has considerably more flexibility than its cousin, cement concrete, and this property facilitates its use in hydraulic works. Water has little effect on it, and it lends no noticeable taste or odor to potable supplies. It may be combined with an almost endless number of other materials to enhance its characteristics and to reduce its cost. A vast myriad of equipment has been developed to convert the asphalt aggregate mixtures into blankets of desired thickness and composition.

On the other hand, there are limitations with which the designer should be familiar before he ventures into the application of asphalt concrete linings in hydraulic facilities. Although this type of lining is not a continuous, impervious, flexible membrane system by definition, it is included herein since, by common usage, it has been placed in a large number of hydraulic facilities where the design intent was, or should have been, to provide seepage control to the system. Its effectiveness at achieving this goal depends on a number of factors which will be examined in this section.

Asphalt is defined as a dark brown to black cementitious material in which the major constituents are bitumens. When heated, the mixture will liquefy. This may occur naturally or be obtained in the refining of petroleum. In the hydraulics field, asphalts are often called asphalt cements; when mixed with aggregates, they become asphalt concrete.

Perhaps the best known property of an asphalt cement is its penetration value, which is also one of its more important parameters. In this test, usually at 77° F, a standard needle placed on the top surface of the sample is loaded with a 100 g weight and allowed to penetrate the asphalt for 5 sec. The unit of penetration is 0.1 mm. Thus a penetration value of 40 designates a penetration of 4 mm and would indicate a harder product than one with a penetration value of 100. The Asphalt Institute recognizes five penetration grades: 40 to 50, 60 to 75, 85 to 100, 120 to 150, and 200 to 300. Many other tests are run on the material, the most important of which are viscosity, flash point, ductility, solubility, specific gravity, and softening point.

In most asphalt concrete mixtures used in the hydraulics field, asphalt cements of 40 to 50 or 60 to 70 penetration are used (see footnote, p. 63). These hard grades of asphalt give better stability to the asphalt concrete when used on side slopes, particularly in hot desert climates.

Mix designs vary, depending on the end use and the experience of the designer. There are literally endless ways to combine the various materials, and experts usually disagree as to the exact proportions of aggregate sizes and asphalt types to use for a particular job. Although each asphalt man has his own ideas as to what is required, in general the final recipe will be reasonably close to the mixtures recommended by others qualified in the field. These variations in design come about because of the complex makeup of asphalt and also its wide range of properties. In general, it does not have sharp values for such things as softening point. Rather, the material becomes progressively softer with increasing temperature, and hence melting point must be carefully defined to prevent its being construed to cover a rather wide range of temperature.

Despite these complications it is possible to list a definite mix design for a typical dense-graded asphalt concrete. The following is representative of one developed by the U.S. Bureau of Reclamation in Denver:

Sieve Size	Percent Passing
¾ in.	100
½ in.	85–100
No. 4	55–80
No. 10	35–60
No. 40	18–30
No. 200	5–12
Asphalt cement, percent by weight of total mix	6.5–9.0

Basically, the mix should have a high degree of workability while hot, be stable enough to support its own weight on side slopes, and have a smooth, erosion-resistant surface. It should be compacted to at least 97%

of laboratory density obtained by the Marshall test method. Aggregate used in the mix should be sound and of a size not more than one-half the thickness of the pavement. Aggregates of low angularity make for easier workability of the mix.

Lining thickness should never be less than 1½ in. If more than 3 in., the lining should be placed in two or more courses, using staggered construction joint techniques. Even though asphalt concrete can be in much greater thicknesses per pass, in reservoir lining work it is recommended that this not be done. For best results here, the 3 in. per course maximum should be preserved.

The total system design must adhere to certain basic principles of lining technology. An important consideration in asphalt concrete work is the side slope. Flat slopes are recommended, and certain restrictions are placed on slopes, depending on the vertical depth of the facility. The following values are recommended by the Asphalt Institute:

Vertical Height of Facility	Maximum Slope
Up to 10	1½ : 1
10–20	1¾ : 1
20–40	2 : 1
Over 40	2½ : 1

In actual practice these slope values are flattened more than the table indicates, particularly in desert climates. The trend by experienced reservoir designers seems to be in favor of slope ratios of 3 : 1, almost regardless of the type of lining being considered. In some cases even flatter slopes are specified.

One of the prime requisites for the use of asphalt concrete as a hydraulic blanket is substrate stability. Although the asphalt concrete blanket has more flexibility than cement concrete, it cannot tolerate any appreciable movements. As a membrane on rock filled dams, where it has been used many times in Europe, it does an excellent job in controlling seepage. Blanket thickness usually runs from 9 to 14 in. and several lifts are applied, each with staggered joints, using special compaction techniques at junctions between blanket edges. But a rock fill dam presents a highly stable substrate, not usually found in cut-and-fill reservoir construction.

Contrary to the generally accepted belief, asphalt concrete is not placed as the prime lining when normal leakage tolerances are specified. Rather, it is used in conjunction with impervious-type clay systems, where it serves as an effective erosion control blanket. Or it may be employed as an overlay wearing surface for thin membrane systems, as in parking deck constructions. Although in both these cases it offers an additional resistance to the passage of water, the extent of its contribution is limited. Publicity releases, however, refer to these types of facilities as asphalt lined. It is this type of misinformation that has led to the confusion with respect to asphalt concrete linings. They have, by this means, been incorrectly credited with excellent performances and also unjustly blamed for bad ones.

Contributing to the confusion are facts developed within the laboratory. Here it is possible to produce asphalt concrete samples with zero porosity. The problem arises from the inability of contractors to duplicate the laboratory results over large areas in the field. Good initial performance from an asphalt lining demands rather careful attention to both mix design and installation details. The control of mix temperatures at the time of spreading, the time lag between this operation and compaction, and the compacting effectiveness itself are three important but difficult parameters to control. In essence the contractor has not been able to exercise the degree of control that will assure 100% probability that these three factors will be the same in the field as they are in the laboratory. For example, obtaining good compaction requires different techniques and is more difficult to achieve on the side slopes than on the flat bottom areas.

To further complicate the problem, sun aging, movement tendencies down the slope, and normal subgrade movements all combine to reduce the effectiveness of the lining just as soon as the contractor has completed his work. After the water has been introduced, the aging process is arrested, except for the areas above the waterline. Nevertheless, all of these factors are constantly interacting, thus hastening the total process of degradation. This interaction is illustrated by the increased aging deterioration as as side slope steepness is increased.

One of the time-proven factors for enhancing the aging properties of an asphalt pavement is kneading. Normally, in a paving composition this working is provided by the tires from passing automobiles and trucks. Abandoned asphalt paved roads degrade at a much faster rate than those in daily use. In a reservoir lining this type of kneading action is missing. It is not very clear as to how important this curtailment of traffic is with respect to the longevity of the asphalt lining blanket in hydraulic use. The effect below the waterline is no doubt of lesser importance, although experts disagree concerning this. Unfortunately, there is nothing in the literature that deals precisely with this unknown entity.

Asphalt concrete does find wide application as a wave erosion control blanket for protecting the slope areas of reservoirs lined with compacted earth. As mentioned, these types of facilities have sometimes been referred to as asphalt lined. Physically, this is a true statement, but the asphalt lining is actually for erosion control; the compacted earth serves for leakage control. News media often do not make this distinction, causing understandable confusion for the layman (e.g., reporting of the Baldwin Hills Reservoir failure).

As a protective slope blanket asphalt concrete is extremely effective. It is vandalproof, economical, and repairable. It can be made in a wide variety of textures, and the degree of voids within the mix can be increased so that reverse hydrostatic pressure does not cause problems on rapid drawdown of water levels.

Asphalt concrete is subject to damage due to icing conditions, which, when occurring within the porous areas of the mix, can cause a spalling effect. Asphalt concrete must be carefully evaluated when certain chemicals are present in the pond, especially oils, which attack the asphalt, or chemicals that attack the aggregates. When the liquids being held are hot, it is well to proceed with caution as asphalt is a thermoplastic material. Naturally, asphalt concrete cannot be used on vertical slope work; generally its use is restricted to slopes at least as flat as 2 : 1.

Because its porosity can be altered, asphalt concrete has had widespread use in the construction of continuous underdrain systems, such as those made popular by the East Bay Water Company of Oakland, California. Here the porous blanket, usually of 3 or 4 in. thickness, is topped by a smoothing course, which prohibits the tendency of the prime lining to extrude into the voids of the open-graded asphalt layer. This continuous underdrain then terminates at the toe in a network of perforated pipes, which carry seepage to a monitoring station. The underdrain blanket then proceeds across the bottom to the opposite toe.

Because of their porous nature asphalt concrete lining systems are prone to penetration by weeds above the waterline, a problem that can plague all linings exhibiting any discontinuity. Here the seeds have access to moisture from above; and when the lining cracks, direct sunlight is also available. The blanket, being black, absorbs heat readily and serves as an incubator for the weed seeds that lie below it. The problem is more or less eliminated if the structure is roofed; otherwise, a soil sterilant is often used, particularly if the facility is built in a location where weed growth is suspect.

Steel

The use of steel as a lining for cut-and-fill reservoirs is a rather novel application for the metal. This is somewhat surprising because steel has been employed for many years in the waterworks construction field. Hundreds of steel water tanks dot the land, and the use of this material in pipeline construction is well known. With these credentials one might ask why this common material has not found application in the earthen reservoir lining field.

There are several reasons why designers have not taken more interest in steel linings placed directly on the ground. Steel, although actually a rather low cost item as structural materials go, is relatively expensive for use as a lining. One of its principal advantages is structural strength, but in a properly designed and constructed cut-and-fill reservoir this is of limited value because the earth itself serves as the structural support. Designers wonder, therefore, why they should spend additional money for a second structural material. Their attitude regarding this point is perhaps the basic reason for the lack of interest in steel as a prime lining.

As mentioned above, its structural strength constitutes a great advantage for the use of steel as a lining. Suppose that something happens to the earthen support. Then the steel can take over and perform this structural duty. It can do this, of course, only if it does not have to stretch so far as to exceed its yield point. In the case of minor earth settlement, however, this will not normally happen.

One of the first reported uses of steel as a lining for a cut-and-fill reservoir took place in 1962 at the Dietz Reservoir in the city of La Mirada, California. The facility contained 7¼ million gallons, and the owner gave, as one of the reasons for the use of steel as the lining, that the reservoir was "a very critical installation, where failure could not be tolerated." Obviously, the steel's structural strength was being used as a backup safety factor.

The entire design and construction procedures were reported in a paper given before the California Section of the American Waterworks Association at its Santa Monica meeting in October 1962.[5] At the time of this meeting there was considerable disagreement among the experts as to the wisdom of placing steel directly on the earth as a prime lining. From the design standpoint the critics were wondering, with considerable justification, about corrosion from the underside. Steel tanks have been in use for many years, and the undersides of these tank bottoms, when they rest directly on earth, have given owners a fair share of maintenance headaches. Even when oil is stored, and when, in the absence of water within the tank no corrosion problems are noticed, there is still considerable maintenance involved in trying to keep the bottoms repaired against the pressures of corrosion from the underside.

Dietz Reservoir has inside plan dimensions of 400 × 128 ft and a depth of 25 ft. The side slopes are 1½ : 1; on two sides there are 8 ft vertical walls, and on the other two sides 3 ft vertical walls. These walls were decided upon when the site grading was half finished. Because of the cost the grading plan was not altered when this change was made. Although the walls did increase the cost, they also added 1 million gal to the capacity of the reservoir.

The underside surfaces of the lining were protected by sealing off circulating air and keeping the area dry. It was the accomplishment of this feat that led to the aforementioned skepticism voiced by some experts. The entire bottom of the reservoir was covered with a 6 in. blanket of ¾ in. crushed rock, and the side slopes were sprayed with SC-2 asphalt.

The 3/16 in. plate thickness was chosen for the lining because it is available in large 8 × 40 ft pieces. A thinner plate would have done as well, but these plates are not made as large and are more costly to weld because they burn through more easily.

Roof column foundations on the sloping walls were poured of concrete flush with the lining and with short pieces of pipe columns cast within them. The lining was welded directly to the pipe and reinforced at these points. Roof columns that rest on the flat bottom areas were attached to bearing plates and bear directly on the lining itself.

The lining was coated after the roof structure was completed. The top 3 ft on the vertical wall was coated with super tank solution. The sides and bottom were treated with hot-applied coal tar enamel placed over a coal tar primer. In all cases the plates were first sand blasted to a gunmetal finish.

On the 8 ft vertical walls the lining terminated at a toe of the wall cast-in-place steel channel, to which it was welded. The owner thought that the lining should have extended all the way to the top of the wall and would design in this fashion on any future installations. It was at the channel that leakage problems were most severe, and later remedial work had to be done there to arrest the problem.

During installation of the plates, which took place at temperatures between 32° and 100°, considerable buckling occurred during the heat of the day, but the panels all laid flat in the cool of the evening. No special expansion joint assemblies were used, as the large area would compensate for any expansion or contraction that might be experienced.

The lining area was divided into 58,000 sq ft of steel and 4230 sq ft of concrete, for a total area of 62,230 sq ft. Over this total area the lining system cost a little over $101,000 or a unit price of $1.63/sq ft. The total cost of the reservoir, including excavation, lining, roof, and landscaping, was $300,000 or a unit price of 4.15 /gal of storage. Had a more conventional, flexible, impervious membrane system been used, over $100,000 would have been saved, thus decreasing the unit price for water storage to less than 2.75 /gal.

Despite the rumblings from the skeptics, the steel lined reservoir in La Mirada is still performing well as of this writing. Time will tell as to how well the facility avoids corrosion problems in the coming years.

Soil Cement

Soil cement is produced from a mixture of soil, cement, and water. Any soil that can be easily pulverized and contains less than 50% silt and clay is suitable for soil cement construction, although a silt and clay limit of 35% is generally most practical and economical for lining work.

Engineers have been using cement mixed with soil for some 40 years to improve the engineering properties of pavement bases and subbases. As the cement hydrates, properly compounded soil cement mixtures become hard, durable, and resistant to freezing, thawing, wetting, and drying.

Soil cement possesses superior stability over natural soils with respect to erosion. It also has much greater shear strength and is less permeable. These properties suggest the possibility of using soil cement as a lining, and in the 1940s several experimental lining projects were undertaken. In 1969 soil cement lining was used in the Cahuilla Reservoir, a terminal storage facility at the end of the Coachella Canal. The water occupies a surface area of 160 acres with a maximum water depth of 12 ft. The lining is 6 in. thick on the floor, increasing to more than 24 in. of the slopes.

A number of soil cement installations for slope protection and dam facings were constructed in the 1950s, beginning with the test section on the south bank of Bonny Reservoir near Hale, Colorado, in 1951. Despite the severe exposure conditions at this site, the dam facing has stood up well, as have 30 similar installations throughout the country, some of which were cooperative projects with the U.S. Bureau of Reclamation.

The Portland Cement Association classifies soil cement in two categories, compacted and plastic types. In the former class installations are made in two ways: *in situ* and central plant mixing. Most of the large installations are made using the central plant mixing technique, as this offers the best overall economics. The smaller installations are more economical if the cement is mixed with the soil and compacted in place (*in situ*). Plastic soil cement requires just enough water to produce a mixture with the consistency of plaster or mortar. This type is used when areas are small or larger equipment is not practical, as in highway drainage ditches. To produce the plastic consistency, more water and more cement are required than for the other construction methods.

For soil cement work, three controls are required: (1) proper cement content, (2) proper moisture control, and (3) adequate compaction.

To determine how much cement is needed, the soils engineer should identify and sample the soils to be used in the mix. Laboratory tests determine how much cement is required for each type of soil present in the project. Test specimens are made up with varying cement contents in a duplicate series. One series is put through ASTM D-559 wetting and drying tests, and the other through ASTM D-560 freezing and thawing tests. A total of 12 cycles is run on each of these tests, and the cement requirement of the soil is based on the loss undergone by the specimens during the cycles. In general, a 14% loss is permissible in granular-type materials, a 7% loss in silt–clay materials, and a 10% loss in soils that lie between these two groupings.

The moisture content and density of the compacted soil cement should be controlled to be close to the optimum moisture content and maximum density as determined by laboratory testing. In the field the object is to mix pulverized soil with the correct amount of cement, as determined from the above tests, with enough water to permit maximum compaction.

There is little restriction on the type of water that may be used in the process, other than that it be free from excessive amounts of alkalis, acids, or organic matter. Even seawater may be used.

In situ construction of soil cement is preceded by fine grading the area to be treated in order to remove any debris, rocks, and the like. Soft subgrade is to be avoided and, if found, is remedied by removing the soft material and replacing it with stable materials or by mixing in cement and recompacting. Specifications generally require that, exclusive of gravel or stone, the soil must be pulverized enough so that 100% of the mixture will pass through a 1 in. sieve and at least 80% through a No. 4 sieve. If the soil is lumpy, scarification may be necessary. Wetting before this step usually helps in this operation.

Weather conditions must be watched rather closely during the placement of soil cement regardless of the technique used. The operation should not be carried out during freezing weather, and any material placed should be protected from freezing temperatures for 7 days thereafter. Rain is not usually a problem, as the mixing water required is generally equivalent to more than 1 in. of rain. The critical period occurs between the spreading of the cement and the compaction. Once the mixture is completely compacted, rain does not bother it.

On small projects the cement may be distributed in bags and spaced according to the cement content required. The bags are broken open, and a spiked tooth harrow or similar piece of equipment spreads the cement evenly. A disker may be used to dry-mix the cement evenly with the soil.

Water is added by means of pressure distributors in amounts as large as the equipment and soil will permit, in order to bring the mixture to optimum moisture or slightly above. Sandy soil cement mixtures need about 5 gal water/sq yd for a 6 in. compacted thickness, while silty and clayey mixes require about 7 gal. A soil cement mixture at optimum water content should make a firm cast when squeezed by the hand and be broken into two pieces without crumbling.

Mixing continues after water is introduced until the soil cement is uniform in color and free from wet and dry streaks. Initial compaction begins immediately. Not more than 30 to 60 min should elapse between the start of moist mixing and the beginning of the compaction step. If this is done, less water will be lost and proper densities will be obtained more easily. Water lost by evaporation should be replaced in light applications. With respect to compaction, specification wording varies. Most specifications indicate a tolerance within 5 lb of the maximum density, but some indicate "at least 95% of maximum density."

In the compaction step the most commonly used pieces of equipment are tamping, rubber-tire, and steel-wheel rollers, but some of the newer types of equipment can also be employed. Plate vibrators, grid rollers, and segmented rollers may be used to compact soil cements made with nonplastic granular soils. For all but the most granular soils, tamping rollers, ballasted to create higher unit pressures, may be employed. Rubber-tire rollers are useful to compact sandy soils with little or no binder material and also sand and gravel of low plasticity. Cohesionless sands may be compacted by the weight and vibration of the tractor alone.

The compaction equipment may be operated directly on slopes of 5 : 1 or flatter. When the slopes are steeper, the use of traveling deadmen on the top of the embankment is often mandatory. When soil cement is used for facing a steep slope or dam, the "stairstep" construction method is used and central plant mixing of the soil cement is recommended, as will be discussed later.

After the soil cement placement and compaction are completed, the installation is immediately covered to prevent too rapid a loss of water. Various materials are used for this purpose, the most common being waterproof paper, plastic, moist straw, moist earth, and bituminous materials. In the latter category RC-250, MC-800, RT-5, and asphaltic emulsions are the usual choices with an application rate of from 0.15 to 0.30 gal/sq yd.

Most large soil cement installations have utilized central plant mixing. Capacities range from 200 to 600 tons/hr, and complete plants are produced by a number of manufacturers. They are made both as continuous and as batch plants. Most specifications call for proportioning devices to be accurate to within 3% by weight of the design mix quantities.

Accurate calibration of the plant is an essential first step before production begins. Soil feed rates will depend on the size, gradation, and moisture content of the stockpile. Although mixing time varies, the usual cycle is around 30 sec. Water is added along with the cement and soil, in opposition to the *in situ* method.

Finished batches are dumped into hauling trucks equipped with protective covers to protect the batch against rain or other types of inclement weather. The hauling time should be restricted to 30 min.

The surface on which the soil cement is to be placed should be firm, clean, and moist. The mix should be laid down by means of a mechanical spreader. If its width is not equal to that desired, a windrow spreader should be used to obtain full width of the loose material. Loose material thickness will normally run from 8 to 9 in. for a 6 in. compacted layer. No portion of the spread layer should remain undisturbed for more than 30 min.

If a multilayered construction is being placed, a sheep's-foot roller can be effectively used to obtain the 96 to 98% maximum density required for slope protection. Pneumatic-tire or steel-wheel equipment can follow up for final compaction of the upper portion of the layer. These are common techniques, but other types of compaction equipment, as discussed earlier, have also been used to good advantage. In any event compaction of each portion to the required final density must be completed within 1½ hr after the material has been spread.

Soil cement for slope protection is normally not trimmed or graded as a final step, as is usually done for road work. Bonding between subsequent layers of the mix is optimum when the existing layer is still fresh. It is not necessary to keep the first layer continuously moist; but if construction is delayed for any length of time between layers, the bonding is improved by prewetting.

As with the *in situ* process, curing is important, and after the final layer is compacted, or no later than 24 hr of this time, a material to prevent loss of moisture must be added. These materials may be the same types as described for the in-place technique.

Inspection and control should cover seven points:

1. Subgrade condition.
2. Cement content.
3. Moisture content.
4. Mixing.
5. Compaction.
6. Bonding.
7. Curing.

The Portland Cement Association has lightened the inspector's workload in the control of soil cement installations by publishing the *Soil Cement Inspector's Manual,* No. SC-16, which is available at any PCA office.

A natural characteristic of soil cement is the formation of transverse shrinkage cracks soon after its completion. No techniques have been developed to eliminate them, although close control of compaction, moisture content, and curing helps to minimize them. The cracks are said not to affect the performance of soil cement as a paving or a paving base. Also, in dam facings, which are installed to prevent erosion of earthen embankments, surface cracking does not appear to be detrimental.

The formation of curing cracks in soil cement prevents the system from being classified as continuous or impervious. This material has been used as a lining in noncritical facilities, one example being the Cahuilla Reservoir, already mentioned. Its use as a reservoir lining, however, is somewhat limited despite its efficient utilization of on-site materials. Even with excellent seepage control data, the Cahuilla facility slope lining in 1969 cost nearly 50¢/sq ft. Adjusting this figure for current costs places it at a competitive disadvantage with modern impervious lining systems, not to mention the lack of a long-term guarantee. However, since most of the Cahuilla lining area was on the bottom, where the lining was only 6 in. thick, the average cost per square foot was lowered to about 12¢, a very competitive overall figure. Thus soil cement will no doubt be restricted to large facilities where the bottom–slope ratio of areas is large and no seepage control restrictions are specified.

References

1. Charles Outland, *The Story of St. Francis Dam,* Arthur H. Clark Co., Glendale, Calif., 1963, p. 175.
2. Portland Cement Association, *Concrete Lined Reservoirs,* Chicago, 1955, p. 10.
3. Russell C. Kenmir (James M. Montgomery, Consulting Engineers, Pasadena, Calif.), "Concrete Reservoir Design," Paper presented at Annual Conference, American Waterworks Association, Cleveland, Ohio, June 4, 1968. Reprinted in *Journal of the American Waterworks Association,* Vol. 60, No. 10, October 1968, p. 1188–1189.
4. *Ibid.,* p. 1193.
5. Hal E. Marron (Chief Engineer, Southwest Water Co. and the Valinda Engineering Co.), "Steel Liner for Earthen Reservoir," Paper presented before the American Waterworks Association, Southern California Section, Santa Monica, Calif., Oct. 25, 1962.

5

Miscellaneous Lining Systems

Bentonite

In the general terminology of linings the word "bentonite" designates what is perhaps the least understood of the many systems discussed in this book. Bentonite has been used in considerable quantities as a means of sealing canals, farm ponds, lakes, and dams, but there seems to be a widespread lack of knowledge as to how to specify this material. There is an even greater ignorance of how to use it and when it should and should not be used.

There are two types of field application, wet and dry. The less expensive wet process introduces bentonite powder or dispersion directly into the water, where it tends to seal up the facility as it settles out. The second, more positive treatment places a layer of pure bentonite on the dry surface of the pond, reservoir, or canal, followed by a protective earth covering. A variation of this is used in sandy, open-type soil, where the bentonite is mixed into the top 3 in. of the soil and then compacted. Protective soil covering is again applied where erosion conditions are expected.

Bentonite had its beginning over 50 million years ago as molten rock in the heart of volcanoes. Massive volcanic eruptions hurled millions of tons of fine ash materials high into the atmosphere. These settled onto the inland seas, which covered much of our country at that time, and then sank to the ocean floor. For eons thereafter, these thick layers of submerged volcanic ash were overlaid by hundreds or even thousands of feet of sedimentary debris that continuously rained onto the ocean floor.

As the fine, glassy ash lay in the ocean muck, a subtle change occurred. It was transformed into a waxy, claylike substance with unusual properties. When the seas receded and the land uplifted, the beds of bentonite were exposed or pushed near the earth's surface, where they could be recovered. Although bentonite deposits may occur anywhere, the most publicized ones in this country are found in Wyoming and South Dakota.

Chemically, bentonite is a hydrous aluminum silicate, technically defined as a colloidal clay of the montmorillonite mineral group which swells in water and carries sodium as its predominately exchangeable ion. Mixed with water, a true bentonite forms a colloidal dispersion, which, by definition, will remain in suspension for an indefinite time.

Although bentonite deposits are millions of years old, commercial interest did not begin until 1910. Even then, it was not until the mid-1940s that significant tonnages were utilized. Bentonite has many applications that dwarf the quantities used for seepage control. It is probably best known for its use in oil well drilling muds, which facilitate efficient removal of rock chips from the exploration hole. Substantial quantities are also used in foundry mold work, in iron ore pelletizing, and as a stabilizing agent in dispersions of all types.

Each of the industrial users of bentonite is interested in certain specific properties of the clay that will most effectively solve his problems. To this end, users have developed meaningful specifications, and purchases are made on this basis. In the case of bentonite's use as a sealing membrane for seepage control, however, specification development has only recently begun, and early users of bentonite as a sealant turned to specifications for drilling muds as a means of bridging the information gap. Since drilling muds require a high-swelling clay, this seemed to be a logical path to follow, but there were two serious drawbacks to this procedure. First, the developed high-swelling clay deposits are restricted to a small area centered near the northeastern part of Wyoming. Thus freight costs play an important role in economic evaluations involving the use of bentonite. The second drawback concerns an erroneous belief that a high-swelling clay is the only type that may be used in sealing operations. It is true that all clays with a high swell index have low permeabilities, but some clays with a low swell index also have low permeabilities. To add to the confusion, a high-swelling clay will not necessarily solve a seepage problem.

Developed bentonite pits are operated in a manner somewhat similar to strip mining. After exploration and classification of the deposit, the overburden is removed to expose the raw clay. Since the clay is usually present with a high moisture content of 40 to 50%, it will often be gummy when first uncovered. It is usually harrowed in place to promote air drying and breakdown of the clay lumps. Then the clay is removed by dragline, crane, or bulldozer and is stockpiled according to the original classification.

The bentonite is removed from the stockpile to obtain maximum mixing and thus minimize product variation within and between shipments. The clay is processed through a ¾ in. screen, which removes the majority of the nonclay materials. It is then ground and packaged for shipment, with a 15% maximum moisture content.

The clay is shipped in bulk trucks with spreading attachments or in 100 lb bags. It is usually available in two or more grades from each mill. The following are typical marketing grades:

80 to 90% passing:
- 325 mesh
- 200
- 20–200
- 40–200
- 20–70 pellets
- 4–20

The 200 mesh and 20–70 mesh (also called 40 mesh) are most commonly used in sealing work. The finer material is usually best for blanket (dry) applications, whereas pellets will do a better job for wash-in (wet) work.

There is no compounding step in bentonite processing that corresponds to the one employed in rubber or plastics work because nothing is added to the natural clay before its use as a sealant.

The effectiveness of bentonite as a sealing membrane depends on a number of factors, the most important of which are as follows:

1. Type of bentonite.
2. Preparation of substrate.
3. Proper design and control of sealing operation.
4. Adequate follow-up maintenance.

The best method of defining the type of bentonite is by means of testing and specifications. As with most natural soil deposits, there is a wide variation of properties in bentonite strata. Since freight is an important factor in the economic picture, it is often desirable to utilize local deposits rather than rely on out-of-state grades. The final decision here will depend on substrate soil tests and the technical specifications for the job.

The most common parameter associated with bentonite is the *swelling index.* This may range from a negative value (shrinks when wet) to over 2000%. Sometimes called the *free swell index,* it is the ratio of the increased volume when wetted in distilled water to the original dry volume. The minimum value required for good sealing is 50%. All the best grades of Wyoming bentonite and some smaller deposits in Montana exceed 600%. Colorado bentonite usually ranges below 300%, while in California the index does not normally exceed 120%. The swelling index seems to be associated with the age of the deposit and the length of time it was resting in the muck on the ocean floor. The older and longer aging deposits have the highest indices.

Layer permeability is defined as the loss in feet of water per day through a loosely packed bentonite layer 0.6 in. thick under a 52 in. head of distilled water at the start of the test. The test is designed to correlate with the dry blanket method of installation. Good clays should have a loss rate of 0.005 ft/day or less.

In cases where the sealing must be done with water already in the facility (wet application), the *filter permeability* test is more useful. Here a standard dry sample is dispersed in 400 ml of distilled water, and the mixture placed in a small laboratory filter press of standard design. Under a 14.7 psi pressure, the water loss rate across the press should not exceed 10 ml/min if a sealing clay is to be effective in wash-in applications.

Another property that is important in the evaluation of bentonite clays is *colloidal yield.* For the yield test a standard size sample is dispersed in 1000 ml of distilled water, using sodium tripolyphosphate as a dispersing aid. After the sample has stood for 24 hr, a hydrometer is used to determine the percentage of the original sample still in suspension. The best sealing clays will show a colloidal yield above 40%, preferably above 50%. A sodium montmorillonite will usually give a value around 80%; its calcium counterpart, as low as 40%. These are referred to as sodium bentonite and calcium bentonite, respectively.

Additional tests such as *moisture content* and *particle size distribution* are run, as well as one called *grit content.* The latter is a measure of the nonsoftening or sand fraction of the sample. For sealing coarse or rocky soil, a grit content of 30% may be desirable. In most cases, however, a grit content of 10% or less is preferred.

For evaluation of clays being considered for wash-in applications, the *mixing index* is a meaningful test. Here a 20 g sample of designated fineness is placed in the test apparatus and washed for 30 sec with an upward stream of water at 3 in. of mercury pressure. The percentage of the original sample lost in the test is designated as the mixing index.

Clays that are particularly effective when wash-in methods are used will have a mixing index of 40% or more.

It should be noted that all of the above tests are performed using distilled water. This places the clay samples in a considerably different environment from that encountered in actual practice. Certain concentrations of salt or chemicals or calcium and magnesium hardness can cause difficulties, particularly in wet applications.

Suitable sealing clays will usually be 10% or more sodium saturated. Generally, the best grades contain the following:

1. Highest cation exchange capacity.
2. Highest exchangeable sodium percentage.
3. Highest pH.
4. Lowest gypsum content.
5. Lowest water-soluble cations.

When a sodium clay is mixed into hard water, it will quickly be changed into a calcium clay, which is much less effective at sealing than its sodium counterpart. A sodium tripolyphosphate dispersing agent is helpful in preventing this change, but it will not stop the action permanently if the water is hard.

Preparation of the site before the bentonite application is important to the success of the operation. The site should be leveled and cleared of vegetation and other debris. Eroding areas at the waterline and elsewhere should be adequately stabilized. Destruction of the seal by erosion is a common problem causing trouble in bentonite sealed projects, especially in facilities located in fine materials, such as sandy to silty soils.

All structures should be checked for watertightness and repaired as required, using waterproof cement or a pack-in filler soil containing 1 part of bentonite to 4 parts of soil by weight. If there are any highly pervious areas, they should be treated separately with the 4 : 1 soil–bentonite mixture.

The application method may be chosen from the following list of four possibilities:

1. Wet application (canals)
   - A. Wash-in. Clay is introduced at head end of canal, and flowing water distributes clay to places needed.
   - B. Multiple dam. Canal is dammed into sections, and clay is added to water in each section. Retention time is increased, thus usually producing better sealing results.

2. Dry application (reservoirs, ponds, and new facilities)
   A. Pure membrane. Facility is overexcavated 6 in. minimum. Clay membrane is laid down, and protective cover placed over the top of it.
   B. Mixed layer membrane. Clay is tilled into top 3 to 6 in. of subgrade materials, followed by compaction.

Much of the bentonite work on canals involves wet application techniques. The wash-in method is particularly useful for sealing canals with steep grades and limited access. The multiple-dam method is generally used when access is good all along the canal. The latter system is really a wash-in method but is more easily controlled. Both methods are highly effective in canals with coarse, rocky, or gravelly materials which are underlain with fines that bridge subsurface void areas. The best clay to use in either case has a high mixing index, a low swelling index, and low filter and layer permeability. The wash-in method is not effective in hard water or in water with a salt content above 400 ppm. In these cases the clay should be placed as a blanket and covered.

When using the wash-in technique, two methods of adding bentonite to the water are available. In the first a dispersion of the dry clay is made in water, and this is added directly to the flowing water of the canal or the still water of the ponded section. The initial strength of the dispersion is between 2 and 8% of bentonite. High-speed agitators, colloid mills, or home-made jet mixers are best for the dispersion work. When the slurry is added to the flowing canal, costs are low, but it is difficult to assess the effectiveness of the treatment. Then, too, erosion problems cause difficulties as the water velocity often carries away the fresh clay particles or otherwise prevents their effective deposition.

In the other method granular bentonite grades are sifted onto the surface of the calm water in the lake, reservoir, or ponded sections of the canal. On large projects a sand blasting gun will do an effective job of clay distribution. When the areas involved are small, hand sprinkling from a boat or raft is often used. In any case the treatment rate runs from about 1 lb bentonite/sq ft water surface to 1 lb/cu ft water ponded. The clay sinks and forms a gel on the bottom, which is drawn into the leaking zones. How much clay to add will depend partly on initial tests of the soil porosity of the on-site soil. One quick test involves the use of a small, bottom-perforated container in which a few inches of the original soil is placed. Different thicknesses of bentonite layers are added by the sifting method, and quantities carefully recorded. When an effective sealing job has been done, the bentonite quantity used is increased by a factor of 50% to account for the variations of the actual large-scale operation.

In both of the wet treatments the bentonite tends to be protected by the degrading actions of freezing, drying, and, to some extent, wading animals. Animals and marine life that burrow into the soil will destroy the seal.

Canal ponding is accomplished by damming up sections of the canal so that the water velocity is reduced, giving the clay more time to settle into the soil voids. Sections up to 200 ft in length are isolated by means of simple earthen dams and the existing check structures.

When the canal soil contains limited fines, it is often necessary to introduce void-plugging agents before pretreatment. These may be sand, silt sand, or even sawdust, when soil is not readily available. In these cases the sealing process is benefited if the bentonite itself contains a fairly high percentage of grit (sand).

For sealing reservoirs and structures by the blanket method, the most suitable clays will have low layer permeability and a high swelling index. This method is most effective for sealing fine materials such as fine sand and sandy silt. As mentioned before, it should be used in sealing applications involving hard or salty water.

To utilize the pure membrane method, the existing soil is overexcavated to a depth of 6 in., a blanket of bentonite is laid down, and the 6 in. soil blanket is replaced. The quantity of clay to apply will depend on seepage test data on the original soil. Again, the simple perforated can will serve as a useful tool in determining the proper application rates. The test results are increased by a factor of 25 to 50% to account for increased head and natural soil and treatment variations. In most soils around 1 to 2 lb/sq ft of bentonite will be required, although there have been instances where as much as 9 lb/sq ft was needed to effect a seal. In such cases the thinner plastic membranes would no doubt have provided a more economical lining method.

The overexcavation may be avoided by applying the suitable clay blanket over the existing soil and then adding the 6 in. protective blanket using additional soil. This method is best suited to the heavier clay soils, where uniform mixing of bentonite into the soil is difficult or impossible. In any event the top protective blanket should be of such a makeup that it will prevent erosion and subsequent destruction of the clay seal.

When the soil is granular (sandy to silty), the mixed layer membrane method is best. In this system the bentonite is first spread on the existing ground. This is done by hand on very small jobs. A grain drill or fertilizer spreader is also effective in spreading the clay, but on very large jobs specially equipped transport trucks are used. The latter equipment can effectively spread 2 lb/sq ft of clay over an area of 6 to 8 acres per day. The clay is then mixed with the top 3 or 4 in. of soil, using a spiketooth harrow, disk, rotary hoe, or similar equipment. Hand raking may be used too, particularly around the edges of the job and at the structures where the larger equipment cannot travel. The mixing operation is followed by the compaction step, using a sheep's-foot roller or other suitable equipment. In some cases the sheep's-foot roller may be used for both mixing and compaction.

Application rates are determined by testing the soil as described above. Tests are run in several locations throughout the job. Usually, good sealing by this method will be obtained if the clay is added at the following *minimum* rates:

Type of Soil	Application Method	Minimum Application Rate (lb/sq ft)
Clay	Buried pure membrane	1.0
Sandy silt	Mixed blanket	1.0
Silty sand	Mixed blanket	1.5
Pure sand	Mixed blanket	2.0

The values in this table are for clay of a powder grade or up to wheat size granules.

The planning of an adequate follow-up maintenance program is a very important part of the bentonite sealing process. Yearly follow-up treatment is the general rule, with each retreatment consuming about one-tenth of the original bentonite usage. In some cases annual retreatment may not be necessary whereas in others higher than average clay amounts may be required when the reworking is done. The best time for the work is in the spring, when the sealed area is dry or the water at its lowest level.

Of the factors working against the bentonite sealing process, bank erosion and undercutting at the waterline are the most bothersome. Either action will destroy an effective seal. When this occurs, extensive earth preparation may be required before further clay treatment.

Other types of activity will cause difficulties, too, such as those due to crayfish, earthworms, muskrats, prairie dogs, and plants of various kinds. Plants are even more destructive when they die, as the voids created by the decaying roots effectively funnel out the water.

Chemical Treatments

Chemicals have been used for some time as a means of stabilizing soils. Although these techniques are aimed at reducing porosity, their primary objective has been to increase soil stability and/or bearing capacity. The literature is ominously silent regarding the use of chemicals *primarily* for the reduction of seepage in cut-and-fill reservoirs and related structures. In this field, chemicals have not enjoyed any appreciable use.

Among the several reasons for the lack of popularity of chemical treatments as linings, three are important: (1) they are not 100% effective—most times they are not effective at all, (2) they are complex and not well understood, and (3) usually they are expensive, although there are a few exceptions.

Soils are complex entities, involving intricate chemical attractions of their own. They are nonhomogeneous mixtures with infinite relationships between moisture content, void ratio, and capillary attraction, to mention but a few. In addition, chemicals themselves are complex, particularly if mixtures are involved. To compound these things together is to develop an extremely complex mixture that is, more often than not, unpredictable in its ultimate properties. Since proper mixing is time consuming and costly, much attention has been given to application by spraying. The key word is penetration; the chemical must penetrate the soil if it is to have much chance for success. If the soil surface is tight, chemicals do not penetrate well. If, on the other hand, the soil is open, they penetrate too much, thereby increasing the cost. It is difficult to obtain the right mixture, the right viscosity. In fact, what is right, when soils are so variable?

The prime consideration in the chemical treatment of soil is to use chemicals that are products of the plant with the lining problem. If they are waste products, so much the better. In any event they are available in quantity, and, regardless of the plant status, their use will undoubtedly result in the lowest possible cost. As an example, sodium brine (salt) waste, when pumped into a lagoon whose soil lining is predominately calcium clays, will usually effect a considerable decrease in the seepage of clean water. It does so by converting the soil lining to a sodium clay, which is more impervious than its calcium counterpart.[1]

Another example is the chrome–lignin process. Here the lignin contained in sulfite waste liquor may be oxidized with a chromium salt. The speed of the reaction may be controlled so that the chemicals can be premixed and added to the soil as a single solution.[2]

Other chemicals that have been tried with some success in clay-type soils are the polyphosphates. Three have been used most often: tetrasodium pyrophosphate (TSPP), sodium tripolyphosphate (STPP), and sodium hexametaphosphate (SPP). Treating rates range from 0.05 to 0.1 lb/sq ft of chemical, depending on the type of clay. This compares to a rate of between 0.2 and 0.33 lb/sq ft of salt. In all cases the chemical gives optimum results only when thoroughly mixed with the soil, followed by compaction to at least 90% of maximum standard density at optimum moisture content. If reservoir water depths exceed 6 or 7 ft, two treated layers should be used. In all cases laboratory evaluation studies should be run before treatment.[3]

Despite the rather sketchy information available on chemical linings, the technique involving seepage control by chemicals is an exciting one that can lead to an almost endless chain of possibilities. Here is truly a virgin field, the surface of which has not even been scratched, and yet a fundamental technique involving chemicals may be the future key to seepage control at a cost that everyone can afford.

Waterborne Dispersions

The waterborne dispersion made its first commercial appearance in 1958, when one was used to reduce seepage in an irrigation canal. Since that time use has spread to ponds, sewage lagoons, streams, lakes, and similar water holding and conveying systems. Although waterborne dispersions are not considered prime lining systems, they are included here in the interest of a more complete discussion of the general subject. Their use presents two advantages: (1) low cost and (2) lining capabilities without removal of the water in the facility. They and bentonite constitute the only two general systems that can boast the latter technique.

The name of the first waterborne dispersion put into general use was Soil Saver 13, commonly referred to as SS-13. It is a mixture of oil-soluble, resinous polymers in a diesel fuel carrier. The material works by decreasing the void space between particles of soil by increasing their ionic attraction to water. The end result is an apparent swelling action as the effective diameter of the particles is increased and the void area is decreased. Less water will permeate a soil layer that has been so treated.

Lakes, unlike reservoirs, do not have a formal underdrain system for the detection and measurement of seepage. The seepage rate must be measured on the basis of water depth recordings. From these measurements a correction factor for evaporation is subtracted. The latter information is obtained from special evaporation pans placed at the site or floating on the water, or it may be a figure agreed upon on the basis of published evaporation rates for the particular part of the country in which the lake is located. In all cases leakage tests should be conducted for at least 7 days to eliminate large variations in evaporation rates.

The seepage loss unit commonly used in lake work is cubic feet per square foot per day (abbreviated as "cfd"). The area referred to is the wetted area of the lake. It will be noted that the unit does not take into account the water depth. This is due partly to the fact that the depth may vary widely throughout the surface area, and partly to the fact that the bottom contour may be unknown. There is a factor that relates this unit of seepage to the one commonly used in formal reservoir work:

$$1.0 \text{ cfd} = 227 \text{ gpm per acre of area}$$

The treatment of an existing lake with SS-13 is a relatively simple process. The chemical is added to the water in a ratio of 1 part of SS-13 to 1000 parts of water in the lake. Effective mixing, which is important, can result naturally or be enhanced by good distribution piping of the sealant or by use of a motorboat patrolling the area during treatment. The chemical migrates to the bottom of the lake and accomplishes most of its sealing action within 48 hr. After installation it is nontoxic to animals, plants, or human beings, but during the actual treatment all fish will be killed. The water is not considered potable for animals or human beings during or immediately after treatment. If the water is to be used for drinking purposes, it should be subjected first to conventional treatment methods for surface water. The SS-13 treatment tends to give the lake water good clarity because of the soil stabilizing property of the sealant.

Results using the SS-13 treatment depend on a number of variables, the most important of which are water quality and soils analysis. Efficiency also depends on the frequency of drying cycles in the lake. The seepage rate often tends to decrease with age after treatment, unless a drying cycle intercedes, in which case the leakage rate will normally increase.

Although the application of SS-13 in itself is not particularly complicated, the pretreatment engineering demands a thorough understanding of treatment chemistry. This phase of the job should not be attempted by anyone unfamiliar with the process and the ways in which the treatment material reacts with various chemicals that may be present in the impounded water.

Since water quality is an important parameter with respect to the seepage of water through soil, percolation tests should be run with the same water that will ultimately be stored in the lake or reservoir. If this is not done, predicted water loss results may bear no resemblance to the final control actually achieved.

Dissolved chemicals in the water can adversely affect the performance of the waterborne treatments. Common salt is one of the materials that can interfere with treatment effectiveness. For this reason the process will not generally be suitable for seepage control in brine pits or other facilities where salt concentrations exceed 400 ppm. Pits that hold other salts or chemicals should be carefully studied by competent chemists with a strong background in waterborne dispersion technology before actual fieldwork is begun.

Joining the ranks of the waterborne treatments in 1960 was Chevron Soil Sealant, later called Seelo W. This material is sold in a number of forms but is essentially an aqueous wax emulsion (nonionic, anionic, or amphoteric) with dispersion size ranging from 0.2 to 10 μ. Like SS-13, it is applied directly to the water of the lake or canal, in a ratio of about 1 part of emulsion to 125 parts of water. Seelo-W forms a thin wax membrane below the surface of compatible soils. The depth at which the membrane forms can be controlled so that the membrane is protected from animals or mechanical cleaning equipment.

Since the formulation of the product depends on the initial seepage rate, soil composition, and particle size distributions, these values must be determined before a proper emulsion can be formulated. In most cases treatment will reduce the seepage rate to a value in the range of 0.1 to 0.3 cfd. Generally, this amounts to a minimum of about 60% control if the original seepage rate is in excess of 0.6 cfd. No manufacturer guarantees complete control for either the SS-13 or the Seelo W system. For this reason the waterborne treatments have been most often used on shallow lakes, canals, and ponds; they have not found their way into prime reservoir leakage control except when the reservoir falls into the classes of facility just named. The treatment would not hold up well under the dewatering and cleaning programs employed by most owners of large facilities. This disadvantage is somewhat offset by the low treatment cost, which ranges considerably below in-place costs for membrane lining systems.

Sprayed Membranes

Linings produced by spray applications in the field defy accurate classification because their final properties are so variable. If they are thin, they usually cannot be produced without pinholes, and thus they are semiimpervious. They may even be noncontinuous. Multiple-coat applications, however, may reduce or eliminate these undesirable attributes. To compound the problem there are many thousands of combinations of materials that may be used. The list is kept small, however, because many of the materials are much too expensive for use as a lining.

The most popular process for developing new linings in the lowcost category has been the one that utilizes a spray application of one sort or another. Heading the list of materials used in this work is asphalt. This comes as no surprise, since asphalt is a cementitious material that is both waterproof and inexpensive. It is the latter feature that has made it the object of so much research and study in the lining field.

Historically, asphalt has had a productive life as a waterproofing agent, dating back more than 5000 years. Early uses were simple and included such things as caulking and cementing rock blocks for forming small baths and similar types of hydraulic structures. These applications took advantage of the thermoplastic properties of asphalt, just as is done today. Although asphalt may be handled in a number of ways, pumping is generally used if large quantities are involved. It is but a short step to place a spray nozzle or jet at the end of the pump line and hence create a spray application system.

After this step has been taken, the easiest route for developing a lining system is to spray the hot asphalt directly onto the ground. This would be the ultimate technique from the standpoint of cost. Unfortunately, however, it is not the ultimate from the standpoint of effectiveness. One reason for this is the natural black color of asphalt, which, in hand spraying, makes it difficult for the operator to control succeeding applications, as each coat is the same color as the preceding one. Although the average rate can be monitored by comparing sprayed areas with consumption volumes, this by no means guarantees that each unit of area is uniformly covered. Use of traveling equipment with a distributor bar is not practical in the irregular shapes and sloping areas of most reservoirs, together with the need for multiple-coat applications.

The black color of the asphalt is an excellent adsorber of radiant energy from the sun. As a result the membrane is quickly heated on a sunny day, causing the material on the slopes to creep downward. Since canals and earthen reservoirs have sloped sides, the warmed asphalt is always at a disadvantage in trying to stay put. As it moves slowly downhill, it tears itself apart, aided by the fact that, when warmed, its strength properties diminish.

The same problem is also troublesome during the application step. The hot asphalt tends to run down the slopes as soon as it is sprayed, and before it has time to cool. This can be counteracted to some extent by the operator's maintaining close control of the application rate or technique. Lower asphalt temperature will also help but can be detrimental to the quality of the coating. Using a higher-melt asphalt is another variation that can be used, at the sacrifice of ductility and aging properties.

Proper spray technique is important, but technical problems are always involved when spraying any material directly onto the ground. As the operator applies the material, he is not shooting from all possible angles. Consequently, some of the small rocks and protuberances that lie in his path are being hit predominately on one side with little coating reaching their back sides. Sometimes no coating at all will get onto the side of the rock that is away from the operator. Even when it does, there is often bridging at that location. Naturally, this type of application is not watertight.

As a final impediment to the use of asphalt directly onto the ground, there is the very serious problem of sun aging. For thin coatings, even an optimistic grader would not give asphalt a very good score in this respect. Most attempts to improve its aging properties are accompanied by degradation of some other property. This phenomenon is not unique with asphalt. Every material known to man exhibits this tendency, which has been the subject of previous comments with respect to the compounding of rubber and plastic materials.

Fillers may be added, and many different ones were tried. Through their use heat resistance is improved, although ductility and tensile strength both decrease. Eventually, all possibilities were exhausted, and it became apparent that asphalt sprayed directly onto the ground left much to be desired in the way of an efficient hydraulic lining. Catalytically blown asphalt was also tried; it was an improvement in some respects because of its great ductility, but its sun aging was notoriously poor. In all cases it had to be buried, and this process itself tends to damage the soft asphaltic membranes.

At about this time, asphalt emulsions made their appearance and interest was renewed. When certain clays were dispersed into the base emulsion, considerable resistance of the fresh coating to sag was experienced. Even after curing (removal of the water by evaporation), the coating did not seem to be much affected by heat from the sun. One big disadvantage to asphalt emulsions, however, is their sensitivity to water immersion, and this had held their use to a minimum. The problem is particularly troublesome in the case of covered reservoirs, where the evaporation process (curing) is hampered. There curing may take an excessive length of time and even then may not reach the optimum point.

Attention was then focused on other types of reinforcement in order to impart sufficient strength to the thin asphalt coatings. In the initial part of this search economy was the primary consideration, and burlap was nominated along with other low-cost cellulose fabrics, including asphalt impregnated roofing felts.

In the long run about the only thing these fabrics did was to increase the cost. They prolonged the life of the membrane to some extent, but not enough to entice many repeat customers. Moreover, soil bacteria and fungi quickly attacked the fibers, and soon all that was left was the original membrane with its original problems.

When fabrics with increased resistance to soil bacteria were introduced, attention focused on woven glass and nylon. Experience proved that the latter can do well when placed next to nonacid soil, but certain solutions of salts cause difficulty. Good design practice is necessary to compensate against wicking tendencies through the edges of the lining, particularly when it is cut to fit around structures.

The performance of glass fabrics was disappointing. Under certain conditions they are affected by water, and they also have a strong tendency to wick water for considerable distances. Alkali soil conditions can cause trouble for glass fabrics. Their lack of good stretch characteristics cause problems too, although nonwoven glass mats showed some improvement in this respect.

There are three key factors in spraying asphalt onto a fabric: asphalt, fabric, and operator skill. These should be blended to produce enough "strike-through"* of the coating for good adhesion, but not enough to completely penetrate the fabric, and all of this must be accomplished without trapping air in the coating. If a hot asphalt is sprayed, it is extremely difficult to maintain the uniformity of temperature control necessary for the proper strike-through. If asphalt cutbacks† are sprayed, their high wetting characteristics produce too much strike-through and lower coating efficiency. Carrier solvent drying rates slow enough to facilitate proper application are too slow to give adequate process efficiency. On the other hand, high-speed solvents cause problems too, as they are sensitive to minor changes in spraying procedure. No one seems to have come up with a solvent system that will give ideal results.

After examining the problems associated with the creation of a hydraulic lining system by spraying asphalt directly onto the ground or on various base fabrics, many segments of the industry turned their attention to other types of coatings. Three general types of flexible materials

*Strike-through" is a term used in the coating industry. Too much strike-through wastes the coating by allowing it to penetrate the fabric. At the same time the tear resistance of the coated material is substantially reduced, although adhesion of the coating to the fabric is improved by strike-through. Too little strike-through will place the coating on top of the fabric, decreasing adhesion.

†An asphalt cutback is a solution of asphalt in a solvent carrier. Mineral spirits and naphthas are commonly used, although slower or faster evaporating solvents may be added to control evaporation to a desired rate.

have been the subject of intense study: the isocyanates, the urethanes (both of which are two-component systems*), and liquid rubber products involving neoprene, butyl rubber, or Hypalon. Sometimes asphalt formulations are incorporated into the liquid rubber systems. The substrate fabric is still the object of interest in these third-generation lining systems, although most of the time the substrates are the same as those used with the sprayed asphalts. There has been some tendency toward glass and glass mats, with the latter predominating.

In the late 1960s one of the major oil companies introduced polypropylene fabric,† the first really new substrate material available since the work on new-type lining systems began. Initial development work dealt with sprayed asphalt of various types, but asphalt emulsions won out, primarily because of their sag resistance. The system that went to market included this type of coating laid down in two equal applications. Initial problems included shrinkage and puckering of the fabric, and there were difficulties in applying the second coat without damaging the initial one. As with spray systems in general, the control of coating uniformity was also a weakness.

In the mid-1970s the oil company took the system off the market for a number of reasons. The lining demanded extremely close control by the applicator, and no specialty contractor firms developed to meet this need. In addition, perfect weather (hot and very dry) was required to effect a desirable cure. Weather much less than perfect interfered with proper curing, and the integrity of the lining was destroyed. Finally, lining costs in place ranged between 25¢ and 35¢/sq ft, rather high for a sprayed system. This factor no doubt hindered market acceptance too.

The polypropylene fabric is still sold to anyone who wishes to use it as a lining base material. The petroleum company, however, no longer stands behind its use as a lining, stating that the purchaser is "on his own."

In parallel with the work on sprayed membranes as linings for cut-and-fill reservoirs, considerable effort was expended to develop systems that could be sprayed on the inside of concrete tanks as a seepage control measure. Despite all of this work, no system has emerged with any degree of acceptance. A few installations have been made, but reservoir owners have not been happy with the results. Hence the materials that have been used are considered to be still in the experimental stage and are not included in this book.

*In this instance "two-component systems" refers to those which utilize two separate chemicals that react to form a chemically cured coating.

†Polypropylene is a cousin of polyethylene, having a unit building block consisting of three carbon atoms instead of two.

In the case of steel tanks it has not been found necessary to line them for the purpose of controlling seepage. They do, however, require protection against corrosion. This type of control is usually referred to as a coating rather than a lining and is covered in the section on steel tanks in Chapter 2.

In conclusion, the newer generation of coating systems is presenting problems that parallel rather closely those experienced in prior work with asphalt coatings. Pinholes within the coating itself is the primary challenge, a problem that seems to be associated with spray technology in general. To eliminate pinholes the coating thickness must be built up slowly by spraying a series of thin coats. Unfortunately, this is not conducive to a low-cost system, which is the major objective toward which everyone is working. Perhaps the pinhole problem will be solved in another way. In any event work continues on the new special systems, and only time will tell which ones will make the grade.

Combinations

Each lining material has certain advantages and disadvantages. No one lining with all of the attributes that would make it the ideal material is available. It would seem logical, therefore, to make use of various combinations that would place each component in the location where its desirable properties could be utilized most efficiently. The best example of this, and one that has been widely used, is to overlay a continuous, impervious lining with a concrete, Gunite, or asphalt concrete topping. Thus a good wearing surface is utilized, while the impervious membrane beneath it compensates for the porosity of the more rigid component, particularly after the composite construction has settled with age.

The major difficulty with a combination involving concrete types and impervious membranes beneath them is that, in the event of leakage through the impervious membrane, there is no easy or economical solution to the problem of closing the breach. For this reason the substrate preparation, the actual lining operation, and the placement of the topping should be accomplished with great care and inspection control. It is advisable to use a good-quality impervious membrane lining that is at least 20 mils (0.02 in.) in thickness.

The above combination has been used in canal lining for many years, although in a great many cases the lower membrane is a compacted clay lining instead of a completely impervious one. Nevertheless, the system works well if properly designed and installed, and this type of underlining is much easier to work on during the overlay operation. When thin membrane linings are involved, there is a greater limitation on the steepness of slope that can be tolerated, since the membrane itself tends to form a cleavage plane with respect to overlay material.

Although the impervious membrane–concrete-type topping combination is highly effective, perfectly melding imperviousness with rugged wear qualities, it is not a low-cost system. For this reason it has given way in recent years to other combinations.

A second general type of combination system that has also been used in the lining field involves the use of two or more flexible membrane linings. Here the most common application utilizes a weather-resistant membrane on the exposed slope areas and a less expensive bottom lining that cannot tolerate weather exposure. For the slope lining, asphalt panels, butyl, EPDM, CPE, or Hypalon has been used. The bottom lining is usually PVC because of its lower cost. This type of combination cannot be used, of course, if the bottom is to be subjected to any kind of mechanical cleaning action. The PVC bottom membrane systems just described will keep the overall price of such a combination at an extremely attractive level. These combinations are usually of interest in large facilities, for in such cases the bottom area is quite large compared to the area that lies on the slopes.

The transition between lining systems of this type should occur at or near the toe of the slope rather than on the slope itself. The combination being considered should be checked carefully for compatibility, as this is one of the problems when dealing with combinations. The best method of evaluation is to subject samples to immersion in the liquid being held and also to exposure in one of the recognized outdoor exposure laboratories (Arizona or Florida). The main point of evaluation is the physical junction between the two materials and the reaction of this junction to weathering and immersion. The various combinations that have been used are tabulated in Figure 5.1.

If it is desired to combine two lining systems, there is one foolproof way in which this may be done. At the transition point a concrete beam may be cast in a trench with the exposed concrete surface flush with the ground and troweled smooth. The linings are joined at the beam with a mechanical fastening device, utilizing appropriate stud anchors, clamping beam, nuts, and washers. A membrane–concrete adhesive system is sometimes utilized together with neoprene or similar gasketing, particularly if the concrete is rough or high integrity is desired at the joint. This method does an excellent job, and any combination lining system may be joined in this manner. Critical installations should be handled by means of this technique. Such a system eliminates guesswork involving the compatibility of membranes and adhesives.

Another type of combination system places a vandalproof lining on

	PVC	Butyl–EPDM	Neoprene	Hypalon	CPE	PE	3110
Asphalt panels	—	12	10	5	—	—	—
PVC		13	—	—	—	3	—
Butyl–EPDM			10	4	—	—	—
Neoprene				—	—	—	—
Hypalon					—	4	—
CPE						—	—
PE							—

FIGURE 5.1. Combination membrane lining systems, showing number of years that each system has been in continuous service as of December 1976.

normally exposed slopes, transitioning at the toe of slope to a lower cost membrane, usually PVC. Asphalt panels constitute the most economical vandalproof system; and since they are windproof as well, they have been utilized many times in this role.

Combination systems also have involved continuous membranes in combination with impervious clays, and on an area basis this type of combination exceeds all others. Here the lining often serves a dual role; seepage control and wave erosion control. This system is really a special case of a cutoff lining, provided that the thin membrane extends into a clay stratum at the bottom of the pit or below it.

In combination work special conditions arise when slopes are very long. To keep costs down to a reasonable level, there is a temptation to transition dissimilar linings on the slope. Although this is not considered good design, transitions or terminations of the lining on the slope areas can be accomplished with the proper technique. To do this, a shelf must be utilized; then the transition (or termination) is made at the toe of the upper slope, as shown in Figure 5.2. This technique is recommended, regardless of the types of linings being married. If the lower lining system is compacted clay, the technique is mandatory from a technical standpoint, in order to preserve the stability of the earth at the termination point.

When two membranes are combined, transitions on the slope are objectionable, primarily because of the physical difficulty of making horizontal joints on slopes. A safety factor provides an added incentive to refrain from effecting horizontal joints on slopes. Although any joint malfunction is bad, one that runs horizontally on a slope will compound

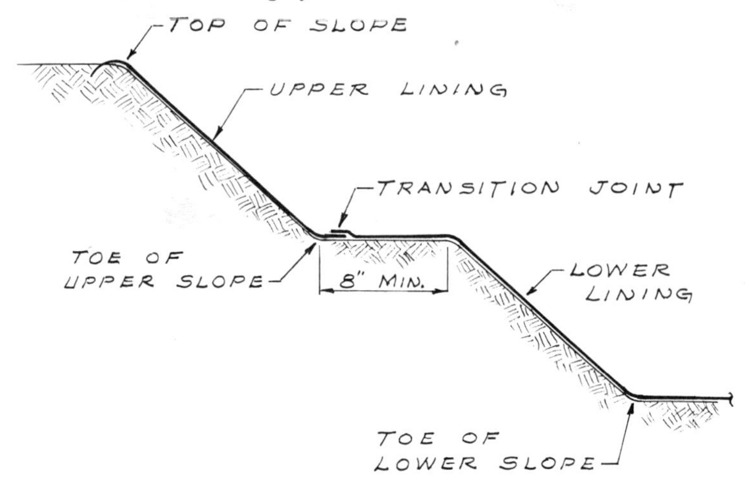

FIGURE 5.2. Lining transitions on slopes. 1. If bottom lining is compacted soil, bury top lining in 24 in. deep trench at toe of upper slope. Make sure that lower compacted earth embankment is stable when wet. 2. Do not terminate or make transition joints on any slope areas.

the problem, both because of its effect on the overall system and particularly with respect to remedial work.

References

1. Gregory P. Tschebotarioff, *Soil Mechanics, Foundations and Earth Structures,* McGraw-Hill Book Co., New York, 1951, p. 329.
2. B. K. Hough, *Basic Soils Engineering,* Ronald Press, New York, 1957, p. 416.
3. Rey S. Decker (Head, Soil Mechanics Laboratory, Lincoln, Nebr.), "Sealing Small Reservoirs with Chemical Soil Dispersants," *Proceedings of the U.S.D.A. Seepage Symposium,* Phoenix, Ariz., Feb. 19–21, 1963.

6

Basic Problems in Lining Decisions

The elements that go into the final decision to build a particular type of reservoir facility are extremely complex and involve detailed appraisal by qualified engineers and waterworks men. This chapter is in no way intended to minimize the complexity of these factors and their interactions; rather, the aim is to summarize some of the more important ramifications of the thought processes involved.

Economics

The size and the type of facility are determined by the engineer in cooperation with the water company or department. Often a survey has been made and is used as the basis for sizing the water holding and conveyance facilities needed. The engineer will first decide between a vertical-wall tank and a cut-and-fill reservoir. Which is chosen will depend on the system that best meets the requirements of elevation, aesthetics, budget, and past experience of owner and engineer.

If either steel or concrete tanks are chosen, basic designs have been worked out by those who market these facilities. In the case of the cut-and-fill reservoir, however, no such standard designs exist, and the engineer is left to his own devices with respect to the design features. This disadvantage of the earthen reservoir is offset, however, by its substantially lower cost. In addition, it can be designed to accommodate any quantity of water without presenting the aesthetic problem often posed by higher rise structures. In fact, many cut-and-fill reservoirs serve to beautify the landscape, in addition to their functional purposes.

The structural strength of the earthen component will be determined by soil analysis. This work will detail the types of soil and the proportions of each to be utilized in building up all embankments. The soils specifications will also include the compaction requirements and will give adequate attention to ensuring that the slopes are perfectly stable, both wet and dry. Such analysis and specifications are definitely a part of the cost of the facility and should not be taken lightly.

The structural design of ancillary items is usually done by someone with a civil or mechanical engineering background and with experience in the design and operation of reservoirs. These support systems are generally of concrete, although metal is sometimes used, along with timber. In all cases great care is given to the soil compaction immediately adjacent to their extremities, for it is in this area that subgrade failures are most likely to occur. Careful planning and control here will save money in the future.

The largest single part of the reservoir cost is the structural roof, together with its columns and column supports. Column design at the points where they pierce the lining is extremely important and should follow the age-old principle that the lining seal is made in the plane of the lining. The coving design universally applied to built-up roof membrane work will not do a good job under a head of water.

To protect the embankments against degradation and loss of earth particles due to water that finds its way into them, an underdrain system should be utilized. The most effective one for critical reservoirs is the continuous underdrain utilizing a "popcorn" concrete or asphalt concrete blanket, although open-graded rock has also been used. The continuous drain is somewhat expensive if one considers first cost only. When the overall effectiveness is considered, however, the price of this drain does not appear out of line.

The lining should be viewed as a component whose prime function is to limit the loss of water from the facility. Hopefully, it will prohibit the loss of any water, but this goal is seldom attained. Some seepage will usually find its way through pinholes, structure or column seals, seams, or other imperfections. Seepage may also originate at unlined structures, such as inlet, outlet, or steps. When seepage does occur, the underdrain is called upon, as the backup system, to lead this water safely out of the embankment areas.

In cases where the lining is laid directly on the earth, it will control soil erosion due to wave action, provided that the substrate has the necessary stability of its own. Many times this is the sole reason for the lining, particularly in sewage pond operations or deep water reservoirs. The lining may be used to control weed growths, and, of course, it also controls erosion caused by rain. All of these functions save future maintenance costs.

The need for a lining will be dictated by any of a number of factors, but the strongest incentive is the profit motive. If a lining will save enough money to pay back the investment cost in a reasonable length of time, it will be installed. Industries differ on what constitutes the proper length of time, however, so this element does vary. Most frequently, the time is set in the range of 2 to 5 years, but longer periods are sometimes used. The choice depends on the size of the investment, the type of benefits derived, and the accounting practices of the owner.

The payback may in the form of value saved, as water and pumping costs, or as product saved by preventing the loss of a valuable chemical solution. These are fairly simple calculations to make and merely match the installation cost and expected maintenance expenses for the lining with the projected annual savings resulting from its installation.

Related to the profit motive is the need for a lining to save the structure. If the reservoir is in danger of destruction due to continual seepage, the owner must act to take care of the problem. In this case the decision is more difficult because it is not easy to accurately forecast the overall effect of seepage on the integrity of the system. To evaluate the need for a lining in this case, soils and structural engineers will usually be required. Sometimes a sizable outlay will be needed merely to study the problem before the cost of a corrective lining can be calculated. If structural work is required as a prerequisite to any lining action, its cost will be part of the overall economics.

It is possible also that the facility could be abandoned and project correction moneys channeled into a new structure of modern design. This may be the end result if the estimated costs of problem analysis and corrective action appear too large.

In studies involving self-destruction of the facility, partial lining systems may sometimes be utilized, such as a slope lining to prevent sluffing. Other examples include special crack treatments and partial linings applied to larger, isolated areas that are contributing most to liquid loss. The dam portion of the reservoir may be lined and sealed off too, if the problem is found to be in this area. These practices, when feasible, can contribute to substantial cost savings.

Should leakage from the facility be detrimental to the life or property of others, a very powerful incentive for lining is presented. Such a condition must be solved as quickly as possible. It makes little difference whether potential loss of life or property damage is involved; either one can be extremely costly and often lead to unpleasant situations. Here the decision to line rests only on potential economic liabilities, (and, possibly, humanitarian considerations). Nevertheless the various factors involved carry much weight because of the tendency of the courts to be quite generous in awarding damages.

A powerful new economic incentive for lining has developed in recent years as a result of various governmental pressures for controlling pollution. Eventually, industries will not be allowed to dump anything that can, in any way, be detrimental to the preservation of minimum quality standards for water and the total environment. New and existing laws are being more rigorously enforced, and penalties against violators can be expensive. The emphasis is on preventing pollution of water, whether it be above or below ground. Even the pollution of ocean waters is being carefully monitored, and strong legislation in this area is already on the books or soon will be. Linings dictated by government action affect process economics too, not only because they entail initial cost and maintenance expense but also because pollution resulting from failure to line is often punishable by sizable fines.

One thing that legislation has accomplished with respect to linings is to greatly enhance their importance in dealing with ecological problems. Linings are being used universally for the containment of undesirable materials in liquid or dispersed form until the potential contaminants can be treated to render them harmless to the environment. In other cases the ponds serve as an evaporation mechanism in which the sun removes the liquids while the solids are left behind to be abandoned after the pit has been completely filled. The same principle, as far as linings are concerned, is present in the case of salt harvesting by pond evaporation methods.

Linings are sometimes justified because they prevent contaminated groundwater intrusion into a reservoir when the stored water is at low levels. Here again the lining is serving to control pollution, but in a reverse sort of way. This is a relatively new application for linings and, although approached with caution, is receiving increased consideration as justifying the cost for lining potable water holding facilities. Again, the problem of reverse hydrostatic forces on the back side of the lining is a limiting factor in the effectiveness of this technique.

Interlaced with the economic needs for lining action is a reason based on psychology. This mechanism is complex, as the study of human behavior is complex. Also, it is usually difficult to separate economics from psychology, engineering, and law. A good illustration is furnished by the California Water Plan and the studies concerning the need for an impervious lining in the large transmission canal. There was a great deal of activity among prime lining suppliers with respect to the proposed 700 mile canal, which would run from Tracy to San Diego. This facility would require over 500 million sq ft of lining. Concrete seemed to be the most logical lining for such a large canal, but seepage losses through it could be considerable. Attention was then focused on the use of an impervious underlining. Such a double-lining system had already been tried on the South Bay Aqueduct and the San Diego Canal, and there was no doubt that the combined system was an effective deterrent to seepage. In addition, the underlining eliminated the hairline cracks often associated with concrete curing. It is these cracks that serve as focal points for the more serious ones that can develop later.

In analyzing the problem, some basic data were needed. First, the expected water loss through concrete had to be known; but since no valid data existed, various assumptions were necessary. Next, the water loss through double-lining systems involving concrete and various underlying membranes was required. Although past data at Livermore, California, on the South Bay Aqueduct project indicated great reductions of water loss through a double lining, no reliable, quantitative seepage data had been developed there either. Again, assumptions had to be made.

Another design criterion was clear; something had to be done in the parts of the canal that crossed subsidence areas. If no precautions were taken there, the canal would slowly sink into the ground if any water percolated through into the unconsolidated substrate.

Naturally, the cost of water was important to this analysis; surprisingly enough, this proved to be a great stumbling block. There seemed to be a rather wide divergence of opinion as to just what cost figure should be used. The state contended that the cost was about $50/acre-ft. This value was (and still is) challenged by many, some of whom placed the cost at $200 or more. Obviously, the value placed on the water is extremely important in the final analysis.

The state engineers showed that the double lining would save money under certain conditions. Since many of the conditions were themselves assumed values, however, it was possible to come up with a variety of answers. Some of these showed that the double lining would not pay for itself, whereas others indicated that it would. As a result of these conflicting opinions, the picture was a confused one.

Because the first need for a lining is dictated by the profit motive, this proved to be the important factor in the final analysis, even though part of the fabric was interwoven with psychological factors. Since water conservation figures were in the vague category, analysts then took a look at the other side of the coin.

They asked themselves, "How much does a lining cost?" Of course, the answer depends on the type of lining and its thickness. In turn, the choice of lining depends on a number of other factors discussed elsewhere in the book. But, in summary, it can be said that an effective lining, including the total design, is not cheap. If the facility is small, the cost of the lining does not represent many dollars. On the other hand, a large lining area like the California Aqueduct is quite a different story. Anything multiplied by 500 million is bound to be a sizable figure. And it was! State engineers estimated that the thinnest effective membrane they could use would cost at least 10¢/sq ft in place. That represented an expenditure of over $50 million. By not spending this money, plus the interest necessary to amortize it over the life of the facility, a sizable fund would be created. This fund could be used to do any maintenance work that might become necessary because no second lining was used. It was a calculated risk, as no one knew how much maintenance expense would be required in the future because of the lack of an impervious underlining system. Maybe the absence of a sublining would not cause $50 million worth of maintenance work during the life of the facility. Another strong psychological point here is that money for capital outlay is usually more difficult to obtain than money for maintenance.

It is difficult to separate the cost of the lining from the other elements of cost that are combined into the overall project. The choice of lining type will certainly influence the cost of the other appurtenances and support facilities. All thin plastic linings must be protected against mechanical damage, and some from degradation by weather as well. Their use also demands finer subgrade texture, more care during installation, and tight supplier specifications to guard against the presence of pinholes or other objectionable defects in the manufacturing process. Any covering required is a deterrent against finding deficiencies in the lining itself, as it blocks the effectiveness of the normal procedures used in leak detection. To find and correct defective areas, the cover material must be removed—a very expensive process.

When all of the above considerations are taken into account, the actual cost is considerably increased over the usual cost associated with plastic-type linings. This same disparity exists in the case of other linings and carries over into evaluations of the different reservoir systems themselves. For example, the cost of the excavation, piping, lining, and roof is not the total cost of the cut-and-fill reservoir, any more than the cost of the steel tank shell is the total cost in its category. Both involve other costs; design, operation, maintenance, and backup systems to protect the structure's integrity.

Some idea of overall project cost range comparisons can be gained by examination of Figure 6.1, which shows actual competitive bid data for the different types of reservoirs in the mid-1960s. It can be seen that the system involving cut-and-fill construction is the most economical in first cost. Although reservoir construction today is more expensive than it was 10 years ago, cut-and-fill systems are still considerably more economical than steel or concrete tanks. Recent figures show that the larger

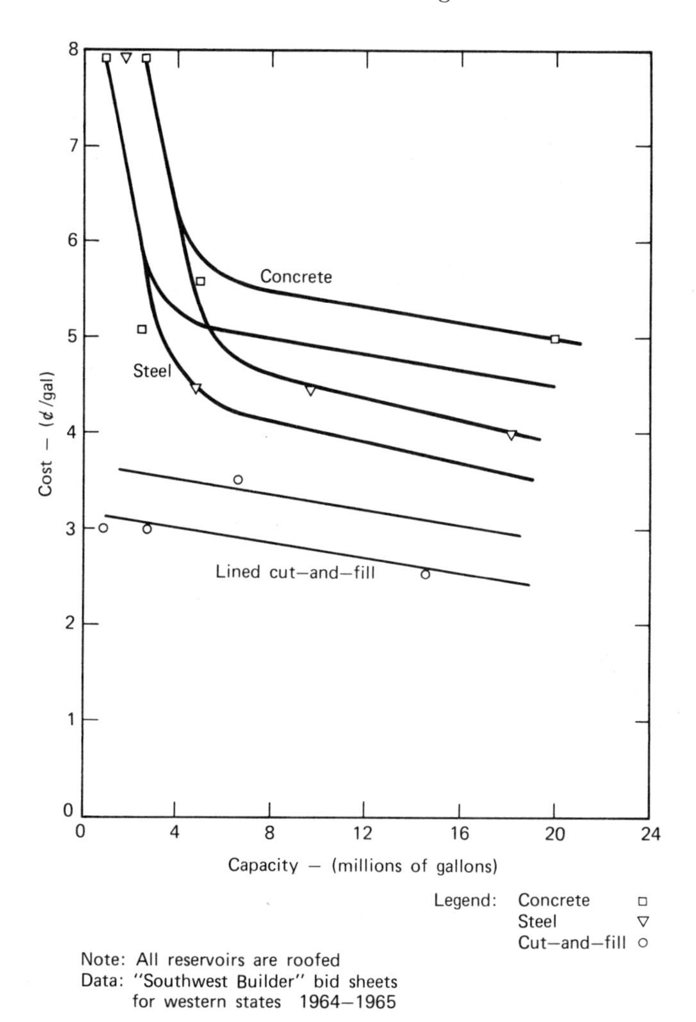

FIGURE 6.1. Comparative construction cost ranges for concrete and steel tanks and cut-and-fill reservoirs. Legend: concrete, □; steel, ▽; cut-and-fill, ○. All reservoirs are roofed (Data from "Southwest Builder" Bid Sheets for Western States, 1964–1965.)

cut-and-fill reservoirs can be built in the range of 2¢ to 5¢/gal, well below half the cost of steel or concrete structures of the same capacity. For reasons already discussed, however, conventional tank designs have many attributes to help balance their higher cost.

Not only must the final design take into account total service life, but also correction factors must be introduced to cover expected outlays for maintenance and safety surveillance. In some designs maintenance can be an appreciable item, often affecting the final decision on what to build.

From the owner's point of view the total cost of the lining system is more important than any other single factor dealing with seepage control. Unfortunately, the cost question is one of the hardest to answer because so many factors influence it. Consider the cost of labor as one of the factors.

Contractors' actual labor costs throughout the country range today from about $3 to over $13/hr. In some areas these costs vary with the season. A job in the northern states will cost more in winter than in summer, whereas the reverse is true in the desert country of the Southwest. If the job is in a remote area, travel and subsistence payments to labor are often required.

The final design of the facility is another key factor. The steepness of the side slopes, along with the ratio of their areas to the area not in slope, is an important consideration. The flatter the slope, the less expensive will be the unit cost of the lining, fine grading, and compacting. As the slopes increase in steepness, it may become necessary to work off ladders and ropes in order to do the fine grading and lining work. Compaction equipment may have to operate by means of steel cables and winches in order to do its work on the slope areas.

The shape of the reservoir also has an influence on the cost of the facility. Particularly in small sizes, circular reservoirs are more costly to construct than ones made in the form of squares or rectangles. Shallow reservoirs are easier to build than deep ones, particularly if the side slopes are steep.

The cost of placing an earth cover on a lining is also subject to rather wide fluctuation. Six inches of cover is the minimum that may be safely placed with heavy construction equipment, such as graders (road patrols), bulldozers, and front-end loaders. The costs depend on the type of earth cover and its availability to the job site, as well as whether or not the contractor applying the cover also dug the pit. Some specifications needlessly restrict the contractor from putting any equipment traffic on the lining at all during the backfilling operation. If it is assumed that earth cover material is available on site, the cover costs will range from about 1¢ to about 4¢/sq ft, 6 in. deep. For a 12 in. protective layer, these figures would fall somewhat short of doubling.

Because there are so many influencing factors, about the only way to discuss lining costs is to work with ranges. On this basis a table of comparisons appears as shown in Figure 6.2. In using Figure 6.2, the reader should keep in mind the many variations in costs of lining work. On extremely large installations the cost may be considerably lower than the chart range indicates, or it may exceed the maximum shown if the job is quite small. The chart is intended only as a guide, and it should be used in that light. For more precise figures the reader is urged to consult a qualified lining contractor.

Design Problems—Risks of Failure

The effort, time, and money invested in the design of any hydraulic structure reflect several things. The initial cost of the facility is no doubt the factor that receives primary attention, particularly from the owner or organization that will eventually foot the bill. "How much will the lining cost?" is the question that is most frequently heard. Before it can be answered, however, many other questions must be resolved.

To simplify the process, the counterquestion is really one that will set the specifications. The situation is like asking a car dealer what a new car will cost. He will want to know what kind of car you want, how it will be used, the safety features desired, and, of course, the color, style, and model. In short, he needs to know your specifications. Part of this information involves the buyer's idea of what he wants and needs, what he can pay in capital outlay, and what compromises he is willing to make to obtain an acceptable package at a price he can afford.

The project engineer is also in need of the owner's specifications. His checklist rather closely resembles the above format. How will the reservoir be used? What special features must it have? What should be its life expectancy? He will also be concerned with the risk of failure, as he is in all his other design work. In the case of the large reservoir, this risk takes on special significance well beyond that involved in nonhydraulic facilities. With other engineering structures the malfunction itself is dangerous but possesses some physical limitations. With hydraulic works, on the other hand, the main danger comes from the sequence of events that follows the initial malfunction. Tremendous quantities of water may be released within a very short period of time, and these add

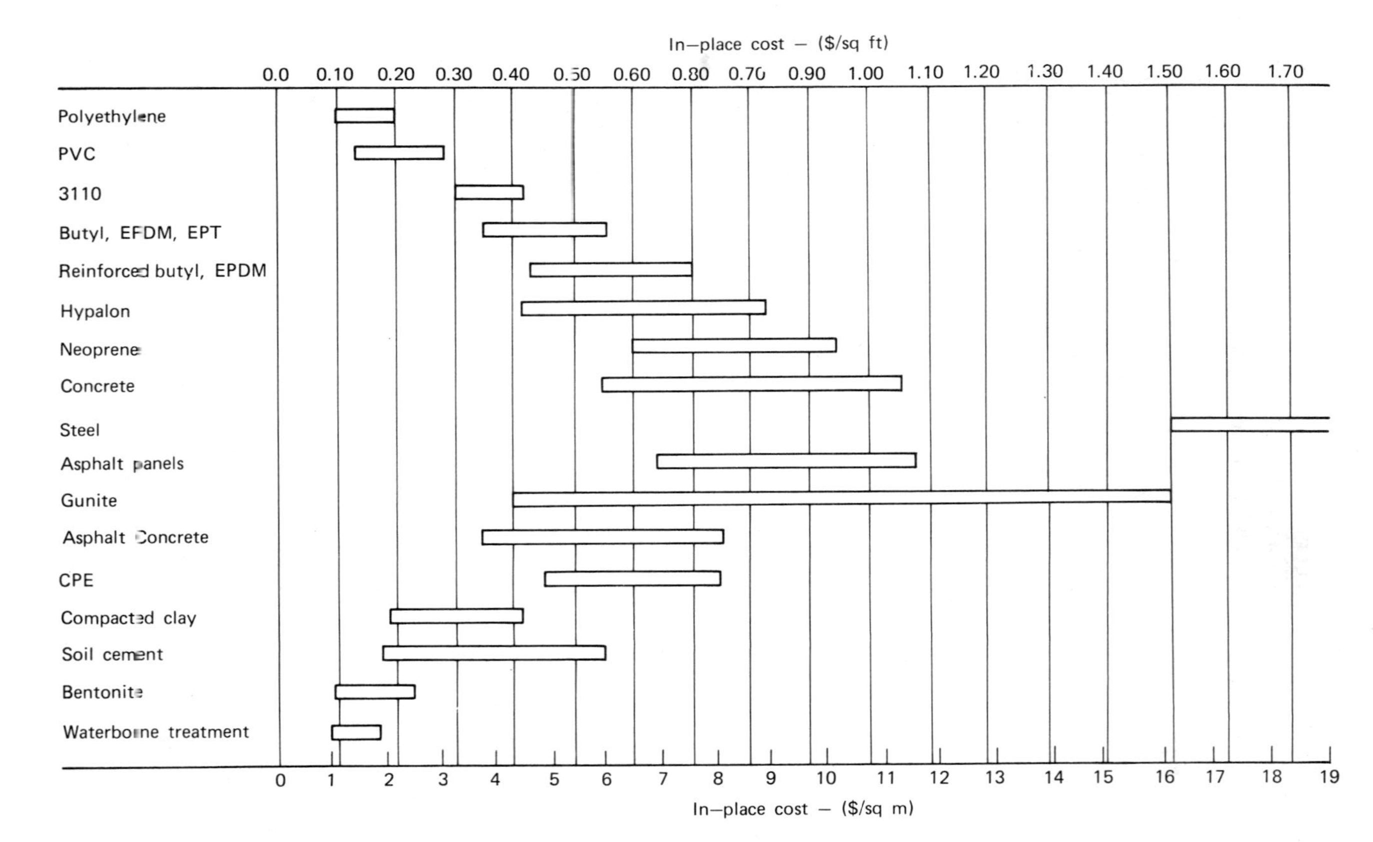

FIGURE 6.2. Cost comparison for linings in the United States.

to loss of life and property. Most people who own large reservoirs of any type cannot help but be appalled when viewing the movies of the Baldwin Hills disaster of late 1963.

Fortunately, many water holding structures are of the low-risk type. The best example of this type is the farm pond, particularly one constructed with no built-up embankments or one with relatively shallow water. Loss from these structures will not normally endanger the lives or property of others, and hence the subgrade preparation and lining may be chosen on a somewhat more relaxed basis.

Contrast this case with reservoirs built in heavily urbanized areas. These often sit on high ground overlooking considerable concentrations of industry and private homes. A breach in this type of reservoir would almost certainly result in great loss of life, heavy property damage, or both. Hence much more care in overall design is required, a heavy-duty lining should be used, and at least one backup safety feature is essential.

Between these two structural extremes, there is a large gray area. It is within this area that experience and cost considerations must be carefully blended with good design and construction practice.

It is well to give careful attention to the choice of an engineering firm to do the design work for large water holding facilities. The paramount item to determine is how much experience the staff has had in designing successful installations of the type being proposed. The situation is much the same as assessing your chances in taking a ride with someone in a small private airplane. You want to know, not how many hours of total flying time the pilot has had, but how many hours he has had in the *type* of plane in which you are going to fly. Like flying, reservoir design is a skill that is learned through proper training and experience.

In recent years we have seen a strong trend toward do-it-yourself engineering of hydraulic construction involving lining work. A goodly share has been beset with problems stemming directly from lack of proficiency in design. This is not to say that well-engineered works will have no problems, but certainly the odds for a successful installation are much better in the latter case.

The principal feature that must be engineered into the popular cut-and-fill reservoir is a stable structure. This means that both embankments and slopes should be stable, wet or dry. This requirement may dictate the use of a well-engineered underdrain system to protect all parts of the earthen members from attack and erosion by water that, for any reason, finds its way into the substrate. The structure design should be compatible with the hydraulic requirements of the facility, and these, together with the substrate texture, should be designed to be compatible with the lining system chosen.

Incredible as it may seem, the do-it-yourself school has blindly tackled the problem of underdrain design, principally on the basis that a trench full of rocks and a perforated pipe should do the job. As previously discussed, the underdrain system is an engineering device whose design should be taken seriously.

The choice of the lining should rest on past performance records in combination with the requirements dictated in the customer's specifications. The lining should be chosen on the basis of job dictates, considering such factors as wind, weather, vandalism, cleaning, maintenance procedures, and the risks of failure. In last place should be budget restrictions, for good total design should not bow to budget pressures, although, unfortunately, this is often done. Such a practice is analogous to building a tall, expensive structure, but cutting back on the money for a good foundation.

In summary, if you must accept the responsibility for designing a water holding facility with but limited experience, *get help.* If it is a cut-and-fill reservoir with definite liability potential, you will be well advised to obtain, as a first step, the services of a qualified soils engineer. If the structure is to be a steel or concrete tank, likewise enlist the services of a qualified engineer who has had experience in this particular field.

Governmental Directives

Governmental agencies are making their presence felt when it comes to linings, particularly in pollution control areas.

Most of the pressure to line waste holding ponds, brine ponds, and related structures is being generated by the individual state health departments with backup from the Environmental Protection Agency (EPA). Both of these groups have been somewhat specific as to lining requirements, although they say less about the actual lining types that may be employed.

In cases involving agriculturally related facilities, such as irrigation water and conservation projects, two agencies are involved in most of the activity. The Soil Conservation Services (SCS) does most of its work through its program of approvals, grants, and partial payments. It is involved in the approval and selection of the lining types that may be used. The other agency is part of the Department of Commerce, although more commonly known as the U.S. Bureau of Reclamation (USBR). Its activities are concerned primarily with water conveyance systems and dam facings, and it has done extensive work in connection with the lining of these facilities. In the majority of cases, the USBR works with the heavy sections of the rigid linings, such as concrete, asphalt concrete, and Gunite, which are utilized both as a lining and as an erosion control blanket. It did considerable early work on the thinner membranes, particularly the butyl rubber and plastic materials.

The U. S. Corps of Engineers specifies large quantities of linings and, like the USBR, deals mostly with the heavy-section, rigid systems. Its interest is usually along the lines of erosion control along river banks, although it has jurisdiction over some reservoirs as well.

State health departments concern themselves with the pollution of open reservoirs by airborne contaminates, but enforcement has so far been rather inconsistent. Some states have been very demanding, whereas others have been complacent, and in some cases efforts at enforcement have been subject to intense political pressures. This situation has caused definite modification and sometimes deemphasis of health department regulations in these areas.

The federal government is affected when drinking water supplies are not up to standard, if the water is to be used in interstate commerce. Thus water supplied to common carriers for potable supplies for the public must come from a U.S. government approved source. On occasion, approval has been withdrawn, and lining membrane materials were then used to remedy the situation.

The Safe Drinking Water Act (Public Law 93-532), introduced in late 1974, laid the groundwork for the establishment of permanent drinking water standards of quality and control as an answer to earlier surveys showing substandard levels of purity in many public water systems, particularly smaller ones. The permanent regulations, which go into effect in June 1977, set minimum requirements to be satisfied by all public water systems. These regulations affect the lining industry, particularly with regard to preventing polluting materials from getting into groundwater supplies, lakes, or rivers that are used as sources of supply, or airborne pollutants from entering potable water reservoirs or clearwells.

The state of California was one of the early leaders in establishing standards for the construction and operation of dams and large cut-and-fill reservoirs. These are generally in the form of design criteria, underdrain protection, and surveillance procedures, all of which are under the jurisdiction of the state dam safety director. Other states have followed this lead, and in the process all of them have come face to face with the thin membrane lining system. Each state is trying to understand this broad realm of technology as it applies to the state's policing operations on reservoirs, both new and existing.

Lining Guarantees

Lining guarantees are becoming more and more a part of the design process. The guarantees were originally applied to the elastomerics (butyl, EPT, and EPDM) and were for a term of 15 years. Later, PVC was sold under special guarantees of 2 to 10 years, but these were rather quickly withdrawn. Reinforced Hypalon and CPE both entered the market with 20 year warranties, but the one on CPE was rescinded in 1975. Most guarantees on neoprene have been for 5 years.

The guarantee is basically nothing more than a merchandising tool employed by early manufacturers of elastomeric linings. It reflected some of their enthusiastic beliefs in their products. At the same time it was a strong inducement to launch these relatively expensive linings in the marketplace.

Generally speaking, all new impermeable membrane lining systems introduced today will require some type of guarantee package, unless they are extremely low in cost. This is necessary to make them competitive with established systems.

Although the lining guarantee plays a role in lining selection processes for most projects today, the term is still not well understood by those who specify and purchase. The word "guarantee" means different things to different people. Many owners have the feeling that the guarantee covers leakage. It does not provide this type of protection, as a careful reading of the document will verify.

What the guarantee generally says is that the lining will not be degraded by outdoor exposure to the elements or by the action of water within the reservoir for a stipulated period of time. If such degradation does occur within the term of guarantee, the manufacturer will furnish replacement material *only*. Usually the replacement is on a pro rata basis, but replacement costs are not included, nor is any contingent liability due to the failure. It should be pointed out that the total cost of replacing a defective lining segment is usually more than the cost of the lining material itself.

The guarantee will not cover mechanical damage of any kind, vandalism, or acts of God. Subgrade or structural failure or settlement will likewise be excluded from the guarantee function.

Most guarantees do not mention the adhesives used in field seaming. If they are not mentioned, it is safe to assume that they are not covered by the guarantee. Malfunctions in this area are extremely difficult to resolve because they usually involve a number of interwoven factors. The analysis of these factors and their relationship to the guarantee coverage constitutes an area of great complexity.

Sometimes the guarantee simply states that the material will meet certain manufacturing specifications, if bought as first-class material. That wording may or may not be coupled with the performance characteristics in outdoor exposure.

When linings are used to hold liquids whose composition may be subject to variation, such as industrial wastes, the guarantee does not cover malfunctions due to chemical attack. This is a very important point to keep in mind when evaluating various linings for possible use in these types of environments. The reason that protection is limited here is that the membrane manufacturer has no control over changes in the chemical makeup of the pit contents.

In summary, the guarantee should be studied carefully by both owner and engineer, and both should be satisfied with what it actually says before the lining contract is awarded. They should keep in mind that this document was written by lawyers, even though the concept may have originated in the sales department.

7

Failure Mechanisms

The general mechanics of cut-and-fill reservoir lining failures have never been scientifically studied for the prime purpose of expanding the available knowledge on the subject. To build a large reservoir solely for study purposes not only would be expensive but also would be of limited practical value. To build a small reservoir for this purpose is not the answer either. Shallow reservoirs do not present the engineering problems imposed by the heavy loads of deep facilities. Reservoirs may function properly when water depths are low, but the same design under heavier impounding may not be satisfactory.

Considerable confusion exists in the industry as to just how effective the underdrain system is in collecting the water that gets through the lining. The answer depends not only on the design of the underdrain but on the characteristics of the soil as well. A major problem, therefore, is that, to study the seepage characteristics of various linings under actual field conditions, the experimental lining must be backed up by a second lining system. The second lining, in theory, would collect the seepage of the first lining, at the same time collecting any loss from the underdrain lines. The experimenter would always wonder if the second lining was doing its job with 100% efficiency.

The earth layer between the two linings has a drastic effect on the seepage loss by the initial lining. To test the effect of variations in this soil layer would be most difficult for obvious reasons. The underdrain system design and construction would also be an important parameter. In effect, the experimenter would really be measuring the effectiveness of the system, rather than the lining alone!

Unlike the effect of the weight of a building on its foundation, the water in a reservoir constitutes not only the weight on the foundation, but a self-contained and highly mobile destructive force as well. Once this destructive force breaches the lining, the support structure and reservoir foundation are under an entirely different environment. It is then that the overall design considerations are given the ultimate test, and any shortcuts or errors in this process will soon become apparent.

The interdependence of the parts of a cut-and-fill reservoir on each other is well illustrated in the Baldwin Hills Reservoir failure, which occurred in Los Angeles, California, in December 1963. A comprehensive report on this failure was issued by the state of California. Although it is the official study of the failure, it discusses also all the important factors in the design of such a facility. It is such a thorough study that all parties working in the design field on facilities of this type would be well advised to study the contents. The basic comments in the report point up very well all of the factors that make up the anatomy of a failure and the closeness of their interrelations.

To approach the problem of failure mechanisms, therefore, recognition should be given to the interaction of all the parts of the total system. These include the substrate, all appurtenances, and underdrain systems as well as the lining. The outline shown in Figure 7.1 illustrates the complexity of the problem and shows that a great many factors are involved in the process.

The actual agent for failing a cut-and-fill reservoir is the water itself. ✓ The triggering mechanism may be earth movement along a fault, seeping water from the reservoir, groundwater, subsidence, or any number of the factors in Figure 7.1 that in some manner disturb the stability of the support structure beneath the lining. It is important to realize that a thin, flexible lining is not a structural member itself. It cannot hold the reservoir together. The foundation and substrate must do this job. And they must do it well regardless of whether the soil is dry, wet, or saturated. This is the very important consideration most often overlooked by do-it-yourself designers of cut-and-fill reservoirs. It has also been overlooked on occasion by more experienced persons who believed that the lining would prevent water intrusion into the soil.

Less attention has been given to the supporting structure for the reservoir lining than any other phase. In over 60% of the cut-and-fill reservoirs built for the industry today, a soils or foundation engineer is not even consulted before the start of construction. He is consulted only if a substantial problem is encountered during the construction or operation phases. Many owners take the attitude that the fee of a soils engineer is a waste of money. On the contrary, his services can save thousands of dollars in future headaches. These same owners would not

Supporting structure problems

- The underdrains.
- The substrate.
  - Compaction.
  - Texture.
  - Voids.
  - Subsidence.
  - Holes and cracks.
  - Groundwater.
  - Expansive clays.
  - Gassing.
  - Sluffing.
  - Slope anchor stability.
  - Mud.
  - Frozen ground and ice.
- The appurtenances.

Lining problems

- Mechanical difficulties.
  - Field seams.
  - Fish mouths.
  - Structure seals.
  - Bridging.
  - Porosity.
  - Holes.
  - Pinholes.
  - Tear strength.
  - Tensile strength.
  - Extrusion and extension.
  - Rodents, other animals, and birds.
  - Insects.
  - Weed growths.
- Weather.
  - General weathering.
  - Wind.
  - Ozone.
  - Wave erosion.
  - Seismic activity.

Operating problems

- Cavitation.
- Impingement.
- Maintenance cleaning.
- Reverse hydrostatic uplift.
- Vandalism.

FIGURE 7.1. Classification of the principal failure mechanisms for cut-and-fill reservoirs.

even question the cost of a soils report if they were going to construct a building; but when it comes to a reservoir, retention pond, sewage pond, or similar facility, they begin by trying to pick the lining! This places the cart before the horse. As discussed previously, a reservoir has a special built-in potential for failure that is more serious in nature than the weight of a building on its foundation. In fact, all things being equal, the building has a better chance of survival than does the reservoir.

The record should be set straight with respect to dams, large water treatment plants, and related engineering projects. There seems to be no reluctance there to do the soil and foundation investigations required and to do them in a thorough manner. It is the gray area between the fish pond and the large reservoir about which we speak, and these constitute an ever-increasing ratio of the total lining business. In like manner large cities generally do a commendable job of planning and study with respect to the anticipitated soil and foundation problems beneath their proposed reservoirs, partly because these reservoirs are large and more critical in nature. Again, it is in manufacturing plants, smaller towns, and poorer water districts that the problem arises.

Obviously, large projects have large funding and small ones are in the poorer category. Unfortunately, this fact seems to become significant in the thought processes of those involved, and the analysis comes out in a manner that relates foundation problems to funding. In other words, if the reservoir is large, there is money available for foundation studies; if it is relatively small, such money is limited or even eliminated. There are many hundreds of small reservoirs with 5 to 20 ft. of water depth that seem to be the victims of this reasoning process.

Another psychological factor stems from the ever-increasing pressure being brought upon industry by all levels of government. The demand is to clean up the pollution, and rightly so! Eventually, no materials that can pollute underground or surface water supplies will be permitted in storage unless the holding facility is adequately lined against seepage. Industries all over the country—indeed all over the world—are rather suddenly faced with an expenditure they never thought they would have to make. Naturally, they are upset, and the same old cry is heard: "Everyone should clean up the pollution mess except me!"

Because industry is forced to accept the cost of the cleanup, its first inclination is to get the job done with minimum effort and expense. The plant that puts in a pollution control reservoir is spending money that will not add a dime to its profit picture. In fact, expenditure will subtract from it, and the figure being subtracted is often a sizable one.

Detection methods for seepage are still not well developed. If a pond leaks, the plant has a pretty good chance that the leakage will not be detected. This is another reason why shortcuts in design and construction are often taken; they represent a gamble that may pay off in handsome reductions of capital expenditures. On the other hand, the reservoir itself is the great equalizer; and, as will be seen, there are few ways in which design and construction procedures may be cheapened without drastically increasing the risks of failure.

The items classed as "Supporting Structure Problems" in Figure 7.1 are linked to both design and construction operations. The prime consideration is not to divide the items into "design" and "construction" categories, but rather to examine each one in the light of how it affects the total performance of the overall system. The subject is complicated because most items are interrelated; malfunction of one will usually have an effect on one or more of the others. Even so, examination of each item separately will shed much light on the subject as a whole. The presentation should be valuable too because most of the data and comments are based on actual field experience. The reader should bear in mind that there is considerable overlapping in classifying reservoir problems, and some items can be categorized in more than one way. Nevertheless, Figure 7.1 attempts to place each item in its most likely position.

The Underdrains

It will be recalled that the underdrain system consists of five parts: the interceptor, collector, filter, conveyor, and disposal area. Each part works in conjunction with the other parts, and the entire system is very much a part of the overall reservoir design. The most common problems occur when the system permits the intrusion of soil particles into the pipes within the drain trenches. This causes a decrease in the volume of the substrate beneath the lining, which in turn places stresses on the membrane. If stresses are too high, lining rupture occurs.

Particles of soil may get into the pipes because of defects in the design or construction of the earthen embankments or as a result of failure of the filter whose duty is to prevent this type of action. Another source of trouble is a broken drain pipe, particularly in the areas where it goes through the embankment.

A serious effect is caused by stoppage within the pipe. The full head of the reservoir is placed on the extremities of the ditch that holds the pipe. Since the underdrain pipe is designed to operate at atmospheric pressure, this places unwanted loads on the walls and bottom of the trench. Seepage can increase through these areas; in addition, soil removal can result if the underdrain trench intersects a pervious lens or passes too closely to a void area beneath the system.

There is another effect of drain line stoppage. If no water issues from the underdrain pipe, the operator may well conclude that the reservoir is not leaking. The same conclusion may result if the backfill material around the pipe is too impervious. In either case such an assumption, if contrary to fact, can have serious consequences.

The actual collapse of the underdrain pipe can come about during the construction operations. As large construction equipment crosses the backfilled trench, the weight can cause crushing of the pipe, particularly if it is made of lightweight or thin-walled materials. This was a troublesome problem at one time but has been remedied by designs calling for stronger pipe.

In cases where soil is used to cushion the effect of the underdrain rock against the lining, collapse can have another definition. Under certain conditions this cushion material may slowly sift through into void areas within the trench rock. This can happen during the course of construction or later, when the reservoir is in operation. This apparent collapse of the trench materials again places undesirable stresses on the lining membrane that can lead to failure. Careful design is the best remedy here, coupled with alert inspection.

If the disposal area does not get rid of water coming through the underdrain pipes, water will back up into the trenches, again placing the full reservoir head at these points. The same situation will also occur if the drain lines terminate in a standpipe. Although such a design has been shown in the literature and in manufacturers' bulletins, it is equivalent to nothing more than a stopped drain or an impairment of the disposal system.

The Substrate

Compaction

It is difficult to state which of the items listed in Figure 7.1 is the most important failure mechanism. All of them are important but in varying degrees, depending on which lining is being used. For example, expansive clays will cause difficulties if the lining is concrete, Gunite, or asphalt concrete, but may go unnoticed if the lining is one of the elastomerics.

Every successful cut-and-fill reservoir is a structurally stable one. The lining will not hold the reservoir together. This is the function of the earth in combination with the underdrain backup system. In every case of ultimate failure it has been defects in the supporting structure that have touched off the final chain of events.

Problems of compaction in earthen reservoirs are concentrated in areas where structures pierce the lining, because it is difficult to obtain good compaction in these locations, even with hand methods. The problem can be minimized with good design (see Chaper 10), coupled with proper inspection and construction procedures.

When differential subgrade settlement occurs, the integrity of the lining may be threatened. The seriousness of the problem will depend on the relationships between a number of factors, the more important of which are as follows:

1. Depth of settlement.
2. Area over which settlement takes place.
3. Speed of settlement.
4. Sharpness of edges of crater.

5. Type of lining.
6. Thickness of lining.
7. Type of reinforcement in lining.
8. Age of lining.
9. Characteristics of subgrade.
10. Liquid being held.

Naturally, a reaction that is dependent on at least 10 interacting factors is difficult to predict. In general, for any particular depth of settlement, the larger the area, the slower the settlement speed, and the smoother the edges of the crater, the better are the chances that the lining will not be affected.

The moderate conditions just described are more likely to prevail in the "open" areas of the facility. Conversely, if settlement occurs at or adjacent to structures, column supports, or other appurtenances that pierce the lining, the depth area relationship will place all lining types at a disadvantage. Since it is particularly difficult to obtain good compaction at structures, these locations predominate as trouble spots in reservoirs. Thus a reservoir that springs a leak should first be examined in the areas at the base of structures, columns, pipes, walls, and the like.

If the integrity of the lining is violated because of settlement, only one thing *should* happen: leakage will increase. In a reservoir not properly situated, designed, or constructed, however, other things may also happen. For example, soil may be removed, additional settlement may occur at other areas, side slopes may sluff, or cracks may open. If these things occur, something is basically wrong with the overall system. The subgrade has not adequately supported the lining. As we saw earlier, the lining cannot hold the reservoir together, heavily reinforced concrete linings excepted.

Subgrade Texture

In addition to possessing proper stability, the subgrade texture should be compatible with the lining system being used. The term "texture" refers to surface roughness caused by such thing as stubble, stake holes, rocks, and other foreign materials. It also takes into account the roughness factor due to the soil particle shape and size relationships.

Rough texture may damage the thinner membranes, particularly the polyethylenes and PVC thinner than 12 mils. The former are subject to damage as they are dragged into position during installation or are subjected to foot traffic before filling. It is especially important that PE be laid on a very smooth surface, preferably fine sand, and any cover material placed on it should likewise be very fine textured. On the other hand, PVC has good abrasion resistance and may be dragged over rough-textured soils (no stakes, stubble, broken glass, or large clods however) without damage. Normal foot traffic will not damage good-quality PVC linings placed on ¾ in. crushed rock if the lining thickness is 15 mils or greater. Despite this attribute the procedure is not recommended, as the lining may fail in use due to "extrusion."

After the reservoir is placed in service, any rocks, hard clods, or foreign materials with sharp edges will probably cut any of the thin membranes, as their cutting and tear resistance is poor under tension. The limiting factors here depend on the size of the object and the degree of sharpness of its edges in contact with the lining. The elastomeric linings are better able to resist these stresses if thicker materials are used. Linings $^{1}/_{16}$ in. or heavier can tolerate more roughness than the normal PVC or PE materials. Naturally, it is much more economical to solve the problem by providing a smooth subgrade texture than to do so by employing thick elastomeric linings.

Voids

Voids are normally associated with subgrade type, compaction, or geological problems. They may be termed "shallow" if they exist at or close to the bottom of the lining. In contrast, they are referred to as "deep" if they are considerably displaced from close proximity to the lining. Examples of the latter situation would be natural subterranean vaults, an abandoned well, or a mine shaft. Coarse crushed rock will provide shallow-type voids into which the lining may be pushed, a situation discussed in the section entitled "Extrusion and Extension."

Reservoirs that have been in service, particularly those with rigid or semirigid linings, sometimes develop shallow voids beneath the lining that are not apparent on visual inspection. These voids are usually the result of subgrade settlement or movement. Slow erosion of soil particles down the slope faces also plays a part. Even though these voids give no outward clues as to their existence, they can usually be detected by tapping on the surface of the rigid lining. A hollow sound will disclose the presence of this type of defect.

Many times, cracking of the rigid lining is associated with the presence of voids. When this situation prevails, the pressure on top of the lining normally is the same as that immediately beneath it, and minimal stresses are placed on the lining. A stable condition may prevail also if the void is some distance beneath the surface, so that, tapping does not disclose its presence. If the roof above the void will adequately support the imposed loads, no problem may result.

It is easy to see what can happen when an impervious lining of any type is placed over a rigid or semirigid lining beneath which there exist shallow void conditions. Either the back pressure is removed entirely, or its buildup is too slow to give the necessary pressure equalization. The lining is placed in the position of trying to hold the reservoir together, a function that is beyond its capabilities. Hence the reservoir system will react by failing the rigid member and rupturing the impervious lining at that point. After this event the reservoir leakage may not be any more than it was originally; on the other hand, it may not be any less either. It is the latter possibility that disturbs the owner, as he went to the expense of lining without reaping any obvious benefits. Previously, even though the reservoir was leaking, it was stable in the wet condition as well as when it was dry. Lining the facility, however, caused a special kind of instability associated with the void condition.

Void conditions are related also to other failure mechanisms, such as extrusion and extension of the lining and problems of the underdrains. This fact again underlines the difficulty of discussing failure mechanisms as separate items, since multiple effects and interactions are the rule rather than the exception.

Subsidence

Certain soils are subject to subsidence activity when they are wetted, and of course they are of extreme interest in connection with reservoir design and construction. The most common type of activity along these lines is termed shallow subsidence, but deep subsidence may also be a problem. The latter is caused by heavy removal of water, petroleum products, or minerals from the ground over a prolonged period of time. The city of Long Beach, California, affords a good example of this type of problem; the ground has been slowly sinking for many years because of intense petroleum extraction in surrounding areas. Shallow subsidence areas of recent publicity are those in California's San Joaquin Valley. Water introduced into some areas of the valley will cause the ground level to sink over 8 ft. with cracks and associated ground movement radiating out several hundred feet from the point of water addition. The sinking is most extreme at the point where water is added, tapering off to nothing as the measuring point moves away from "ground zero."

Shallow subsidence is caused when silts are washed down from mountains over a period of thousands of years and are deposited in areas of sparse rainfall. Therefore they never attain natural consolidation. Later, when large quantities of water are spread on this ground, the consolidation that failed to occur by natural means will take place at a relatively rapid rate. If a structure is built on this type of soil, water reaching the foundation can seriously damage or destroy the facility. The action has a destructive effect on all lining systems too, and they should not be laid on ground that has subsidence tendencies.

The subsidence problem may be solved by replacement of the unconsolidated soil with well-compacted material. Common practice is to utilize the "collapsing" soil itself, as it is usually capable of being compacted. How much work must be done in this area to guarantee stability depends on the type of structure being built and the water problems associated with it.

Shallow subisdence may also be eliminated by artificially introducing water into the area before construction is begun. This process, known as "ponding," was extensively used in connection with the construction of the California Aqueduct when it crossed known areas of shallow subsidence. The ponds were kept flooded for periods of 2 years or more, in attempts to remedy this unstable condition. Ponding does not remove all of the potential settlement hazard. Only by ponding for an infinite time can this be accomplished with 100% effectiveness. Therefore it is necessary to compromise on a time interval that will reduce the hazards to as low a point as is possible, without consuming so much time as to make the project schedule completely unrealistic. For water holding and conveyance systems built over areas of bad shallow subsidence, the ponding interval required will range from 1 to 3 years. Even though 100% effectiveness is never achieved in so short a time, ponding on this moderate schedule will remove subsidence for all practical purposes.

Deep subsidence is a phenomenon that must be carefully studied before a deep water* reservoir is placed on such an area. It is possible to design structures to compensate for this site deficiency, but the task is certainly one that should be delegated to the soils or foundation engineer with experience in this type of problem solving. It will be remembered that any type of void condition beneath a hydraulic facility can provide the "sink" necessary for the disposal of fines that seepage water may remove from the embankments or floor of the facility.

"Subsidence" is sometimes used to describe soil settlement due to improper or poorly compacted lifts in a man-made embankment, but

*"Deep water reservoir" is an arbitrary classification for reservoirs having over 40 ft. of water depth. Medium depth facilities hold water from 20 to 40 ft. while those below 20 ft. are termed shallow depth.

usually the term is applied to the natural condition described above. The sinking of man-made embankments is usually referred to as settlement.

Holes and Cracks

How smooth must the subgrade be? This question is asked by the earthwork contractor, whose work must precede that of the lining contractor. The answer is that the subgrade should have no rapid changes in elevation such as would be caused by rocks, bottles, cans, stubble, cracks, or stake holes.

Although a subgrade that is free of holes and cracks would seem to be an obvious prerequisite to the placement of any thin lining system, it is amazing how many times this elementary requirement has been violated. The subgrade should be firm and smooth. Generally it is good practice to roll the soil with an appropriate piece of equipment that will render a smooth texture on which to lay the lining.

If the lining has to bridge any appreciable void, such as a crack or hole, it is subject to failure by extrusion. Some failures of linings due to this cause have been attributed to the lining itself, instead of to the subgrade which is supposed to do the supporting. Again, it should be emphasized that no lining system is going to hold the structure together or to bridge an impossible gap. Therefore it becomes of utmost importance to make certain that the substrate is free of all cracks and holes. In the case of ½ in. prefabricated asphalt panels cracks up to $^{3}/_{16}$ in. in thickness can be tolerated, but even then experienced lining contractors demand a better subgrade.

Groundwater

At various levels below the surface of the ground there exist, in many parts of the land, water deposits called groundwater. This water is sometimes in stationery "pools" but is usually on the move. The pools of water are not pools as we normally think of them; rather, the water is dispersed in various types of formations. These may be fractured rock, sandy or gravelly strata, or porous-type rocks that hold the water much like a sponge. The upper level of the groundwater tends to rise during the rainy season and to recede when it is dry.

The reservoir site may be underlaid with groundwater. When this water is a considerable distance below the bottom of the reservoir, no problems are usually encountered. Sometimes, however, groundwater is at or near the bottom elevation of the reservoir, or at certain times of the year its level may actually rise to a point above the floor of the reservoir.

When the reservoir physically intercepts groundwater deposits, or is close to them, a number of special problems arise that are of great significance to the designer and owner of the facility.

The effect of this external water on a reservoir depends to a large extent on the details of when and how the groundwater problem arises. If the reservoir is constructed in dry weather, and the groundwater level is at a considerable distance below the bottom elevation of the reservoir, no problems would be expected. This assumes, of course, that there are no foundation problems that would cause the facility to be unstable. The situation will be examined first on the basis that no underdrain system is present.

If the reservoir is completed under these dry conditions, and it is filled with water, the gradual rise of groundwater will not usually be a serious problem as far as the lining is concerned. It may trigger other reactions that may play a part in the failure mechanism, and these are discussed in other sections of this chapter. For the moment, then, no complicating interactions of any sort will be considered other than the groundwater itself.

If the water level in the full reservoir is then lowered to the point where it is below the groundwater elevation, the reservoir passes into a danger zone where a great deal of trouble is possible. This condition is referred to as reverse hydrostatic pressure. Next to the various types of subgrade stability problems, this probably ranks as the second most serious problem of cut-and-fill reservoirs. The reverse pressure begins to push on the lining system from the back side. The lining being essentially impervious, water cannot pass through in either direction, so the lining will begin to bulge inward. The seriousness of this condition will depend on what happens next. If the water in the reservoir is lowered further, the reverse hydrostatic problem will tend to get worse, particularly if the level is lowered quickly. Raising the reservoir level, on the other hand, does not mean that the problem is solved. Again, the situation is a complex one and depends on a number of factors such as the porosity of the soil beneath the lining, the proximity of structure or column seals (if they exist), and the lining type. In all fairness it should be pointed out that lowering the reservoir water level during conditions of high water tables does not automatically entail 100% assurance of disastrous results. This action may effect the groundwater flows in a way that is beneficial to the reservoir. However, the odds favor the creation of serious problems.

If the reservoir is filled for the first time, and the groundwater level is below the elevation at the bottom of the reservoir but very close to it, another set of conditions arises. As the reservoir fills, the weight of the water will affect the groundwater deposit in some way. If the water is moving through the soil, the direction and velocity of movement may be affected. The groundwater level may rise as a result of a damming condition caused by the increasing weight being applied to the strata. As the reservoir is dewatered, pressure is decreased on the formation and normal groundwater flow may resume without causing the expected reverse hydrostatic problem.

An unusual phenomenon can occur if groundwater rises above the bottom elevation of the reservoir while the facility is empty. Strange as it may seem, filling the reservoir at this time will not cause the water beneath the lining to recede into the substrate in a way that will prevent a problem. Instead, the water being introduced will begin to "pond" in one or more locations. The ponding will block some of the normal escape routes for the trapped water and, at the same time, will force it to be confined in a smaller space. This will increase the pressure on the encapsulated water, and the lining will be subjected to increased back pressure. As a result the lining will stretch and a large bubble will be formed. As more water is added to the reservoir, more pressure will be placed on the water beneath the bubble. The lining will expand still further, and the bubble will rise higher into the air. This condition, in an asphalt panel lined facility, has been known to produce bubbles over 150 ft. in diameter and greater than 7 ft. in height. Obviously, the lining must be of the flexible type to take this sort of treatment, and even then damage of some type will usually result. For one thing, the freshly made joints are jeopardized, and any small nicks in the thinner membranes will be focal points for a tear.

When this type of bubble condition develops, introduction of water into the reservoir must stop immediately. The bubble should be cut and water removed from it. If the reservoir is emptied, pumping may be required. The lining is then laid back into its original position, excess material removed, and a thorough inspection made for damaged areas. Particular attention should be focused on the structural seals, since the lining is anchored at these points.

In the above situations, it is important to know what type of substrate lies just below the lining. It is also important to know whether any type of impermeable formation underlies the substrate and, if so, its type and elevation. When the reservoir is built on a well-drained type of soil such as sand, it has a good natural underdrain system. Groundwater may not cause any appreciable problems under these conditions.

If a natural impermeable stratum lies just below a well-drained surface, the problem becomes more complex. In this case the reservoir itself can be a source of the groundwater flow. Water is collected in the natural basin, and the groundwater level rises. A reverse hydrostatic condition can be generated by this type of action. It may take longer to develop, however, because the initial water fed in is taken care of for a time by the natural basin beneath the lining.

Even though the reservoir does not leak, groundwater can be a factor with respect to the structural integrity of the facility. Foundations can be saturated as well as large areas of the subgrade. Groundwater movement may also dislodge soil particles beneath the lining. These may be removed from the site because the natural underdrain may not have a natural filter built into it. Eventually, enough soil may be removed to cause the lining to rupture, thus exposing various areas of the soil to the direct action of the water from the reservoir. Fluctuating reservoir levels will hasten the failure of more supporting subgrade by means of a "pumping" action. As water heights move up and down, water is pumped in and out between the lining and the subgrade. Each cycle loosens a few more particles, and this process of slow degradation is, of course, a cumulative one.

Although we have spoken of the natural underdrain possibilities beneath the reservoir, the best way to solve the problems associated with groundwater or reservoir seepage is to construct an adequate underdrain system. When the reservoir has a formal underdrain system, the problems of groundwater are pretty well eliminated, provided that the disposal area is not affected by this underground water flow. In cases of heavy groundwater flows and critical facilities, it may be desirable to construct a special drain system that underlies all of the lining on the slopes as well as on the bottom. In such a system the continuous underdrain intercepts the perforated pipes, which serve as a manifold to carry the water to the disposal area. The East Bay Water Company in Oakland, California, was the first to develop the continuous underdrain for reservoirs, and we are indebted to them for doing a great deal of fine work in this area.

Expansive Clays

Some clay materials have the property of expanding when wet. If confined, they can exert tremendous pressures on foundations, slabs, and linings. These expansive clays, as they are called, will usually cause cracking of rigid linings such as concrete and asphalt concrete. Fortunately, soils engineers can detect their existence in advance of the construction step.

When an existing facility containing expansive clay requires lining, there are two methods of handling the problem. One is to remove the clay and replace it with select compacted fill. The other is to utilize a flexible, continuous membrane lining and leave the clay undisturbed. The latter alternative is normally a good one provided that the clay is not bothering any appurtenances of the reservoir. If the first alternative is chosen, the lining choice is then widened to a considerable degree, in that the rigid systems may then be used.

The existence of expansive clays in a reservoir facility has a special meaning to the design engineer. He should be working on the premise that water will get into the subgrade, either from the reservoir itself or from some other source. Thus the clay will be very likely to exert expansive forces. If this possibility is known in advance, a great deal of future expense can be eliminated.

There are some clay materials that shrink when wet. This negative expansion can also cause difficulties with rigid systems by creating bridging conditions beneath the lining. Flexible membranes can tolerate either shrinkage or expansion of the soil provided that (1) it does not occur adjacently to structures and (2) transitions to undisturbed soil are gentle.

Gassing

Continuous, flexible membrane linings are sometimes damaged or failed by gas pressure beneath them. Gassing can produce mammoth uplifts and will generally cause the lining to fail in a short time.

Gas formation is common to oil production, and lined pits in these areas can suffer from what may be termed reverse gas pressure. Although caused by gas instead of liquid, it will produce the same effect. Unlike reverse hydrostatic conditions, gas formation beneath the lining is serious whether the reservoir is full or empty. The gas has the effect of making the lining lighter than the liquid it is trying to contain. There have been many instances in which a large area of the lining has risen to float on the surface of the liquid as a result of buoyancy conditions caused by gas. This causes high-stress conditions in the lining and is particularly damaging if any structure seals occur in the vicinity of the bubbles.

Lining that is forced to float on the surface of the liquid is at the mercy of the elements. Wind is the big problem by adding continually fluctating stresses and impact loads to the material. If the situation is not remedied quickly, the lining has little chance to remain intact.

In reservoirs not situated in or near oil producing fields, gassing can sometimes occur if organic materials are present beneath the reservoir. When these materials are attacked by certain bacteria, gasses are released. Although the rate of gas formation may be slow, and the gas may not cause a wholesale uplift problem, it can still be troublesome in small, local areas. The problem can be prevented in many cases by clearing out all organic material before building the reservoir.

When gassing occurs. the first remedy is to remove the liquid from the reservoir as quickly as possible. This will relieve the effects of the gas by decreasing its effective pressure. The lining may then be punctured to let the gas escape, and the hole can be repaired. This procedure may solve the problem if the gas producing mechanism tends to be passive in character. If gassing occurs before the liquid is first introduced into the facility, the filling operation should not be started until the gassing problem has been solved. Adding liquid to the reservoir before removing the gas will make the situation worse, causing more pressure and higher rise of the uplifted lining.

In Gulf Coast locations and other areas where marsh lands and oil deposits are found, gas producing mechanisms tend to be active. Hence, these areas should be carefully surveyed to determine this situation before the pit is built. If continuous gas evolution is detected, the reservoir should be built with a bottom slope of about 3%, and a continuous, pervious underdrain should cover all bottom and slope areas, the high side slope lining being equipped with built-in gas vents located just below the top of the berm on the slope just above the higher elevation of the bottom area. Spacing may vary for vent location but is usually about 50 ft. on centers.

In some cases the lining has been ballasted to prevent gas uplift problems. Whether or not this is a good solution will depend on economics and technical considerations. With respect to the last point it may not be possible to cover the lining if the covering material will interfere with the operation of the pit. This can be the case particularly if the pit is holding an in-process chemical solution. For pits already in service and afflicted with gassing, ballasting may be the best and quickest solution.

Sluffing

In a later section on weather, wave action is shown to be responsible for sluffing problems on a reservoir's unprotected side slopes. Although this is the cause normally associated with sluffing, other things can also precipitate this kind of difficulty.

Sluffing can occur when the shear stresses over a relatively continuous surface exceed the shear strength of the soil. This does not necessarily mean that the soil mass as a whole is unstable. Many things can cause increased stresses. Removal of lower supporting soil material by wave erosion is one possibility, as discussed later. Other factors may be such things as increased weight of water content, seepage erosion, water pressure in earth cracks, or shocks caused by earthquakes or blasting.

Decreased soil strength may play a part in the mechanism. A number of things can bring about this condition, such as swelling of clays by adsorption of water, breakdown of loose soil due to vibration or seismic activity, or thawing of frozen soil.

From this discussion the need for an experienced soils engineer to analyze and cope with the problems of the soil's sluffing tendencies is apparent.

One of the peculiar aspects of the tendency of a material to sluff is that such action is not arrested by merely placing a lining on the slope. This is best illustrated by a typical example of a reservoir constructed in sand. The side slope was 1–1, which corresponded to an angle greater than the natural angle of repose for the particular material in question. When the sand was dry, it could not remain on this slope for any period of time. A thrown rock or gust of wind would cause many particles to trickle down the slope toward the toe. Anyone walking on the slope would cause a much greater disturbance. When the sand was wet down, however, it exhibited greatly increased stability. A thin membrane placed on this surface would retain the subgrade moisture.

If, immediately after lining, this reservoir was filled with water quickly, it would be assumed that the stability of the side slopes would be assured, particularly if the reservoir water level were not reduced. Unfortunately, such would not be the case. The reservoir side slopes would continue to sluff to the bottom until the natural angle of repose at the existing moisture content of the material was satisfied. This situation would not be greatly altered by placement of a rigid lining; sluffing would continue in most areas under the lining. For this reason, if linings are placed on cohesionless (sand) slopes, the designer should be certain that the slopes are considerably flatter than the natural angle of repose.

Slope Anchor Stability

There have been recent strong tendencies to design a lining system as a wave erosion blanket to protect slope areas. If the liquid level will remain constant, there is a theoretically limited area of concern, namely, at or near the waterline. To save money, some designers have terminated the lining at a point a few feet below the normal water level instead of running the lining down to the toe of the slope for its lower terminal anchor point.

If wave action on the slope is the only concern, the entire slope should be lined. If this is not done, about the only really safe and workable alternative is to anchor the lining in a trench on a shelf at the point where the shelf meets the slope. This will ensure a stable anchor point but will cost more in earthwork charges. However, this so-called more costly system may be more economical in the long run.

To terminate the lining at a point on the slope is extremely hazardous and from this standpoint is not an economical solution. The risk involved is that the lower anchor point is highly unstable to the actions of water and wave, even though the water level is not expected to be at this location for very long, perhaps only briefly as the facility is filled. Nevertheless, the ground at the anchor point will soon become saturated. This condition is coupled with the fact that a natural cleavage plane is formed at the point where the lining slices into the anchor trench. The situation thus created is a natural one that favors the sluffing of the lower lip of the trench down the slope. The lining is exposed when this happens, and it becomes highly unstable to the action of wind generated water currents below the surface. An unanchored lining at that point is more likely to fail than to remain in position.

Even stabilization of the lower anchor of the trench by pouring cement into it is of questionable value, since the concrete itself also depends on the trench stability for its foundation. Here, again, the concrete does not stabilize the lower trench lip, and the same condition exists as before. In fact, the weight of the concrete anchor may hasten the destruction process; and when the anchor starts to skid down the slope, it will certainly pull the lining apart somewhere.

The only really successful design in this case will run the lining all the way down to the toe of the slope to a stable anchor point. Although the lining cost is higher, the maintenance savings will more than pay for the difference, probably in less than a year.

Mud

Lining systems laid on a base of soft mud do not have a history of good performance. Once again, the principle of providing a firm and stable subbase beneath each and every lining prevails. In other words, the subgrade must be stable, *wet* or dry.

The first problem connected with lining on a soft mud base is the probability of getting moisture in the seam area. Under these conditions the risk of making poor-quality seams is appreciably higher than when the subbase is dry. This is particularly true for butyl and asphalt panel-type linings.

During the placement operation there is a certain amount of foot traffic across the new lining. Any foot traffic that hits on or near freshly made seams can sink into the mud, placing undue stress on them. Since butyl and asphalt panel-type seams are weak right after assembly, they are especially subject to damage and eventual leaks.

If mud is on the slope areas too, its sliding movement down the slope is a definite possibility. Such movement usually causes trouble by pinning down the lining in folds as the substrate material slides down the slope. The best solution to this problem is to return to the basic concept of providing a stable subbase on which to place the lining.

If the mud problem does make an unexpected appearance, methods for stabilizing the mud by adding certain chemicals to it should be considered. The chemicals used depend on the type of soil, the depth of mud, and the availability ot equipment for application. It is not within the scope of this book to recommend specific materials but it is well to know that chemicals do exist which may be able to solve this problem when it occurs unexpectedly.

Frozen Ground and Ice

The placement of linings on frozen ground and ice should be avoided when possible, although this situation does occur from time to time. The problem is threefold. First, frozen ground conditions usually mean sharp ridges of hard ground, which may damage the thinner membranes linings or, in the case of the heavier asphalt panels, can cause bridging conditions. Sccond, the ground has expanded; and when it contracts later on during the thaw cycle, ground displacement may cause difficulty due to extension and extrusion or other factors that generally come into play with unstable substrate conditions. Third, if these movements occur in the proximity of structures, the lining may tear.

If no structures are present in the system, there is a good chance that frozen ground will cause no trouble to the lining upon filling if it is a type which has fair elongation characteristics. The prerequisite here is that no mechanical damage occur to the lining during its placement. Another condition is that the lining be of sufficient thickness and strength to be able to handle the stresses placed on it.

Another problem with frozen ground is that the lining may cause some thawing beneath it during the day, particularly during sunlight; then at night the lining will freeze to the ground. It is some time before thawing will permit the lining movements during installation which are usually necessary for proper alignment of the joints. This results in delays on the job; even more importantly, it may sometimes cause areas of joint stress even while the joints are being made.

After the lining is in place, ice can cause damage, particularly when the spring thaws come, or if the water level is continually raised and lowered during ice formation periods. There is some discussion of this subject in the section entitled "Weathering."

In retrospect, if the substrate is properly constructed and has proper drainage, ice damage will be limited. Water that gets into the substrate is effectively removed by the drainage system, thus limiting the amount of water that can freeze.

The Appurtenances

Every reservoir has a number of auxiliary parts required for its operation. These include such things as the inlet, outlet, drain, overflow, columns, roof, and transmission lines. Grouped together in a classification called the structures, they may be associated with failure in two general ways: (1) structural inadequacy, and (2) transitional problems.

Structures involve the use of a combination of steel and concrete, with the latter representing the major constituent. The concrete itself may be defective because of voids (improper vibration), cracks (improper curing), mixing (nonhomogeneous), or an error in the recipe. The completed structure may malfunction for any of these reasons alone or for two or more in combination. Inadequate reinforcement may also cause problems for the structures, in which case they may crack because of excessive loading.

The location where a concrete structure joins the earth is termed a transition area, and it is in these areas that many problems lie. Although the difficulties actually relate to the soil and its compaction, the resulting malfunction is closely associated with the structures, and such problems are often referred to as structural problems. Indeed, the presence of the structure makes the achievement of efficient compaction very difficult. This point is strongly emphasized in experience records, which show that over 70% of malfunctions involving subgrade compaction occur at the transition points. In his 1968 paper H. R. Cedergren summed up the situation so well that his words are worth repeating for special emphasis (see Chapter 3, section entitled "Compacted Earth"):

> Many of the failures of earth dams have been caused by piping along the outlet pipes, beside or under spillway walls and slabs, or along other contacts between natural or compacted earth and rigid structural members. If the designers and builders of dams give sufficient attention to fundamental principles of seepage analysis and control and the work is carried out under carefully prepared plans and specifications, seepage problems can be virtually eliminated.[1]

Although the above quotation refers to dams, it applies equally well to cut-and fill reservoirs and related earthen facilities. The seepage control (lining) of the reservoir must be backed up by a method which safely removes any water that gets through the lining. No lined structure of any type should be built that does not adequately take care of breaches in the lining. If soil or rock pieces start washing out of embankments, failure is inevitable unless the process is quickly controlled.

Occasionally, the concrete roof columns have been the source of some leakage because of their porous nature. This is particularly true of older facilities, where the concrete quality as cast may not have been of the highest order. Although it is not normal procedure to line the roof supports, this has been done when they appear to be possible trouble spots for leakage. The two continuous lining systems that have been most frequently used for this purpose are butyl rubber and PVC. In reservoirs that have concrete steps, there are additional possibilities for leakage. The most likely spots occur at construction joints, but if any other cracks develop, these may also let out water.

Inlet–outlet structures, whether in the form of separate castings or as one combined design, are usually not lined. It is customary practice to construct these as one monolithic pour, which tends to eliminate any problems with cracks at the corners and wall joints. When older reservoirs are being considered for relining, the structures should be examined carefully, as many have been cast with construction joints. Such joints are often the source of trouble.

Points where the pipes join the inlet–outlet structures and places where they make 90 degree bends in their vicinity are other likely spots where leakage may occur. This may be speeded up because of the vibrations set up by rapid movement of the water at these locations. When they occur, leakage in underdrains will usually increase during the pumping cycle and subside when no pumping is being done, although the same behavior can occur as a result of settlement at the inlet structure. The latter is an example of a transitional problem, where subgrade settlement occurs adjacently to a structure.

The Lining

An erroneous impression has persisted throughout the years that a lining system will ensure the success of a reservoir. Many feel that the mere designation "an impervious membrane" grants the reservoir some sort of magical insurance that all will be well. Actually, the term "impervious membrane" has been much overworked. It stems from advertising claims that linings are impervious; and for all practical purposes, they are. Any salesman of linings will proudly admit as much, emphasizing that the material he sells does not leak. These comments and cleverly worded advertisements have lulled the designer into thinking that he need only pick an "impervious" lining and his problems are over. If there is one thing that has contributed greatly to reservoir problems, it is without doubt this early attitude on the part of owners, designers, and manufacturers.

The preceding paragraph indicates that the lining is not the key to the solution. Yet, if it is impervious, why will a problem arise? If the lining does not leak, no problem should result! This statement may or may not be correct, depending on how well the support structure does its job. Instability in the latter is more cause for concern than lack of integrity of the lining itself, but the situation is actually of the teamwork type. Each element must do its job.

A serious malfunction always brings up the question of which came first, a lining failure or a substrate failure. The problem is an interesting one and has been debated frequently by contractor, designer, owner, and manufacturer. It may appear that the question has no answer, and yet the answer is a very simple one. It goes back to Murphy's Law: if there is any possible way that the lining can leak, it probably will! So the first thing that the potential user of a lining must consider is that it may leak. If he can make this assumption, his faith in linings may be shattered, but his thought processes will be on the right track. This is the very first step that he must take if he is to approach the subject of linings from the practical standpoint.

To illustrate how thinking must be changed in regard to linings, the writer was called into an engineer's office a few years back to comment on plans and specifications for the lining of a 20 million gal. reservoir. The project was a rather complex one, involving over 200 pages of drawings and specifications plus the usual contract forms. There were several phases of the project, and the appurtenances were complicated, as were the piping and the general design itself. Soils and foundation analyses were very complete. No detail had been overlooked. Under the guarantee clause was the statement, "A bottletight job is mandatory. There shall be no leakage through the lining." For any contractor who has built a reservoir before, this is one statement that will make him nervous. He may do good work and may be proud of his record, but "no leakage" is a rather binding clause. To be absolutely sure of the results, the specifications and design must be faultless and every subcontractor must do his work in a perfect manner. The general contractor must also perform flawlessly, and there should be zero defects in any materials furnished to the job. Since this chain of events smacks more of heaven than of earth, the engineer was asked what leakage tolerance was allowed for the reservoir. "None," he replied. "That's why I'm specifying a lining."

The best approach to this problem is to relate the contractor's work to that of the designer who puts out the specifications. When the specification calls for a reservoir with absolutely no leakage, this means that a rather complicated chain of events must occur with 100% probability. First, the manufacturer of the lining must produce his product without a single defect or pinhole. Then he must package it and ship it to the job with absolutely no mishap during any step of the shipping process. He has excellent equipment, pinhole detection apparatus, thickness gauges with amazing accuracy, and a top-flight quality control department ready to reject anything really bad—if they see it! Then the contractor unpacks the material. He does not have a laboratory for checking the quality. Many specifications reserve this right to the owner, but it is seldom exercised. The material is installed assuming zero defects, although, as any quality inspector well knows, no manufactured product is entirely free of defects.

The installation of the lining involves several more tricky processes. The lining should be handled very carefully to prevent mechanical damage. The contractor will do his best, everyone is sure. The lining is then put together. Here a few variables are involved such as heat, wind, moisture, dust, subgrade imperfections, inexperienced people, and an inspector who may not have seen this type of material before. Also, the contractor may be using some type of adhesive system with which he is totally unfamiliar. Each of the field operations must also be carried out flawlessly. When the water is placed in the reservoir, the subgrade must be stable and properly support the load. No settlement may occur at columns or structures. All of these things must happen for zero leakage to be realized. Unfortunately, the probability of all these things happening in a perfect manner is almost zero. If something goes wrong, the contractor's money is held up and arguments develop. Who is right?

Let's look at the engineer's operation. He must prepare plans and specification. These are complex too; but if any mistakes are made (regardless of who makes them—printer, engineer, owner, etc.), an addendum may be put out. That is usually why addenda are issued—mistakes are made. The producer of the lining may make mistakes too, and they can be made by the contractor on the job. All of these people have other people working for them, just like the engineer. Whereas a mistake in a document can be corrected by merely putting out an addendum, mistakes in construction are not as easily corrected. Sometimes they are not noticed, even after repeated inspections. Therefore a tolerance is needed. The leakage clause is that tolerance and it is to the contractor what the addendum is to the engineer.

Another very practical reason why a leakage clause is an important part of the specification involves the detection of leaks. This is a really serious problem in cut-and-fill reservoirs. After the reservoir is competed, there is absolutely no known, practical way to look for leaks. Electrical measurements, like those for detecting pinholes in steel tank coatings, cannot be used in cut-and-fill reservoir designs as the substrate is not an electrical conductor of the type that will lend itself to these techniques. A leak detection system that will pinpoint defects of any type in the lining before water is introduced is badly needed, but no one has developed it as yet. The situation is not much improved by using an underwater detection method. This is strictly an art and not a science. The technique is extremely limited as to the size of leaks detectable.

Leak detection before filling is difficult unless the defect is quite large or obvious. It is made even more difficult by the fact that a leak which will pass as much as 1 gpm at 20 ft. of head may be easily missed, even when the location of the leak is known in advance. The cost of looking for and detecting such a small leak would be prohibitive.

In 20 ft. of water depth a leak passing 20 gpm of water will be rather easy to detect. Between the 1 gal. and the 20 gal. leak, there is a gray area. But obviously somewhere a line must be drawn. The line is called the leakage tolerance, and the industry is indebted to the East Bay Water Company of Oakland, California, for its work and analysis of the problem. The company was the first user of linings to conclude that an absolutely perfect, bottletight cut-and-fill reservoir is impossible to attain. Therefore the contractor should have some latitude with respect to the leakage. A leakage tolerance figure was put into the company's specifications. Leakage below this figure was acceptable, but leakage rates above it must be fixed at the contractor's expense. The formula that defines the leakage tolerance figure is given as

$$Q = \frac{A\sqrt{H}}{80}$$

where Q = maximum permissible leakage tolerance (gallons per minute), A = lining area (thousands of square feet), and H = maximum water depth (feet). The value of Q may not be less than 1.0.

Although there is definite need for leakage tolerance in modern reservoir design practice, this does not mean a relaxing of vigilance in the lining step itself. Much can be done to maintain high-quality levels in design and installation. One important step in this direction can be accomplished if both designer and installer possess a thorough knowledge of failure mechanisms.

Since the original concept of the leakage tolerance by the East Bay Water Company, the denominator in the formula given above has been subjected to some experimental maneuvering. In some cases it has been increased to a value of 160, which in effect reduces the tolerance by 50%. It should be remembered that the tolerance applies only to the lining installation; leaks in the structures, steps, or other appurtenances are not included. Pending the accumulation of further data, the formula shown above gives the generally accepted leakage tolerance for facilities that are lined with any of the continuous, impermeable, flexible membrane lining systems.

The preceding sections have dealt with failure mechanisms developed through man-made problems within the earthwork and its appurtenances. The following subsections will look at deficiencies that can be attributed to the lining itself or to its field fabrication.

Field Seams

For each lining system the manufacturer has developed a special technique for making seams in the field. In the case of the cured materials, such as butyl, EPDM, and neoprene, the factory joints are made in a different manner from those in the field. The factory may employ a vulcanization system, whereas the field crews must resort to adhesives of one sort or another. Some make use of a special tie gum tape to be incorporated in the joint by means of adhesives, while others use adhesives only. Adhesives are of two general types; (1) cold-setting, one-part systems, and (2) two-component systems.

In the factory PVC and CPE are put together electronically, whereas one-component cold-setting adhesives are used for fieldwork. For Hypalon, a one-component cold-setting adhesive is used in most cases, both in the factory and for field work. In some respects CPE differs from all others in that it has been fabricated by means of a solvent mixture only, although this method never achieved widespread use. The 3110 lining is field fabricated by heat welding techniques.

It is often said that the success of the lining project lies in the adhesive joint system. Anyone can spread large sheets of rubber or plastic on the ground, but it takes considerable experience and know-how to be able to put them together successfully so that they will stay. It also requires a manufacturer with a strong background in adhesives technology to develop and manufacture a good workable joint adhesive. The jointing of all lining systems in the field demands considerable attention to details, such as proper procedure, accurate measurement , and effective policing of the clean, dry surfaces required during field seaming operations.

Failure of field seams is due to a number of causes:

1. Improper adhesive system.
2. Defective adhesive.
3. Surface contamination, including oil.
4. Adhesive contamination.
5. Improper tack development before closure.
6. Moisture or excessive humidity.
7. Too low or too high a temperature.
8. Lining contractor inexperience.
9. Inproper lining layout.
10. Solvent evaporation from the adhesive.

There are also other items, but these represent the most common problems. Careful examination and study will develop 10 remedial actions, one for each cause. For example, the proper adhesive must be used for each particular lining system, and only the lining membrane manufacturer can recommend the right one. If he cannot recommend an adhesive, his lining system should not be used, for no lining contractor is qualified to work in this area.

Seam performance is closely related to subgrade and/or underdrain performance, and malfunctions in either are often blamed on the others. Unfortunately, it is usually very difficult to evaluate the overall problem after troubles have occurred. It is even more difficult to reach a conclusion to which everyone will suscribe. It is well to point out again that subgrades which have the necessary stability, both wet and dry, will do no more than leak water should a lining seam separate. No massive subgrade settlement should occur as a result of seam failure. If it does, the problem is more basic than that involved in the field seaming process. This simple axiom is probably the least understood of all the facts involved in the technology of linings and lining systems.

Fish Mouths

With the advent of the butyl rubber family of linings, a special new type of defect gained recognition and deserves special mention. Although "fish mouth" defects are possible with any type of membrane seam, the probability of their occurrence was greatly increased with the advent of the butyl membrane and its cousins, EPT and EPDM.

Basically, the problem stems from the fact that all of these cured membranes have excellent resistance to the solvents normally used in the rubber industry. At first this would seem to be a blessing, as everyone likes to see a membrane lining that is not attacked by anything. Unfortunately, it works to a disadvantage in the case of the cured rubber-type linings.

When two butyl-type cured sheets are joined together in the field, the adhesive is extremely slow to develop its full strength. Although the joint in tension on the slope or bottom of the reservoir is not affected too much within an hour or two after it is put together, the joint in "peel" does not do well. Peel tests generate extremely high unit stresses at the bond interface. Theoretically, they approach infinity, since the bond area under stress is actually only the thickness of a line. Since a true line has infinite thinness, it can readily be seen that this can be a problem.

If the membrane is stretched, particularly if one piece is stretched more than the mating piece, one sheet will have more material in the joint than the other sheet. When this condition exists, a segment of the joint may pop open to relieve the unequal tensions in the two sheets. The open area of the joint is termed a fish mouth. It may occur very quickly, but, most often it happens a number of hours after the joint is put together. The only remedy is to cut out the defect and apply a patch.

Fish mouth conditions, even though they may not traverse originally through the full width of the joint, will do so in most cases after water pressure is placed on the joint. If the joint lies in a fold of the material, the possibility that fish mouths will pop up is increased.

Structure Seals

Lining seals to structures are not well understood in this industry, and improper technology here has contributed to a considerable amount of difficulty throughout the country.

Most failures of structural seals are caused by nine factors, although it can be seen that there are some interrelationships that tend to cloud the issue:

1. Coving of the lining up the side of the structure, resulting in a "bridging" condition.
2. Subgrade settlement.
3. Failure of the membrane-to-concrete adhesive.
4. Concrete texture too rough.
5. Concrete curing compounds used.
6. Moisture in the concrete.
7. Cracks in the structure at the seal.
8. Water velocity (turbulence) at the seal edge.
9. Elimination of mechanical anchor strips.

Very little evaluation work has been done in the area of membrane adhesives to concrete in underwater environments. Thorough evaluation tends to be rather time consuming, so as a safety factor it is well not to rely on adhesive integrity alone. If adhesives are used, they should be given added security with a mechanical anchor device of some type.

Bridging

Pressure of water on linings that overlie a crack may cause failure by a process called extension and extrusion.

There is a similar way in which the lining may fail, but it does not depend at all on a subgrade condition. It is possible to build-in the problem during field fabrication operations, or during the installation of a simple one-piece drop-in lining. The problem, termed bridging, is such an important factor in lining installation work that it rates a separate section of its own.

The problem occurs at the corners of vertical or steep-walled structures, which already possess a rigid-type lining, as well as at all intersections between the floor and vertical walls. Where structures intersect the existing lining the same potential situations are present.

When bridging occurs, the lining will generally fail at that point when water is introduced into the structure with any appreciable depth. In addition to being built-in by the installing contractor, bridging can result from shrinkage during or after the installation work. The best remedy is to remove the bridging tendency by the insertion of additional material so that the lining may lie relaxed into the wall or corner intersection areas. Preformed 90 degree corners and pipe boots are often used, particularly in butyl-type systems, where joints are not as strong as the material itself.

Factory fabricated or field fabricated pipe boots are also utilized in Hypalon systems as a matter of convenience when no sealing pad is available.

Porosity

In regard to porosity, lining systems fall into two classes: (1) the continuous, flexible, impervious membrane types, and (2) the noncontinuous types. In the first category are asphalt panels, butyl, CPE, EPDM, Hypalon, neoprene, PVC, 3110, PE, and related plastics and elastomers; the latter class includes cement concrete, asphalt concrete, Gunite, soil sealants, chemical treatments, bentonite, and so on. Systems in the first class do not contribute to leakage because of their natural porosity, whereas those in the second class do.

No controlled measurements of leakage from reservoirs due to porosity have been made because, as mentioned elsewhere, the phenomenon is exceedingly complex. Since tests on very small samples are misleading, a very large prototype facility is needed for proper evaluations, and to date this has not been built. Thus concentrated study of the problem is still waiting for someone to initiate a program based on scientific evaluations.

The reader should be reminded that MVT (moisture vapor transmission) or perm ratings* do not indicate the porosity of a lining to liquids. They do indicate its porosity to gases (such as water vapor, a gas), which is a separate phenomenon and which, incidentally, is not a factor in seepage control mechanics. The driving force for leakage through linings is gravity, and not humidity, the latter being the driving force for the passage of gaseous water vapor or moist air.

Porosity comparisons between the different linings are difficult. Not only is porosity a function of the lining, but it also depends on design, construction, and field measurement techniques. One approach is to utilize coefficients of permeability, the term applied to compacted earth, concrete, and other noncontinuous lining systems. One difficulty with doing this, however, is that all the continuous, flexible membrane systems have values of zero for this measurement. Another difficulty is that the coefficient is derived from laboratory data under ideal conditions of measurement. Perhaps the best method of comparison is to utilize seepage rates that have been experienced by different linings on actual con-

*The term "perm rating" refers to the permeability of a thin membrane to the passage of water vapor under controlled temperature and humidity conditions. The units are grains of water transmitted per square foot of sample per hour per inch of mercury difference in vapor pressure between the two sides of the sample. These readings, when multiplied by the thickness of the sample in mils (thousandths of an inch), provide a basis for comparison of membranes of different thickness. In this case the units are called perm-mils.

struction jobs after the facility in question has been in service for 1 year. From the author's experience data, unpublished records, and industry case histories, the data of Table 7.1 have been assembled. These values will serve as a rough guide, but the reader should bear in mind that actual ones can vary appreciably from those shown.

Holes

There are three sources for holes in membrane lining systems. First, there are those developed in the manufacturing process. Secondly, holes may be developed during shipment, unloading, or installation. In the third category are those that occur after the facility is placed in service.

Although the thin membrane lining systems have limited thickness, there is not as much trouble with mechanical holes as one might expect. Holes incurred in the manufacturing process do crop up in PVC membranes below 15 mils in thickness, but butyl and EPDM types 45 mils and over have little or no difficulties in this respect. Hypalon, neoprene, asphalt panels, and CPE have no problems in this area either, as they are all made in thicknesses of 30 mils or heavier.

TABLE 7.1. Seepage Rate Comparisons[1]

Material	Thickness (in.)	Minimum Expected Seepage Rate at 20 ft of Water Depth after 1 yr of service (in./day)
Open sand and gravel		96
Loose earth		48
Loose earth plus chemical treatment*		12
Loose earth plus bentonite*		10
Earth in cut		12
Soil cement (continuously wetted)	4	4
Gunite	1.5	3
Asphalt concrete	4	1.5
Unreinforced concrete	4	1.5
Compacted earth	36	0.3
Exposed prefabricated asphalt panels	0.5	0.03
Exposed synthetic membranes	0.045	0.001

[1]The data are based on actual installation experience. The chemical and bentonite (*) treatments depend on pretreatment seepage rates, and in the table loose earth values are assumed.

Most mechanical holes are caused during the shipping, unloading, or installing operations. In regard to shipping, most success has been achieved by utilizing two or three-ply asphalt impregnated burlap wrapper over corrugated paper, which surrounds the rolled goods. The linings also demand a reasonable amount of care during handling and unloading operations. The biggest problems have been the damage caused by forklift operations and the snagging of the lining on nail heads or other sharp objects protruding from truck floorboards or shipping aprons. Nail heads working up through the bottom of the skid have also caused damage to lining systems that are shipped on pallets. These nail heads should be isolated from the lining itself by the use of buffer strips or extra-heavy-duty exterior roll wrappings.

During the installation step the lining is in by far its most vulnerable position. Not only is it exposed without benefit of a protecting package, but also much more of it is available to damaging blows by tools, heels, or rocks lying beneath or on top of it. The problem is really not so much that the lining may be damaged as that the person who caused the damage may not report it to his foreman. Although damage does not occur often, and it is usually not difficult to take care of; the foreman does need to be made aware of the problem, as damaged areas not repaired contribute to overall leakage values.

Holes also may be caused by a number of other agents, such as birds, animals, and those intent on mischievous damage. Large holes from any cause should be patched as quickly as possible; otherwise, they are subject to further damage by wind. After the lining is in service, holes below or near the waterline can cause damage to the substrate, especially because of wave action. This damage, in turn, may cause further lining damage.

If vandalism is anticipated at the site, it may be well to choose a vandal-resistant type of lining, such as prefabricated asphalt panels. These lining elements are, for practical purposes, resistant to almost all types of aggressive action by vandals. The panels are tough and present to anyone wishing to vandalize them a relatively smooth, hard surface, highly resistant to cutting by razor or knife blades. Only a very heavy and sharp object, hurled point blank at high speed, will penetrate the ½ in. membrane. Even cleated golf shoes, which are lethal to other thin membranes, do not have a deteriorating effect on the panels. Some of the other lining systems, even though carrying long-term guarantees against degradation on exposure to weather, have had to be covered to protect them from attack by vandals.

Like defective seams, mechanical holes in the lining should not cause a holding facility to fail if it was properly designed and well built initially.

Pinholes

The pinhole defect is reserved for the manufacturing or processing part of the business. Although it is possible to produce pinholes on the job during the installation, this is a very remote contingency. Any damage done to the lining by the installation contractor is much bigger in size than what is classed herein as pinholes.

Pinholes, by definition within this book, are essentially invisible to the unaided eye. They may be visible under a very strong magnifying lens, or upon examination with a strong light (such as sunlight) placed behind the membrane. Even then, sometimes they are visible only when the material is stretched. Although pinholes are processing oriented, they may, in the case of some resins, be tied in also to the resin production stages.

In the case of PVC materials, pinholes may develop from problems involving plant cleanliness. Foreign particles in the air may become mixed into the plastic, or small particles of burnt resin may drop into the batch in the mixing process. Pinholes can result from this action, particularly if the burnt particles are well carbonized. The original powder resin may also contain contamination, not the least of which can be due to small particles of a much higher molecular weight than the average molecular weight of the batch. These higher molecular weight particles do not plasticize or soften as readily as the rest of the resin, yielding "fish eyes" in the material. When viewed under strong light, these appear to be holes in the sheeting. Actually, they are not holes, but they are focal points from which holes may develop, as the "fish eyes" may pop out, particularly when the sheeting is stretched and subjected to water pressure at the same time.

With the cured rubber-type materials, pinholes can develop also from contamination, as well as from defects in the curing cycle or from moisture in the batch. With Hypalon materials, thermal breakdown during processing is one of the causes responsible for pinholes. Each type of membrance has, therefore, certain inherent conditions that may contribute to the production of pinholes.

Material with pinholes was a problem in early PVC production, and even today all 10 mil and some 15 mil films carry pinhole tolerances. Generally, as the thickness increases, the pinhole problem decreases. Some films are built up of two laminated membranes, and this of course almost eliminates the pinhole problem caused by the factors just mentioned. Although a purely mechanical situation can produce the defect, such as a small metal burr on a calender roll, these types of pinholes are relatively scarce.

In unsupported Hypalon some manufacturers experienced considerable early difficulties with pinholes, to the point that the Hypalon sheet specifications stated that "pinholes could be repaired after the lining installation was completed." This placed the material in the pervious category and certainly did nothing to promote the healthy growth of the product. In fact, experienced, high-quality lining contractors and design engineers shied away from this type of material during its early years. The introduction of reinforcement fabrics into the Hypalon technology, thus mandating two-ply membrane construction, generally eliminated pinholes, but the unsupported membranes were a problem for several years. Even today, design specifications tend to call for reinforcement, although multi-ply Hypalon nonreinforced sheet is now produced with no pinhole problems.

Pinholes in unsupported membranes are focal points for the beginning of the tearing process, particularly if the membrane is under any kind of stress at the point of the pinhole. Since all of the unsupported membranes have poor tear resistance, the pinhole problem can have far-reaching effects on the eventual quality of the lining job. Since pinholes also pass water, the integrity of the installation suffers from the very beginning. If tearing results from the pinholes, leakage is further increased.

In cut-and-fill reservoirs the leakage rate from pinholes can be appreciable if many are present and heads are high. In concrete tanks pinholes can allow damp spots to exist on the outside of the walls. Sometimes this is the major reason for lining a concrete tank, so in such cases the pinholes are extremely troublesome.

The pinhole type of defect is difficult to remedy, since it is extremely hard to detect visually, particularly after the installation is complete. Repair, if many defects are present, is extremely difficult and costly. Psychologically, the owner is not happy either if he is aware that *any* pinholes are present in his job. Remedial action may well mean the relining of the facility, or at least that part of it in which pinholes are suspect.

Prefabricated asphalt panels do not suffer from pinholes, although they do have a manufacturing defect that produces large holes caused by a section of asphalt material that has not been heated to the fluid state. As it hits the extruding roll, it "drags," producing a long and narrow void that is easy to spot. Most such defective panels are easily detected on the assembly line, and those that get through are spotted by the sealing or unloading crews. Since they do not occur very often, and repair techniques are quick and easy, these holes have never constituted much of a problem for the lining contractor.

Tear Strength

It should be recalled that all thin membrane systems, including asphalt panels, have very poor tear resistance. Reinforcement by nylon or Dacron fibers will enhance this property, but without the presence of auxiliary fibers the linings are very susceptible to tearing.

The lining systems vary in their resistance to this type of action. The asphalt panels are most resistant, and because of their ability to flow or surround a sharp object they have a better chance to do well on a subgrade of less than ideal texture. The unsupported vinyls, whose tear resistance mil for mil exceeds the values for many of the cured membranes, can take considerable abuse; but when tension is applied at the point of nick, they tend to tear. Hence they must not be subjected to tension stresses in combination with sharp objects in the subgrade. All of the cured membrane systems exhibit more resistance to the formation of focal points (nicks or punctures) within the lining system; but when such points do develop in areas subjected to tension, these systems, too, fail quickly because of tearing.

The best defense against failure by tear is to utilize top-quality reinforced linings and keep them out of situations where tension stresses are high.

Tensile Strength

Technically speaking, thin membrane linings do not fail because of poor tensile strength. Whether a lining fails due to tensile or tear deficiencies is dependent on definitions. We know that, as the lining thickness decreases, the tensile strength increases. Therefore, if tensile strength were really an important issue, we would strive to produce linings of infinite thinness. And yet we know that this would not result in good or even usable linings. Why, then, is tensile strength given such an important role in all manufacturers' bulletins on linings?

The reason is probably associated with habit to a large extent. No doubt it also has something to do with the ease of running the test, coupled with the fact that everyone understands tensile strength better than any other parameter that can be determined. But what does tensile strength have to do with successful lining performance in the thin membrane systems? Not much.

A high-speed photograph of a membrane failing in tensile strength would show a progressive and very fast tear shoot across the material at failure. The tear must start at a focal point within or at the edge of the sheet or test specimen. Because the test is run relatively quickly (usually at a jaw separation of 12 in./min.), the effect of tension on the lining in developing a failure by tear is eliminated or shielded from view. When the sample fails, we merely say it has failed in tension, rather than use the more descriptive term of "failure by tearing."

In the case of the thin membrane systems, it is doubtful whether any lining has ever failed because of poor tensile strength! There is overwhelming evidence that tear resistance is the dominant factor here. Some reports of lining failure do mention "failure in tensile strength," which in reality is a technical misnomer.

In the case of prefabricated asphalt panels, asphalt concrete, or cement concrete, on the other hand, lining failure may well be due to deficiencies in tensile strength. These lining systems do, in fact, fail in tension. The panels are subject to tearing failures; but when stresses are more or less evenly applied, a true failure in tension will result. Again, the speed at which this occurs is important in determining just which action takes place. Sometimes both actions occur in combination.

To reduce lining problems with respect to "failure in tension," slopes should be flattened, subgrade texture should be given more attention to be sure it is smooth, and there should be no bridging of the lining.

Extrusion and Elongation

All of the flexible membranes boast a high degree of elongation, particularly high in comparison to such materials as concrete, steel, and asphalt concrete. At first this property was thought to ensure the success of the lining because the material could "give with the blow." It is now known that great extensibility of a material is not closely related to its ability to perform as a lining membrane. Other properties are more important, as we have already seen.

As users, designers, and manufacturers of these membranes became aware of this fact, they attempted to design the layout of the lining so that excess folds were left in the toe areas after the installation was complete. It was thought that, if the subgrade was depressed at any point in the facility, folds from this "reserve" would "feed" enough excess material to prevent overstressing the membrane itself. If the membrane was not excessively stressed, it was believed, the material's poor tear resistance would not be highlighted.

The proponents of this thinking were surprised to find that the facts did not support the theory. Underwater inspections of linings by divers confirmed what engineering logic decreed. The coefficient of friction between the lining and the rough texture of the ground will not let the lining move except in a very limited way. It cannot "borrow" excess or elongated material from another area unless that area is infinitely close. The heavy weight of the water pushes down so hard that the lining is literally embossed into the ground surface. The principle may be easily demonstrated by trying to move a section of a plastic swimming pool lining under a mere 4 ft. of water head. The fallacy in thinking comes about from the fact that the lining, when first spread on the ground, may be pulled around to move wrinkle patterns from one area to another. This ease of movement is eliminated, however, when the extreme pressures of contained liquid are applied on top of the lining.

Because of the lining's inability to redistribute wrinkles and excess material to areas of need, the large elongation percentages extolled in specifications are of limited value. In fact, they are not even available during installation. Even a 10 mil, 100 ft. square PVC lining will not cover an area that is 110 ft. on a side. This represents a mere 10% overage in each direction, whereas the material's ultimate stretch is well over 200%. The problem is that the amount of force required to stretch the material even 10% exceeds the limits of practicability. Also, the force must be applied uniformly, or ripping and tearing of the membrane may result. The plastic's "memory" effect directs the material to return to its original dimensions, and it does so with great authority.

The elongation properties of the thin membrane cause a problem that could be described as the reverse of the situation just discussed. As these flexible linings lie on the slopes, they slowly begin to elongate. Most of this action takes place before filling, or when water levels are low. There is some indication that this slow creep, or cold flow, continues at a reduced pace even when the reservoir is held full. Whether or not this happens depends on the length and steepness of the slope, the type of lining, and its unit weight. Such action is also facilitated considerably if water levels fluctuate.

Elongation will sometimes accumulate folds of excess material at the toe of the slopes. Theoretically they cause no damage, but actually there may be some problems at seams, particularly on elastomeric linings. Folds in PE linings will usually cause cracks at the points of 180 degree bend. Asphalt panels will cold-flow slowly on very steep slopes until they have been set by the weight of the reservoir water. Once set, they maintain good stability, but aluminum-type reflecting coatings are often used as an added safety factor. They are recommended when water will not be placed in an unroofed facility for some time after the lining operation, particularly on slopes steeper than 3 : 1. The rigid linings also display a tendency to slip downhill. In these cases the problem manifests inself in the form of tension cracks.

For the flexible linings, columns and structures on the slope areas introduce another difficulty. At locations just above the protuberances, the elongation is interrupted and wrinkles and folds collect. They can cause some difficulty with the seals to the structures at these points. Since the the wrinkles usually cross vertical field seams, there is a chance that seam integrity may be affected, particularly in the case of elastomeric or cured lining systems.

Extrusion of the lining into cracks or holes is merely another manifestation of the extension problem just discussed. The hole may be only ½ in. in diameter, infinitely small compared to the total lining area. In these cases, too, it is not the ultimate elongation that is important, but the local situation around the hole itself.

At the immediate area of the hole the lining is being placed in a set of testing jaws, so to speak. The clearance of the jaws is ½ in. at the start of the test. As water pressure is applied, the material will begin to extrude into the hole. It cannot "borrow" much lining from outside the hole, as we have seen. Of course, the thickness of the original lining is important; and if it is thick enough, extrusion may not occur because the force required for this is more than balanced by the resistance of the material itself. As a thin lining extrudes into the hole, the 200% elongation mark is quickly passed. If the hole is deep, the lining will eventually rupture, since 200% of ½ in. is only 1 in.—and that is not a very deep hole. This explains why linings of this type should not be laid on uncushioned crushed rock: they may well be able to withstand the tendency of the rock to puncture them, but the extrusion process will be the ultimate deciding factor. The mechanism described for holes is similar to that for cracks in the subgrade.

Rodents, Other Animals, and Birds

The rigid linings, such as concrete and asphalt concrete, stand up well against animal traffic of all kinds, but the thinner membranes cannot make this boast. There are two types of hazards involved; large animals, which do mechanical damage to thin linings because of their great weight and sharp hoofs, and smaller ones, which cause damage associated with their search for food and water.

It is good practice to fence all reservoirs, even when not required to do so by law or health regulations. A fence is not 100% insurance against trespassing by animals, but it will keep out the majority of them. A 6 ft. fence is not high enough to prevent deer from jumping it. They are very curious by nature and will vault over the fence to check the reservoir from time to time.

The hooves of the cow, deer, or horse are damaging to PE, PVC, or butyl-type membranes. Just how much damage they will do, depends on the slope of the side walls and the firmness of the subgrade. Cows, in particular, are attracted to reservoirs because they smell the water. In one recent case they walked over a mile for the privilege of drinking from a new, unfenced reservoir. The 12 mil PVC lining suffered moderate damage. Although it was quickly repaired, the incident could have been an expensive one.

Smaller animals such as the gopher, beaver, rat, muskrat, prarie dog, and mouse will attack the linings for two reasons: (1) they may be attracted to them because they smell or taste good, or (2) the lining may be blocking their natural path to food or water. In the latter case the behavior of the animal will depend on the accessibility of alternative routes. If no alternative path exists, most animals of this class will cut through any lining system for survival. For instance, rats trapped from their food source can cut through concrete, glass, or aluminum; of course, thin plastic or elastomeric membranes offer absolutely no resistance to them. Damage from these causes is not common, but it does occur occasionally. Adequate soil studies will usually alert the designer to these potential problems.

Small aquatic animals often have their burrows covered by lining systems, particularly in old canal or reservoir installations. In these cases they will cut through the lining if this is the easiest means for reestablishment of some particular route that is important to their routine. It is almost impossible to forecast this type of event.

In regard to rigid linings, there are no known instances in which small animals ate through them. It is not that they could not get through, but rather that the work required to do so was too great, and so an alternative route was found that involved less effort.

Some linings have a pleasant odor and taste to small animals. The most notable example of this is the PVC linings. Their plasticizers or softening oils, called esters, constitute a class of chemical compounds not very far removed from perfume. They also have a somewhat sweet taste, and the combination of these two properties apparently makes for good eating. Ground squirrels have eaten profusely of some PVC membranes. The practice is not habit forming, however, because the animals die from digestive difficulties. The resins are not digestible, and hence the animal's vital processes are clogged. This action is not fast enough to benefit the owner of the facility; it occurs only after the animal has eaten a moderate sized hole in the material.

Although PE does not have an odor, it likewise has some difficulties with small animals because of the blocking of access routes to food, water, or shelter. With butyl rubber compounds, there have been some isolated instances of rodent ingestion. Neoprene has had no problems in this respect, although its use is still so limited as to make any statement in this regard somewhat premature.

Birds have caused some damage to the thin PE and PVC linings, but they have not seemed to bother any of the other systems. In the former cases it is not known whether they were picking bugs off the lining or whether another type of driving force was involved. Some experts think that birds like to peck at the shiny surface exhibited by the plastics. All of these surface attacks by birds or small rodents have been countered by the almost universal tendency to bury this type of membrane beneath earth in areas where it would normally be exposed.

Burial is also a help in protecting elastomeric and plastic membrane systems from mechanical damage by hoofed animals. When membranes on slopes are covered with earth, however, the problem of sluffing is reintroduced. With the membrane forming a slip plane, the tendency of the wet earth to slide down the slope is greatly enhanced. In short, one problem is solved and another one is created. Thus lining system utilization, like lining compounding, is a study in compromise.

Insects

Insect attack on lining systems is minimal. Insects have been reported to have been involved in the destruction of PE systems on a couple of occasions, but this is somewhat surprising, considering the inert character of this polymer. The ants that were reported in and around the disintegrated lining membrane could have well been there for other reasons, and not for the purpose of eating the lining itself. The situation here was never well explained.

Insects have not been reported as attacking PVC lining systems, as far as is known to this writer. Also, butyl, EPDM, Hypalon, neoprene, 3110, and CPE have had no record of attack, with the same inertness reported in the case of asphalt panels.

Although PVC is affected by some microorganisms that attack the plasticizer system, insects themselves have not found it very palatable. Microorganisms have also attacked the original felt outer layers of asphalt panels, but his problem, as reported earlier, was solved by replacing the cotton felts with glass fabrics.

Weed Growths

Whether weeds alone can destroy a lining system is subject to some debate, but they can do considerable damage. When a lining system is abandoned, weed growths will eventually take over, just as they do when any man-made structure is abandoned. It is not so much that the weeds destroy the lining as that their presence is merely a result of the abandonment.

Weed growths do cause problems with lining systems, however, because in the ground, before the lining is laid, there are dormant seeds in many soils. This is particularly true if there were weed growths in any of the areas of excavation before the lining operation. After the lining is installed, the sun gives increased warmth to the soil, and the lining "draws" moisture to condense in this area. Hence two of the three important requirements are present for weed growth; only sunlight is missing. This brings up the subject of quantitative measurement. Just how much sunlight is required to initiate growth? Also, just how much sunlight is actually getting through? Perhaps there is a minute pinhole in the area. If there is, it will admit enough sunlight for some weed growths, and the initial shoots will penetrate in this area.

Once weed growths begin, the hole through which the first shoots penetrated are enlarged, and more weeds force their way through. Wind then begins to enlarge the hole thus produced, so more weeds can grow, and so on. In addition, weeds may grow on top of the lining in areas where silt has been blown in along with weed seeds. The roots of the plants growing on top may, if they lack sufficient moisture, penetrate the lining in search of more, a process that is enhanced by the presence of pinholes or other defects in the lining.

Weed growths may be arrested by the use of weed killing chemicals placed on the ground just before the lining operation. They may also be controlled with an effective maintenance program that makes sure they do not get started. In fact, good maintenance practice is usually the more important of these two ingredients.

When pond linings are buried to protect the upper portions against sun aging, wind, or vandalism, the weed problem becomes more prevalent. This is particularly true when the ponds hold sewage or high concentrations of other nutrients. Here first-class growing conditions are present, and the weeds respond accordingly. Where sewage-type materials are being held, it is usually better to choose a lining that does not need burial for protection against the elements.

The Weather

Weathering

No lining system is completely immune from the weathering process. Sun, wind, hail, ozone, rain, and freeze–thaw cycles take their toll. Although all lining systems are affected by these forces, those that suffer the most are the PE and PVC membranes. Both are immune to ozone effects, which afflicted early rubber materials. It is difficult to say which of the elements are the most destructive, but sun aging and heat seem to be the worst enemies of the PE and PVC polymers. Only Hypalon and elastomeric linings seem to be able to weather these agents rather well; at least, the effects are slower to develop. The 3110 system appears to possess good aging qualities, although it is still too early to draw meaningful conclusions here.

Polyethylene is notoriously poor against sun, and vinyl, which rates much better, is still a long way from ideal. The performance of the PVC membrane depends to some extent on where it is placed. In the northern states, the northeast portion of the country, and the Pacific Northwest, aging does not appear to be quite as serious as in other parts of the country. The desert areas of California, Arizona, Nevada, and Texas, along with the southern and midwestern states, all seem to be hit hard with respect to sunlight degradation. Western low desert country, with its twin forces of intense heat and high winds, is particularly hard on these materials. There PVC materials generally show serious degradation within 3 years on exposed berm areas.

Degradation from sun aging does not occur as quickly in the case of the rigid linings. Asphalt concrete is affected because the heat of the sun slowly distills off the volatile components of the asphalt. This causes the binder material to lose some of its strength, hastening the cracking process. Both asphalt concrete and the cement concrete materials suffer another effect at the waterline. Here, in summertime, the lining is cool just below the water surface, but extremely hot just above it. At this transition point the forces of expansion and contraction, acting in close proximity, tend to cause hairline cracks to form. Although these cracks are small, they are a part of the degradation process that afflicts concrete during the winter months of freeze–thaw cycles.

The custom of burying PVC and PE lining membranes as a precaution against sun degradation and mechanical damage will protect them also from the wind uplift effects described later. Even so, wind will cause problems by generating waves (see Table 7.2), which in turn begin to erode the earthen cover. The size of the reservoir, the velocity of the wind, and its direction are all important factors in determining the maximum height of waves than can be generated. Placing the long axis of the reservoir perpendicular to the direction of prevailing winds will reduce the intensity of the waves. The erosion effects may be further reduced by flattening the slopes. Flatter slopes have more natural stability and hence do not erode quite as badly as do steep ones, but they do erode. Erosion can also be lessened by using soils that are well compacted and stable. Sometimes soil cement has been used to provide stability here, along with riprap and asphalt concrete blankets. In all these cases, careful design is essential, as none of the methods automatically guarantees against erosion by wave action.

Hail is very destructive on thin PE membranes and will also damage PVC membranes if it is heavy enough or the pellets are of large size. The latter membrane is also subject to much greater damage if it is old or if the weather is cool at the time of the hail's impact. Since both of these materials are normally covered with soil on the berm areas, however, hail damage is not considered an important problem. The butyl elastomer family has held up well against hail in most cases, and of course the rigid lining systems are not bothered by this type of pelleting.

Rain does not affect the flexible membranes, asphalt concrete, Gunite, or concrete to any appreciable degree. It does not normally cause damage to compacted earth facilities or reservoirs lined with soil cement. Bentonite lining systems may be affected by erosion of rainwater on the side slopes.

Freezing weather will cause PVC and PE membranes to stiffen; but if they are not subjected to impact loadings during this time, there are no degradatory effects from the cold weather alone. As the weather moderates, the membranes will again become pliable. The cured elastomerics are not affected by cold or freezing weather, as they are not thermoplastic.

The effects of freezing weather and freeze–thaw cycles are felt almost entirely by the rigid linings: concrete, shotcrete (Gunite), and asphalt concrete. If these linings are improperly installed, the mix design is not right, or curing is not carefully controlled, water can enter the body of the lining. When this water freezes, the lining is damaged by the expansive action. Even if no water gets into the lining, this same expansive action exerts an outside force against the lining when the reservoir water freezes, and this can be damaging also. If there are any voids beneath and adjacent to the lining, or if there is differential settlement of the substrate, the process is speeded up considerably. The speed at which the rigid linings are affected by freezing problems depends on the interaction of the above factors. Although problems may occur within a very few years, generally the action is slower, requiring 10 or 15 years or more before remedial action must be taken. In any event the spalling action continues to affect the lining in this manner; and unless it is stopped, the lining can be entirely destroyed. As is mentioned in Chapter 4, air entraining agents will help concrete to resist this action, but they will not entirely eliminate the problem.

Although earthquakes are not strictly a weather phenomenon, they are discussed here for convenience. At least, like wind, rain, hail, and freezing, they are beyond the control of man. Several million earthquakes occur annually throughout the world, according to some estimates. About a thousand of these (probability about 0.0002) may be classed as strong, capable of causing considerable damage in the areas where they occur. Fortunately, most earthquakes occur beneath the sea, where they cause little concern except when seismic sea waves (tsunamis) are generated. Although the probability of an earthquake affecting a reservoir is rather low, in recent years in the western part of the country several dams and reservoirs have been damaged, a few severely. In most cases the quake registered above 6.0 on the Richter scale (VII on the Modified Mercalli scale). If the reservoir lies at the epicenter of the earthquake, and the intensity of the latter is high enough, it really does not matter what the lining is. The same applies with respect to other high-energy disturbances such as tornadoes and hurricanes.

Wind Effects

Rigid linings are not affected by wind itself, but in the case of thin membrane linings wind is a problem in two respects: first during the installation, and again after the lining system is in place. During the installation care must be taken to keep the wind from getting under the edge of the lining. Generally, winds up to 10 mph do not cause appreciable trouble. As the wind increases from 10 to 20 mph, however, installation problems begin to appear. Above 30 mph, it becomes a real problem to lay and seam thin membrane systems of all types. It is generally better to stop all operations during this kind of wind activity. Dust is blowing through the air at these wind velocities, and this interferes with proper joint assembly.

The most common method of preventing wind damage during construction is to place sand bags or automobile tires at intervals across the lining, particularly at exposed, unseamed edges. After construction the weights are removed before the reservoir is placed in service. Another method is to roll or fold the exposed edges of the lining back on themselves during construction and then, in addition, place weights in these areas. This method is much more effective than placement of bags or weights alone. As the wind blows, additional edge material may be folded over on itself, but each such action further stabilizes the membrane.

On occasion, asphalt tack coats have been used to hold thin membrane linings down against light to medium winds. Another method is to anchor sections of the lining in trenches throughout the installation. As each large sheet is added, it is merely sealed onto the existing section at the points where they emerge from the trenches. Should the wind get under a section, damage is limited to that section only. The method is effective if the sheet size is not too large, but is a slower and more costly installation procedure.

Some installations have been overlaid with "boots" or "sausages" filled with sand and running from the top of the slope to the toe. The tubes were made 8 to 12 in. in diameter and placed on 20 ft. centers. They did stabilize the sheeting; but when winds reached 40 mph or more, the spacing had to be reduced to 10 ft. It is not necessary to run the sand filled sausages to the toe of the slope. Sausages placed just below the berm and parallel to it also help. In addition to their weight, about 50 lb/linear ft, they also disrupt the air flow pattern.

Suction equalizing vents have also been tried, and these seem to do a better job for less money. At first they were just slits in the lining membrane near the berm. To prevent wind from getting beneath the lining at that point, the slits were covered with a flap that was cemented to the lining on three sides of the slit. This method was later modified by fabricating vents or tubes, which pierced the lining at points on the berm. At first these were constructed of pipes or tubing, with flap-type covers. Later they were fabricated from the same type of material as the membrane; by means of a flange, they could be adhered to the membrane itself. Today these units are made to adapt to various slopes, so that the emerging tube is vertical. The openings at the top are hooded and enclosed with stainless screen material. The vents are installed on 50 ft centers about 1 ft below the top of the berm.

All of the above methods have been used, together with a number of variations. Quite often, during the installation process the contractor requests that water be placed as quickly as possible on these portions of the lining installation that are complete. Filling the reservoir with water has always been thought to be a good deterrent against wind damage. Unfortunately, thin membrane lining systems are not necessarily stabilized against wind effects merely by filling the reservoir with water. In fact, of the stabilizing methods described above, only the vertical tube vents on 50 ft. centers have done a 100% effective job.

If the lining is properly fabricated on the bottom and has no holes or large wrinkles in it, most wind related problems occur near the top of the slope. At this location air leaving the reservoir creates a suction effect on the lining as a result of increasing its velocity at that point. If air is present in the soil interstices, and if it is able to migrate behind the lining to points at which the suction is applied, the lining will balloon upward.

As the bubble forms, it alters the points where the wind velocity creates pressure and uplift forces. This, coupled with the nonuniform nature of the wind (gusts), causes these points to oscillate as the forces create unstable conditions. The result is that flutter or buffeting of the lining occurs. This buffeting causes the lining to vibrate against the substrate and can result in its eventual failure due to chaffing and abrasion. Fatigue may be involved too, as the lining is alternately under stressed and relaxed condtions. In some observed cases, the lining rise above the berm has exceeded 6 ft. Regardless of the spacing of the various weight systems on the lining, the buffeting occurs until the weight pattern is almost continuous.

Another method of breaking up wind forces is to construct a short wall at the top of the reservoir slope. As far as is known, no walls have been constructed in reservoir installations specifically for this purpose. Some reservoirs have been built with walls at the top of the slope for other reasons, however, and it appears that these walls have a desirable stabilizing effect on the lining. The Army Corps of Engineers[2] has experimented with various short vertical and sloping wall segments to break up wind velocity patterns produced by downwash from heliocopters. These baffle studies confirm the theory that short walls help in controlling the wind uplift problem.

The effect of the slope ratio on the degree of the problem is another question. Theoretically, the steeper the slope, the greater will be the expected velocity differential, which should translate into a greater problem with suction effects at the berm. There is an offsetting effect, however. As the slope becomes steeper, the velocity pressure component against the lining increases, until it reaches a maximum with a vertical wall. Superimposed on these effects is the fact that the wind is highly variable, shifting in direction and magnitude in unpredictable ways. Ballooning tendencies of linings hanging on vertical walls have been observed on both the leeward and the windward walls. In actual practice no one has been able to predict ballooning tendencies with any degree of certainty.

Reinforcement of the thin membranes with low-stretch fabrics will decrease the ballooning tendency of a lining, but will not eliminate it until the lining weight and the weight of the fabric are increased beyond the point of economic acceptance.

The conclusion, based on many hours of "berm watching," is that there are three effective ways to deal with the problem of wind uplift on lining membranes:

1. Install vertical tube vents.
2. Cover the upper slope areas of the lining with Gunite or concrete.
3. Use a wind-resistant lining.

Ozone Cracking

Some lining systems are subject to ozone cracking. The effect can best be illustrated by placing a natural rubber band with about 30% stretch in an atmosphere of pure ozone. The results are dramatic in that the rubber band will break within minutes. This is one of the poorer properties of natural rubber and resulted in cracks in tires, windshield wiper blades, and auto door extrusion gaskets, which formerly were made exclusively of natural or reclaimed natural rubber.

With the introduction of butyl rubber linings amid a fanfare of "ozone-resistant" claims, the ozone problem was said to have been solved. It was not long, however, before users found that this was not the case, for in areas where the lining was sharply folded and exposed to ozone laden atmospheres cracking did indeed occur. The effect was startling and extremely rapid. Even "soft" folds could trigger cracking problems in certain industrial areas.

In this respect the introduction of butyl rubber seemed to be a step backward in the search for an improved lining system. Even the lowly PE and PVC linings have never been affected by ozone in any concentration.

Initial countermeasures adopted to combat the cracking problem included the manufacturers' insistence that linings be installed in the "relaxed" condition. This indeed lessened the incidence of cracks. At the same time the manufacturers embarked on a program of better compounding in an effort to stop the cracking or at least to slow it down so that customers could live with the materials.

Later, EPDM was introduced, and its main advertising claim was that it gave industry a more ozone-resistant material. Indeed, compounds of this material seem to offer much greater resistance to outdoor aging of all types, although some rubber men claim that butyl itself can now be made as ozone resistant as EPDM compounds. On which side of the fence one stands on this issue depends on the rubber company to which he owes his allegiance.

Even today, part of the reason for the installation of "relaxed" membranes is the early unfortunate experiences with taut lining systems of butyl rubber. If there is even the most minute tendency for ozone sensitivity, relaxed systems will give the user improved performance.

Hypalon, CPE, and 3110 are the newest members of the lining family that definitely are not affected by ozone. Likewise, the rigid systems, concrete, Gunite, and asphalt concrete (as well as the flexible asphaltic panels), are not ozone susceptible. Although ozone attack can be eliminated by burying the lining system, this is never done. Rather, an ozone-resistant material is chosen in lieu of one whose resistance is questioned.

Wave Erosion

Wave erosion works on the banks of all unlined earth facilities. Wind generates waves that beat upon the side slopes, patiently nibbling away at the soils and causing continual sluffing problems. In time this action will cause sections of the embankment to fall into the water as the material sluffs to the toe position. If this action continues long enough, the entire bank can be eroded away to the point where reservoir failure occurs.

The severity of wave erosion is dependent on the soil type, the size of the waves (which in turn depends on wind velocity), and the "reach" of the facility parallel to the wind direction. In other words, the longer the distance the wind can blow across the open water of a reservoir, the larger the waves that can be generated. The stronger the wind blows, the higher will be the waves, and the more severe the erosion can become. Thus a long reach or fetch and a high wind can do considerable damage to unprotected slopes in a very short period of time.

One of the factors that affect the tendency of water to erode the slopes is the slope ratio itself. The steeper the slope, the more vulnerable it is to the action of waves. It should be emphasized that, within practical limits, it is not possible to make the slope flat enough to give it complete immunity from erosion problems. Even beaches, where the slope may be extremely flat, are victims of wave erosion. It should be recognized that the stability of the slope materials, too, is important. For cohesionless soils the closer a material is to its natural angle of repose, the more likely it is to suffer heavily from erosion attack. This is the reason why loose sand banks generally make for poor design even though they are protected by a lining. The safety factor here is very low; once the lining is breached, the waves begin their steady erosion of the sand particles beneath.

There are a number of ways of preventing or reducing the hazards of wave action. Riprapping, or the placing of rocks on the side slopes, is one method that is used universally. The rocks may be dumped or hand placed. They should be hard, dense, and durable, should resist long exposure to weathering, and should be at least as large as the maximum wave height expected at the slopes. This height may be calculated by any number of empirical formulas[3] or charts.[4] Some examples of wave heights expected under various conditions are listed in Table 7.2.

The thickness of the riprap blanket will vary with the characteristics of the reservoir and the wind velocities. A 3 ft. blanket is usually ample for most reservoirs, although this may be reduced for small facilities.

TABLE 7.2 Wave Heights Expected to Be Generated by Winds in Reservoirs

Wind Velocity (mph)	Fetch (Ft)	Expected Wave Height (Ft)
50	1000	1.0
80	1000	1.6
50	2000	1.4
80	2000	2.3
50	3000	1.6
80	3000	2.8

Concrete and Gunite have been used for erosion control, but in the words of the U. S. Bureau of Reclamation[5] "the number of failures is tremendous." On the other hand, some concrete installations have withstood the test of time, and this fact has led engineers to continue specifying this type of construction. Embankment protection by this method is not cheap, however, and because of the efficacy of riprap it has been economical to use the latter in preference, even if the rocks have to be hauled over 150 miles.

Grass plantings of various kinds have been tried on the side slopes of reservoirs in efforts to halt wave erosion of embankments. Their value is questionable. Maintenance is time consuming and the grass does not normally grow well beneath the waterline, thus reducing its efficacy. Moreover, fluctations of the reservoir water depth cause problems. Hydrophylic plants have been tried, but unless careful control is exercised, they produce dense growths and get out of hand.

Brush mattresses have also been used. These consist of bundles of saplings, 1 to 2 in in diameter, of willow, tamarisk, or cottonwood. The bundles are tied with heavy wire. As placed on the slopes, perpendicular to the toe, they range in size up to 18 in. in diameter and 20 ft. long. They are very effective, but again the maintenance problem precludes their use.

Because of its greater flexibility and lower cost, asphalt concrete has been used for the control of erosion due to wave action. This material, properly applied, does a reasonably good job, although it suffers from a few problems such as resistance to freeze–thaw cycles, expansive clays, and poor sun aging. Although these cause maintenance headaches, asphalt concrete seems to present fewer problems in general than most other materials, if initial cost is not an overriding factor.

Earthen reservoirs, properly compacted with the right soil mixture, do a much better job of resisting erosion than might be expected. The only problem involves uniformity because of the difficulty of perfectly grading, placing, and compacting the soil to a highly uniform consistency. Earthworking equipment in skilled hands can do a remarkably good job, however, and there are soils that can stand up for long periods of time.

Although the thin, flexible linings have been used in good volume for wave erosion control, they cannot boast overwhelming success in this area. Naturally, some of them are relatively low in initial cost, but those that possess the required weather resistance, such as Hypalon and the EPDMs, are in the same price range as asphalt concrete and often cost more, particularly in the heavier thicknesses. Because of their poor aging qualities, PVC and PE materials are seldom used for this purpose, unless on a temporary basis. Open exposure, high winds, and vandalism limit the utility of all of the thin membranes. Wave action and the strong currents generated by the wind have caused difficulties, too, by pushing the thin, light lining materials back and forth. Not only does this increases the chances for mechanical damage from below, but the pumping action beneath the lining works on bank stability as well. For best results the bottom edge of such linings should be buried into an impervious clay layer at or inside the toe of the slope.

Another problem with the thinner membranes for this use is that they are being utilized as partial linings, and this in turn presents some unsuspected difficulties. Water will get behind linings used in this manner, unless they are thoroughly sealed at their extremities. Since this design feature is seldom given much attention, these areas are often the main access routes used by the water to get behind the lining.

One of the problems involved in the evaluation of materials for wave erosion control is their resistance to mechanical damage. In many large reservoirs this problem occurs in the form of floating logs and other debris that can be propelled into the slope protection material with considerable impact. This is a case where riprap has a decided advantage; no impact force is too big, if enough riprap of the proper quality and size is used.

Seismic Activity

It goes without saying that reservoirs can be failed as a result of earthquakes, but there are a number of ways in which this can happen. In the first and most common case the embankments are further consolidated as a result of the shaking action. In other words, certain of the earthen segments of the system occupy less volume after the earthquake than they did before it. Local faulting is an even larger source of displacement. Since the movements are not necessarily uniform from place to place, settlement is variable. This causes uneven stretching of the lining, which may be great enough to exceed its ultimate elongation. Adjacent areas may move in different directions with respect to each other. Linings may be sheared or mechanically damaged by abrasion or chafing on the rough textures below. They may also be mechanically torn in areas near or adjacent to rigid structures.

In another action related to earthquakes the mechanical movements manifest themselves in the form of cracks in the supporting earth. In these cases the lining is asked to bridge the cracks; and if they occur below the waterline, reservoir integrity may be seriously affected. Although some minor bridging can be accommodated, excessive bridging, as we have seen, leads to failure of the lining. In addition to cracks in the earth, the concrete structures may also crack as a result of ground movements. Any lining anchored to the structures will be ruptured at these points.

Accompanying the earthquake activity is a wholesale movement of water within the facility (seiche). The movement is in the form of waves not unlike tsunamis. These rapid "tidelike" waves in a large reservoir or lake may continue for hours or days, and on occasion they have been so great that water actually sloshed out of the basin. Although the waves are normally not of great height in a reservoir, in a recent quake in California nearly 10 ft. waves were generated in a 120 ft. deep, 1 billion gal. reservoir.

A properly designed and compacted earthen reservoir is quite stable to earthquakes. Many earthen structures built before 1925, however, were constructed by "puddling" the embankments. In this method the embankments are built up by pumping water and soil mixtures through pipes until the sides of the reservoir attain the desired heights. Embankments constructed by this process, never dry out and are prone to sluff when shaken by earthquakes. When these structures are lined, the lining will be damaged or destroyed by the tendency of the embankments to flow during the shaking process. Today most states ban the construction of puddled embankments, and those so constructed are being abandoned or reinforced.

Reservoirs built on sandy or cohesionless soils are prone to "liquefaction" when subjected to shock waves generated by an earthquake or other exterior force. In such cases the soil may lose its ability to support a structure with resulting failure of the lining at the transition points between structure and soil.

Seismic activity in itself generally will do no harm to thin, flexible membrane systems, including asphalt panels, because all of these have a great deal of flexibility. There is no record of earthquake damage to any of these lining systems. At the Baldwin Hills, California, reservoir the lining was compacted earth, considered a flexible, high-quality lining. Although earthquake activity in the area did precede the failure, it did not seem to be the prime factor in the process. Rather, deep subsidence appears to have triggered the malfunction. This was a case, it will be remembered, where the reservoir was said to have had an asphalt concrete lining, although in reality this blanket served only as an erosion control membrane over the compacted earth lining.

Operating Problems

Cavitation

Cavitation plagues the rigid lining system, but its effect on concrete is perhaps best documented. The problem can occur whenever the direction or velocity of moving water is suddenly changed. Such action may happen in discharge flumes, in dam spillways, in canals around structures, and near inlet or outlet structures of reservoirs and tanks.

High-velocity water flow in large conduits and canals, especially near inlet, outlet, or other structures, is never steady, but is turbulent and subject to continuous variations in velocity and pressure pulsations.

There exist points near structures, pumps, or other protuberences where the relative velocity is particularly high. In these locations the flowing stream parts from the surface of the structure and leaves a void filled with eddies. When the absolute pressure is reduced to the vapor pressure of the liquid, this void or cavity becomes filled with water vapor, air, or other gases. As the flow moves downstream, the static pressure rises again and exceeds the vapor pressure. At that instant the vapor will suddenly condense, producing a collapse of the cavity and an explosion—or, more accurately, an implosion. This action is not confined to the larger cavities but extends into the pores of the lining material itself. There it bangs into the bottom surface of the pore, where a water hammer action takes place. It is equivalent to hitting the surface of the structure with millions of tiny hammers. Each collapsing bubble can generate pressure intensities that exceed the tensile strength of the rigid linings, including metal. Failure is by fatigue, and small particles are ripped away, giving the surface a spongy appearance. The action is called pitting and in the case of metal was originally thought to be chemical in nature, such as rusting or galvanic corrosion. It also occurs in wood and glass, and the latest thinking is that the effect is a mechanical one, as described above.[4,6,7]

The best defense against cavitation is proper hydraulic design. In addition to control by good design practice, it is also possible to eliminate cavitation effects by use of some of the impermeable membrane lining systems that are designed for bonding to concrete. The elastomerics are best for this use. Considerable attention must be given to mechanical anchoring systems. They must prevent the turbulence associated with cavitation from tearing the lining off the surface to which it is bonded.

Although cavitation is usually associated with a destructive mechanism against concrete, the same force also attacks steel, asphalt concrete, and, of course, Gunite. The more resilient the surface, the less is the chance of attack. When the surface can absorb and dissipate the shock without dislodging particles of itself, no degradation will take place.

Fortunately, the cavitation effect does not often appear in reservoirs. This is partly due to the lower velocities associated with water moving through the large inlet pipes and partly because the point of inlet is usually below the water surface. The energy of the incoming water is thus dissipated by the inertia of the water already in the system. Cavitation can be a problem, however, if the inlet water is allowed to splash directly onto a concrete surface or if it falls into an area of shallow water. Since the impact of water against a membrane lined surface is hard on the lining because of vibrations and sheer mechanical damage, inlet designs of this type should be avoided.

Impingement

Though not well known, there is an enemy of lining systems that has caused some malfunctions in recent years. The problem arises when incoming solutions impinge directly on unprotected thin membrane linings. They may do this by falling from entrance pipes or other overhead conveying systems.

Impingement produces problems due to vibration of the lining and its abrasion on substrate particels. In addition, evidence points to consolidation of soil substrates as a contributing cause.

When solutions impinge on linings which lie on a substrate that does not consolidate or displace, such as a heavy section of concrete, the texture of the substrate becomes of vital importance. The impinging solution induces rapid vibration, and failure is due to mechanical abuse only. If impinging solutions are hot, the problem is worsened. The heat contributes to faster degradation, even when heat-resistant polymers are used.

In some cases liquids are introduced into a pond by means of a spillway or are simply allowed to cascade down the slope of the lining. At the point where the incoming solution enters the pond a hydraulic jump is produced, which manifests itself in the form of turbulence. The degree of turbulent activity depends on the volume and velocity of the incoming stream. If both these factors are large, the turbulence may be extremely violent. Although no formal studies have been made in which hydraulic jump was analyzed with respect to its effect on flexible lining systems, we know that rigid linings such as Gunite or concrete are rapidly degraded by the "boiling," turbulent actions of the water at the jump. This is the cavitation condition just described.

Although cavitation "hammers" as such do not degrade the flexible membranes, the violent turbulence will usually cause failure due to mechanical damage. The mechanics may first involve subgrade settlement or displacement if the substrate is not stable. Here, again, positive proof is difficult to obtain because of the lack of scientific study. Vibration induced mechanical chafing seems to be the most likely failure mechanism if the substrate is Gunite, concrete, or asphalt concrete.

The best solution to the impingement problem is the use of inlet structures placed on the bottom of the reservoir, where the water in the facility will act as the energy dissipator, and it is strongly recommended that designers use this technique on all types of hydraulic facilities equipped with thin membrane lining systems.

Maintenance

Into this category of failure mechanisms falls a conglomerate of items, not otherwise classified. One of the more important things in this classification is damage due to cleaning operations.

Most reservoirs are cleaned from time to time. Quite often, fire hoses are used for some part of this cleaning, but little attention is given to the care of the lining during these operations. Fire hose couplings make use of heavy brass lugs. When new, these lug handles are smooth and rounded, but most of them seem to acquire points and burred edges as a result of continually being dragged across pavement and other abrasive materials. The lugs have a history of damaging the thin membranes, which often are not patched before being placed back into service.

Another damaging practice is to allow the cleanup crew to shovel debris directly off the top of the thin membrance. Again, the thinner linings just will not take this kind of abuse. The use of wheeled motor vehicles on top of these materials will also result in damage in some cases, particularly on rough-textured soils or in areas of unstable substrate.

If cleaning operations are frequent, shoveling should be done on a concrete pad designed into the facility, or a lining should be chosen that is resistant to mechanical abuse. This problem is sometimes solved by use of an asphalt panel lining system on the bottom or in the areas where cleaning is prevalent; this is then transitioned into a thinner, less expensive lining on the slopes, where there is no mechanical abuse. Sometimes the thinner membrane linings have been overlaid with protective pads that guard against mechanical damage. Again, asphalt panels have been used for this protection.

Mechanical damage can also be prevented by overlaying the thinner lining with other materials. The most common cover is soil, followed by riprap and then Gunite, but concrete and asphalt concrete have also been used. All of these materials, however, preclude the possibility of search and repair operations with respect to the overlaid lining. The loose asphalt panel overlays offer a good solution to this problem, in that they can be easily removed from the lining to facilitate its inspection and repair.

Reverse Hydrostatic Uplift

Those with experience in operating reservoirs have no doubt witnessed the chain of events that can occur when water gets behind a rigid lining during a time of rapid drawdown, particularly during wet weather. When this happens, lining slabs or segments can be uplifted, as there is no balancing water pressure on the top side. The problem is no less severe with respect to the thin membrane systems. Their seals around columns or structures are usually pried apart by uplift forces, and tearing occurs at the mechanical fastening points.

In addition to the uplift phenomenon, the effect of heavier saturated soil on rapid drawdown can create another spectacular result. This extra weight of the soil can cause downslope movement in which the lining ends up at the toe, neatly wrapped around a large segment of soil that formerly occupied the slope area.

There are two ways in which the reverse hydrostatic problem may be attacked. The first is through design, and the second is through good operation. Usually, both methods are used.

In the design area the best protection against the dangers from hydrostatic uplift is the use of an underdrain network. As already discussed, however, protection is not necessarily forthcoming if the underdrain design is not correct. Certainly, a "ditch full of rocks and a perforated pipe" do not insure optimum performance and can, under certain conditions, contribute to the reservoir failure. The underdrain system must be designed properly and must be installed under the supervision of a diligent, well-trained inspector.

The reservoir operator is a most important factor, too, in combating uplift problems. Most owners insist upon a policy that does not permit draining the reservoir during the rainy season, except under emergency conditions. If the operator follows this procedure, problems of lining uplift from reverse hydrostatic forces are minimized.

Vandalism

Every water company is familiar with vandalism, and it would do no good to list all the possible forms that destructive damage to lining systems can take. Perhaps the most common is cutting the membrane, either for fun or for the removal of a large section to serve the vandal as a tarp or similar type of waterproof covering. This practice in some sections of the country has led owners to cover even 20 year guaranteed materials in order to prevent their mass destruction.

There are two methods of attack on vandalism that have practical merit. The first is to keep the lining out of sight and away from vandals, and the second is to use a vandalproof system. The first method will utilize a cover material of earth, rock, Gunite, or concrete. The second will utilize a rigid-type lining or a flexible, continuous vandalproof lining.

Another approach has been used in connection with keeping the lining out of sight. It is an indirect method, but it has considerable merit. If the fence that surrounds the reservoir is placed down the outside berm instead of at the top of it, the facility will be more or less out of the sight of vandals. It therefore does not display an attractive pattern for those who have intent to do damage. The fence should be a sturdy one, preferably with three strands of barbed wire on top of it. This should be done in connection with any potable water holding facility, in order to keep vandals from throwing anything into it. The bottom of the fence, in all cases, should penetrate into a concrete sill to prevent entrance by tunneling beneath the fence. The road around the top of the berm should be inside the fence and not outside it, as is the case with many installations.

References

1. Harry R. Cedergren, "Control of Seepage in Earth Dams," *Proceedings of the Second Seepage Symposium,* Phoenix, Ariz. Mar. 25–27, 1968.
2. U. S. Army Transportation Research Command, *Heliocopter Downwash Blast Effects Study,* Vicksburg, Miss., 1964.
3. American Society of Civil Engineers, "Review of Slope Protection Methods," *Proceedings,* Vol. 74, June 1948, Subcommittee on Slope Protection, Soil Mechanics and Foundations Division.
4. C. V. Davis and K. E. Sorensen, *Handbook of Applied Hydraulics,* 3rd edition, McGraw-Hill Book Co., New York, 1969, pp. 4–20.
5. U. S. Bureau of Reclamation, *Design of Small Dams,* U. S. Government Printing Office, Washington, D. C., 1960, pp. 201–204.
6. T. Baumeister and L. S. Marks, *Standard Handbook for Mechanical Engineers,* 7th edition, McGraw-Hill Book Co. 1967, pp. 9–198.
7. James H. Potter, *Handbook of the Engineering Sciences,* D. Van Nostrand Co., Princeton, N. J., Vol. 1, 1967, p. 996.

8

Pollution Control Linings

Pollution control linings constitute the fastest growing segment of the lining business today and are destined to be the biggest area of lining use for some time to come. The contamination of our land must be halted if we are not to suffocate in our own waste. As President Johnson said in his February 8, 1965 message to Congress on natural beauty, "Every major river system is now polluted. Waterways that were once sources of pleasure and beauty and recreation are forbidden to human contact and objectionable to sight and smell." Many others have echoed the same message. And the problem is in no way confined to the United States. The problems of pollution control and the complex disciplines likely to be required for their eventual solution may well serve as an unexpected incentive for nations to band together for their common survival.

Only recently have medical experts begun to see relationships between air and water pollution and serious health problems such as cancer. Recent newspaper articles have stated that one-half of the people in this country are drinking contaminated water! This is really a fantastic declaration to make of a country that prides itself on cleanliness. But it is true, according to government surveys of a representative sampling of water supply systems throughout the country. In the summer of 1970 hundreds of persons in California's Squaw Valley area suffered an epidemic of "sewage poisoning," the third incident in 5 years within the state. A short time later, on the beaches of Pensacola, Florida, there was another mammoth "fish kill" due to contaminated water, the fifty-second such incident in only 18 months.

There are a number of possible solutions to the problems of contamination, but most involve a considerable amount of time. One technique that is quick and positive is the use of linings to segregrate polluting solutions from potable water supplies (Figures 8.1 and 8.2). Although linings for pollution control are rather closely related to lining technology in general, there are enough differences to warrant a separate chapter.

The psychology of linings, as we have seen, plays a part in lining design and choice, and in the control of liquid pollution it is often a very important factor with respect to overall design. This is particularly true of pollution control linings planned for already existing plants.

In such cases a mandatory directive has probably been handed down (or soon will be) by one or more governmental agencies. The plant, faced with an expenditure not previously budgeted, is not likely to look with favor on an expense item that does nothing but decrease the profit picture. The result of these circumstances has been an unmistakable trend toward getting the job done as cheaply as possible. This means not only a minimal lining, but also minimal attention to support facilities and other related expenses in connection with the project. Although this attitude may get the job done at the time, relieving the pressure from controlling authorities, it can well result later in an additional (and sometimes greater) expenditure to correct the built-in problems. This is a

FIGURE 8.1. A Texas installation uses blue Hypalon lining on the slopes and black Hypalon on the bottom of this cut-and-fill brine pit.

FIGURE 8.2. A special black 30 mil reinforced Hypalon pollution control lining for a wastewater treatment plant in the Southwest.

situation encountered much too often with pollution control lining systems.

Fortunately, in new plant construction pollution control facilities are not regarded with the same contempt as in existing plants, because, more and more, this segment of the plant is considered as part of the original design and construction budgets. Nevertheless, it is an area that is subject to more casual analysis than the profit producing areas of the plant.

In some facilities, particularly in sewage lagoons, there is a common belief that almost any lining will tend to seal itself in time. Nothing could be further from the truth, unless the word "seal" is used very loosely. Even the so-called sealant chemicals developed specifically for this purpose do not make this claim. And yet, to the objective view, perhaps nothing is impossible. It could be that, in an occasional isolated case, a facility might actually seal itself 100%. But it is not the type of thing to bet on, particularly with today's ecology-minded citizens. It is doubtful whether any pollution control authority would endorse this theory.

An almost perfect parallel is to state that a typical commercially built backyard swimming pool does not leak. It is not that the builders could not achieve good results here; rather, no one would pay the extra expense required to meet this type of specification. Moreover, except in special cases, leakproofness is not critical for this type of backyard pleasure spot. But any owner who has tried to stop permanently the leakage from his pool has no doubt been surprised to discover how difficult this really is. And the maximum depth with which he works is a mere 8 ft!

There are isolated cases where a concrete lined facility gradually decreases its leakage (as measured in underdrains) to zero or near zero. The prerequisites are small initial leakage and hard water or water that has been treated with lime. If we can make the assumption that leakage really did stop (i.e., none got by the underdrain), well and good. But an owner could become old waiting for this to happen. For every facility where this pleasant experience has occurred, there are a hundred where the leakage stayed the same or got worse. Unfortunately, it usually gets worse.

Normal sewage is not generally an enemy of the newer lining systems. But what is "normal sewage"? As one operator has said, "Normal sewage is getting hard to find anymore." In today's economy sewage may contain quantities of oil, solvent, petroleum derivatives, acids, alkalis, and a host of other chemicals. Some municipal sewage systems are tied into lines that may introduce these diverse agents into their treatment plants. Today the possibility of these occurrences must be reckoned with if any lining systems are being considered for the treatment plant. More and more, chemical plant owners and other producers of these off-brand wastes, who were formerly discharging them into sewer lines, are being forced to discontinue this practice and are being asked to pay for their own special disposal arrangements. This, of course, is one of the major objectives of pollution control activities.

With some of the chemicals listed above, the use of PVC, PE, CPE, butyl rubber, neoprene, Hypalon, 3110, concrete, asphalt concrete, or asphaltic panels may be questionable. Table 8.1 may be useful in the proper selection of a lining system, but it should be considered as a guide only. On-the-spot tests of potential linings and their adhesive systems are recommended as added insurance.

Somewhat related to the usage discussed above is the entry of linings into sanitary landfill operations. Here the membranes serve to prevent pollutants from being leached into high-quality groundwater supplies. Care in selection of the proper membrane is important in this applica-

TABLE 8.1. Lining Selection Guide Chart[1,2]

	Type of Lining										
Substance	PE	Hypalon	PVC	Butyl Rubber	Neoprene	Asphalt Panels	Asphalt Concrete	Concrete	Steel	CPE	3110
Water	OK	OK	OK	OK	OK	OK	OK	OK	CP	OK	OK
Animals oils	OK[3]	OK	ST	OK	OK	Q	Q	NR	OK	OK	OK
Petroleum oils (no aromatics)	OK[3]	Q	NR	NR	SW	NR	NR	OK	OK	OK	OK
Domestic sewage	OK	OK	OK	OK	OK	OK	OK	OK	OK	OK	OK
Salt solutions	OK	OK	OK	OK	OK	OK	Q	NR	NR	OK	OK
Base solutions	OK	OK	OK	OK	OK	OK	OK	Q	OK	OK	OK
Mild acids	OK	OK	OK	OK	OK	OK	OK	NR	NR	OK	OK
Oxidizing acids	NR	NR	NR	NR	Q	NR	NR	NR	NR	NR	NR
Brine	OK	OK	OK	OK	OK	OK	OK	Q	NR	OK	OK
Petroleum oils (aromatics)	Q	NR	NR	NR	NR	NR	NR	OK	OK	NR	NR

[1]OK = generally satisfactory, Q = questionable, NR = not recommended, ST = stiffens, SW = swells, CP = cathodic protection suggested

[2]It is recommended that immersion tests be run on any lining being considered for use in an environment where a question exists concerning its longevity. Consult the lining manufacturer or an experienced testing laboratory when in doubt.

[3]Must be a one piece lining.

tion. Not only is it necessary for the lining to be continuous, but also it must stay that way despite the irregularly shaped particles of the backfill and possible subgrade settlement as landfill constituents decompose.

Pressure against thermal polluters is now on the upswing. The problem stems basically from the fact that power generation requires enormous quantities of cooling water today, because economy and efficiency dictate that each plant be of extremely large size. After the water has done its cooling job, it is considerably increased in heat content. It is the dissemination of this heat that poses the problem.

Lining systems, as we know them today, can make but a limited contribution to the solution of thermal pollution. Some of these thin membrane systems, however, can contribute to the solution of the problem in a rather unique manner, as will be seen later in this chapter.

In addition to pollution by liquids and by heat, airborne pollution also causes trouble with potable water supplies. Included here are such things as radioactive fallout from nuclear activites, the natural trash of man and nature, and the intrustion by insect and beast. Although conventional roof cover systems are placed over potable water supplies in order to keep out all of these things, their cost effectiveness is limited. Later in this chapter, we will see how thin membrane lining systems have been adapted to solve the problems of airborne pollution.

Ponds for Waste Products

There are a number of special considerations in regard to lining ponds holding waste products. Many of these are obvious, but some are not.

In the waste pond lining category, the designer is faced with some very sobering risk versus cost factors. One potential risk is the contamination of groundwater should the pit leak. This can be a very serious mishap. Today, the government takes an extremely dim view of any action that contaminates the evironment. In fact, the potential cost of such an occurrence is usually of such magnitude that it dwarfs the entire cost of the holding facility.

Other risks are related to the one just discussed, such as fish kills resulting from leaks, damage to agriculturally productive lands, and, should the facility fail, lack of a place to put the continuing flow of waste product from the plant. Superimposed on thse items are the other previously discussed dangers associated with leakage from any facility.

Analysis of the risks of failure will generally establish the criteria as to the design standards for the entire facility, including the lining type and thickness. Generally, the closer the pit is to the lake, river, or stream or to the groundwater level, the more important it is to make the pit design foolproof with respect to leaks. The emphasis is even greater if the plant has already been warned about contamination by pollution control authorities.

Naturally, the lining must be chosen to be compatible with the pit contents. The rule here is that the lining be resistant to *every* chemical whose presence is expected in the pit, regardless of its concentration. The rule is a confusing one for those specifying lining systems.

Of all the chemical materials, two types tend to attack linings in general: (1) oxidizing acids and (2) solvents. The former will tend to attack all the linings in common use today, particulary if there is any combination of fluid velocity, heat, or impingement on the lining. Although solvents are somewhat more selective, most are suspect. Among the hardest to contain are the aromatic compounds (benzene, toluene, the xylenes). The chlorinated chemicals (carbon tetrachloride, chlorobenzene), turpentine, phenol, and ketones (methyl ethyl ketone, acetone), to name a few of the basic ones, can cause a lot of trouble, too. A problem that arises quite often these days is to choose a lining that will resist minute quantities of these chemicals dissolved or dispersed in some sort of slurry, waste stream, cooling water discharge line, and so on that is to be contained in a lined pit.

The answer to this lining question is difficult to predict with any degree of certainty. Laboratory tests at best are only a guide. If carefully run, they may be useful in the decision making process, provided that the contents of the pit will never vary in makeup, temperature, or concentration from the sample. For most industrial processes this is something no manufacturer can guarantee. Generally, the processes vary from time to time and are always subject to a concentrated dose or "slug" of an offending chemical. Depending on the density of this component, its solubility in the stream, and a host of other factors that affect chemical or solvating activity, the slug may pass through the system unnoticed, but it is more likely that irreversible damage will be done to the lining. Destruction of part of the lining, either on the bottom or on the slope (or both), can occur, rendering the pit lining ineffective and subject to complete annihilation by outside influences such as wind or turbulent fluid flow conditions.

Another way to look at the situation is that linings in an oil refinery should be oil resistant, even though no oil is supposed to be in the pit. There is always the possibility that an oil spill, the opening of a wrong valve, or some other such incident will place a concentration of oil in the pit that will destroy or seriously damage a non-oil-resistant lining. It is sound logic, then, to install a lining that is resistant to every component expected to be in the pit.

Next on the list is to check the adhesive system. The fact that the membrane itself is resistant to the waste liquids does not mean that the adhesive used for the field joints is similarly endowed. Both lining and adhesive should be checked if there is any doubt about either component's performance. Again, pure concentrations of the individual components should be evaluated, as well as the mixtures normally encountered. If the field seams are to be made by one of the heat welding processes, the need for the adhesive checking step is eliminated.

Elevated temperature of the pit contents is another factor that can cause difficulties. Again, evaluations of small lining samples in the actual pit solution will yield some qualitative data. Samples should be placed in the pit in areas of highest velocity and greatest expected temperatures, for it is in these locations that the weakest link in the performance chain is to be expected.

After these evaluation steps have been taken, the techniques of lining installation are similar to those employed in other types of jobs. There should be a continuous underdrain system intersected by perforated pipe toe drains. There is one additional design requirement with respect to these drain lines: the drain pipes should terminate in a heavy-wall concrete or appropriate metal lined sump, equipped with a pump and a return line to the pit. In this way any product that escapes for any reason may be returned to the pit instead of eventually contaminating some segment of the environment.

Unless the pit is constructed in sand or gravel, it is not usually necessary to install a cutoff lining beneath the continuous underdrain blanket or the trenches that hold the pipes. The resistance to flow within the blanket will generally be less than that offered by the substrate. If the substrate is coarse sand or gravel, however, the preceding statement will be incorrect and an impermeable lining must be installed below the drain pipe trenches or the underdrain blanket or both. The latter construction is known as a double-lining system. The common technique here is to enhance the impermeability of the area beneath the underdrain blanket by means of a thin, compacted soil blanket. Since the normally constructed underdrain has a very low resistance to flow, it is not usually necessary to have an extremely dense, impermeable membrane beneath it. In the trench below the perforated pipes, however, continuous membrane films are used in relatively critical cases.

Another type of pollution control lining or barrier is required in connection with landfilling operations. Many cities handle their garbage disposal problems in this manner. Seepage of rainwater into the ground, together with underground water, can and does leach from these areas potentially dangerous chemicals that may find their way into aquifers or other water holding facilities. These landfill areas also generate gases such as carbon dioxide and methane, which may at times cause additional problems for the lining system.

Three general types of membranes have been used for landfill anticontamination purposes; polyethylene, polyvinyl chloride, and thin reinforced asphalt sprayed membranes. Reports of test installations indicate that all three membranes are of questionable value over the long term. Mechanical damage seems to be a big factor. It can occur during installation, particularly in the case of a curtainwall member.

Although landfill operations involve compactive effort, it is not done with the same degree of intent as is required in a reservoir construction project. Therefore settling can occur, as the earthwork is not generally designed as a foundation for any structure. The settling is further magnified because decaying items reduce their volume and thus contribute to subgrade settling. These settlements can place undesirable strains on the side portions of the unreinforced membranes. If reinforced membranes are used, the reinforcement must be a material not affected by soil decaying mechanisms. With any buried membrane, particularly in a landfill operation, there is no practical way to locate or repair a ruptured area. Nevertheless, proper choice of materials and special pit operation procedures could do much to minimize the problem.

Butyl rubber would be a good choice as a lining material for landfill application, as far as its general chemical resistance is concerned. Nylon would be the preferred reinforcing fabric because of its stretch characteristics and good resistance to biological attack. Unfortunately, the present state of the art prohibits the production of high-strength nylon reinforced butyl membranes at an affordable price. The top tensile values run in the neighborhood of 125 to 150 lb/in., and tear values are quite low. For landfill membranes these values would not be strong enough.

The next obvious choice of material would be PVC with a nylon reinforcing fabric. Here high tensile constructions are comercially available. Tensile strengths can range from 300 to 1200 lb/in. with correspondingly high tear strengths. Although the vinyls are not generally as resistant as the butyl compounds, the higher strengths makes them a more attractive choice. Large sheet capability is another advantage to this type of material. With either of these materials, the principal objection is price. Hence neither has enjoyed much sales activity in this area. Again, budget considerations seem stronger than technical correctness.

Cooling tower operation requires high-quality water. Nevertheless, many towers operate with water containing some dissolved materials, which are slowly concentrated in the towers as the operation proceeds. These solids must be "blown down" from time to time. If good-quality groundwater is present below the site, pollution control regulations will probably require that pits holding these concentrated salts be lined. As already described, the solids are usually left to build up in the pit until its capacity is consumed, at which time the pit is abandoned and a new one dug.

The philosophy of pit discarding makes more sense than meets the eye. It may well be the answer to the obvious question of running out of space to contain the solids extracted from the above process. Most people believe that either the contamination problems will be solved by other means or the process will be discontinued before salt buildups within pits bury the owners. Most pits, therefore, are designed to hold accumulated buildup for a limited period of time, say 5, 10, or perhaps 20 years. Probably, by that time other factors will have been brought into play to lessen or eliminate a further need for the pit.

Power plants or other industries that operate large-scale boilers will experience more serious problems of salt buildup. With boilers the concentrated salts from the water may contain other undesirable boiler chemicals so that the total waste package may present a particularly bad situation in the eyes of pollution control experts. In these cases boiler blow-downs generally fall into the growing list of things that must not be allowed to contaminate groundwater supplies.

Holding Ponds for In-Process Materials

Ponds holding liquids in process within a plant or otherwise holding liquids with basic value to the process are termed holding ponds. Although technically, holding ponds can hold any liquid, those that hold liquids that can pollute were discussed in the preceding section. Liquids of profit value to the plant may often be contaminants too, but primarily they have considerable worth to the pond operator. Because of this fact the pond, its lining, and its general operation are of special concern to plant management.

Generally, holding ponds contain a water solution of some chemical, although oil may also be the liquid. In any event the liquid is considerably more valuable than water, and hence the design and construction material choices should reflect this fact. It is surprising how seldom this obvious axiom is followed. In fact, in consulting work on linings in general, the trend seems to be toward a lining choice influenced by budget considerations alone. This has always seemed odd. The situation is almost like building a $100,000 home and then roofing it with newspaper. If the holding pond is part of the operation and is holding a valuable product, its failure will be costly. Not only will the valuable product go down the drain, but also the plant may be forced to shut down temporarily at a cost that can amount to several thousands of dollars a day. Yet, in a great majority of cases, holding ponds are built on unconsolidated material with no attempt to bring in a reliable soils engineer to study the substrate problems before the work is done.

Another common problem should be mentioned again. The time to discuss linings and their integration into the total project is in the initial design stages. Too often, this is not done. The facility is designed and sometimes actually built before attention is given to the lining. Linings are a part of the initial design, and the choice of a proper lining depends on other construction features. If lining intent is known early, the cost to adapt the facility for lining is minimal. If, on the other hand, too much time goes by before a lining decision is made, preparing the surface to receive the lining may cost more than the lining itself.

A quick reference to Table 8.1 will provide a general indication of what type of lining may do a particular job. In holding ponds, however, the proper choice is of prime importance with respect to the profit picture. If unusual solutions are involved, particularly ones containing oil, solvent, or acids that may have oxidizing capabilities, a careful testing program should be initiated, and it should preferably be run by a firm competent in this field. Usually, general design engineering firms do not possess the time, facilities, or know-how to evaluate effectively this chemically oriented area. Often it is very wise to conduct latter phases of this program on the actual material that is to be held in the pond.

Another factor that should be considered is temperature. If elevated temperatures are present in the solution, the lining may undergo fairly rapid changes that will affect its performance. Generally, PVC and PE systems should be kept below 125 or 130°F in sustained operations, despite some manufacturers' bulletins to the contrary. Butyl materials were originally rated by most manufacturers at over 250°F. Later, the recommended operating temperature was reduced to the 160 to 180°F range, and even then the manufacturer wants to examine carefully all operational details of the pit. At elevated temperatures evaluation of factory and field joints should be made, and substrate sensitivity toward the chemicals at operating temperatures should be evaluated also. The designer should never overlook the possibilities of interaction of the lining with components within the liquid, particularly when elevated temperatures are involved.

Low-temperature use also demands some attention to lining choice, although more latitude is available here than in the high-temperature area. Butyl again has an advantage, being good at -40°F or lower. Although PVC linings are operating well at 0 and at -10°F, they could be risky at -40°F if localized high-impact loadings were involved. Local settlement of the substrate could be a factor in the successful performance of the thermoplastic materials at low temperatures. Because of their tendency to stiffen as temperatures drop, subgrade movements can cause critical problems.

Since berms of PVC lined facilities should be buried anyway, this requirement will help to shield the plastic from the ravages of depressed temperatures. Unfortunately, not too many data are available in this area, except that PVC installations have been put in place at temperatures as low as 0°F. No cracking problems occurred during any of these operations. Pit operators have said little about PVC linings cracking at low temperatures.

Asphalt panels are also brittle in cold weather, and high-impact loading at depressed temperatures should be avoided. This type of lining is also poor in high temperatures and except under special circumstances should not be used in solutions over 125° F. Even then, slopes of 3.1 or flatter should be utilized.

Holding ponds containing oil present special problems in two areas. Hypalon, CPE, and neoprene have been used in some pits containing oil, and all have showed physical effects. In general, they "pucker" because of swelling action; this is a little less than ideal performance, even though manufacturers state that the puckering is of no concern to the owner.

The second problem is the effect of oil on the field adhesive system. Limited data are available in this area, but at this writing none of the manufacturers is pressing this point very much, as shown by a considerable reduction in the tenure of the guarantee, if it is included at all. In addition, almost all manufacturers want to submit samples to the pit owner for *his* evaluation.

When the oils contain aromatic derivatives, there is no economical, proven lining system on the market that can effectively be used, except for steel and the nonferrous metals. A very significant void in flexible lining technology exists with respect to this problem, although several firms are actively engaged in developmental efforts.

When a pit is known to contain trace quantities of oil, it has been assumed by many that a potential problem exists only at the liquid surface. The theory here is that all the oil is floating on the pond surface and therefore should affect the lining only where it comes in contact with the slope. Although the assumption is a reasonable one, the facts do not support it. In such cases any non-oil-resistant lining used in the pit will undoubtedly be degraded not only at the waterline but also on the bottom of the pit as well. Pit bottom lining degradation has happened with less than 150 ppm of oil in the water. The mechanism of failure involved here is twofold: (1) emulsification of the oil, in which the emulsified product takes on the characteristics of the oil itself (probably some water-in-oil emulsion), and (2) interactions between the oil–water emulsion (or the oil itself) and other chemicals in the pit to form a sludge that is heavier than water. This sludge settles to the bottom of the pit, where it takes on the characteristics of oil, to attack the lining. If there is *any* oil in the pit, even trace amounts, an oil-resistant lining represents by far the safest choice.

Harvesting Ponds

Harvesting ponds are in common use where chemicals (usually salts) are to be recovered from water solutions. In many cases this recovery method is less expensive than the one requiring the use of heat exchangers, boilers, and evaporators. Since solar heat is free, the heat for the evaporation process is available on a no-charge basis. The offsetting entries are the cost of the land, pit construction, and lining. In most cases desert sunshine conditions and low annual rainfall are required to make the process economical. Unfortunately, when these attributes are present, the soils are generally porous and good clay deposits from which impervious compacted linings can be made are absent.

Since compacted clay linings are not usually feasible here, the economics of evaporation ponds will normally dictate the use of the thinner, less expensive membranes. These are then specially protected against mechanical damage during installation and harvesting operations. Polyvinyl chloride linings have been utilized in the greatest volume for harvest ponds, the largest being installed for Texas Gulf Sulfur at Moab, Utah. Here over 23 million sq. ft. of the PVC material was used in the evaporation ponds to prevent pond solutions from contaminating the Colorado River.

As evaporation proceeds, salts are precipitated out of solution and fall to the bottom of the pond. After these deposits build up to a certain depth, overlying liquids are removed and the product crystals are "harvested" by the use of scrapers, front-end loaders, or similar equipment.

A technique to prevent lining damage during harvesting operations is required. Two methods are in general use. One is to control the level of equipment operation by elevation sightings, using surveying equipment. The other method is to allow original deposition of salt crystals to a 6 or 8 in. depth. Overlying liquids are drained off, and the deposit is allowed to dry; then a red oil paint is applied, usually in broad stripes, and allowed to dry. Further salt deposition is then allowed to continue until it is time to harvest. Equipment operators know when they see the colored paint that they are within 6 in. or so from the lining, and therefore harvesting operations are stopped at this point.

It is possible to control lining damage during harvesting by burial of the lining beneath earth or rock. Neither of these methods is used much, however, because the vast areas involved are too expensive to cover with anything but the cheaper linings. Then, too, soil coverage can introduce contamination problems.

As in any lining situation, the membrane choice will depend on the solution being impounded and on the value of the end product that is being obtained. Water solutions of salts will work with all common linings, but for cost reasons PVC or PE linings are most often used. Because of its big-sheet capability, moderate cost, and puncture–abrasion resistance, PVC, in particular, is on the best-seller list for this application.

In harvesting ponds, high lining integrity is required if the process is to be economically feasible. Underdrains are not used because of their expense in the large areas involved. The efficiency of the lining may be accurately charted, however, by continuously monitoring the percentage of total solids in the pit contents. Leakage will show up as a failure of the total solids to increase. After operation experience is gained, losses from leakage will also be revealed as unusually high feed rates for the maintenance of liquid levels within the pond.

If the pond leaks through large holes in the lining, heavy rains may show the locations of these holes, as fresh water flowing through the salt cake will cause it to dissolve. The appearance is subtle, however, unless the rains are extremely heavy. Thus the owner is faced with the tough problem of locating and repairing any leaks in the lining. For this reason PVC is utilized more than the less expensive but puncture-prone PE lining.

Because of the great size of chemical recovery ponds, combination linings are often used. Here, as discussed in Chapter 5, the idea is to place the less expensive lining on the large bottom areas, transitioning to a more weather-resistant type on the smaller slope areas. Most combination possibilities have been used, the most popular ones being those that involve clay as one of the linings.

Another type of harvesting pond growing in popularity is employed on fish ranches. Since the pond is often built on cheap, pervious soils, a lining is required to conserve water. Single lining membranes are usually utilized here, as areas are not as vast as in chemical salt evaporation ponds. Again, PVC is the preferred choice, although PE is used too.

Ground and Groundwater Contamination

Health agencies throughout the country are concerned with groundwater contamination in two situations: when the groundwater is being contaminated and when the groundwater is doing the contamination. In this section the latter possibility will be discussed insofar as linings are concerned.

In the preceding sections we examined the subject of pollution control linings. In all cases the objective was to prevent the pit contents from contaminating the goundwater. Health officials have asked on various occasions, however, that unlined reservoirs be lined to prevent their being contaminated by adjacant groundwater flows. This is quite a departure from the usual reasons for lining to control pollution, as we have just seen.

To locate a potable water reservoir in an area where groundwater pollution is prevalent is a risky business. On the other hand, reservoirs are designed to hold contaminants immediately above known groundwater supplies, and the latter are not polluted. It sounds very much like one and the same problem, but there is a difference. When contaminants seep into the ground, there is some possibility that they may be adsorbed or destroyed between the point of contamination and the point where they enter high-quality groundwater supplies. On the other hand, when a reservoir is contaminated by inflows of undesirable materials, the reservoir water has nowhere to go but into the feed mains to be served directly to the water user. Such a condition involves a risk out of proportion with respect to costs and alternative solutions.

There may be situations in which the reservoir is already present and the groundwater contamination problem arises later. Or, as some health authorities state, they want to "insulate" the potable water of the reservoir from contact with the ground. This will prevent any movement of soil microbes, insects, or worms into the reservoir. In addition, the water will not be muddy in appearance. Lining to prevent this type of "ground" contamination is not at all uncommon, although the owners are undoubtedly motivated by the other advantages of lining at the same time.

If the lining is to be used as a barrier against the intrusion of contaminated groundwater, some important basic principles are involved. First, there is the problem of reverse hydrostatic forces. To guard against this problem, the reservoir must be protected by an underdrain system, as already discussed.

Operation and testing procedures for a reservoir located in a contaminated environment will obviously be quite different from those used for a reservoir located where this problem does not exist. Groundwater flows and elevations should be charted regularly, and water quality tests run on a very frequent basis. At the same time serious thought should be given to building a new reservoir in some other location. Fortunately, since cut-and-fill reservoirs are normally built on higher elevations, their contamination by inflowing groundwater is a relatively rare occurrence. In any event the handling of water flows within cut-and-fill reservoir embankments is the same whether the water comes from the reservoir (leakage) or from some other source (groundwater flows or rain).

Airborne Pollution

The preceding sections have discussed the use of linings in controlling pollution when the pollutants are liquids. This section will deal with the situation that exists when the pollutants are solids. Generally, for solids to contaminate a reservoir, they must be airborne at some time or another.

Many types of contaminants fall from the skies, and over an extended period of time the volume of these substances can reach sizable proportions. For some time health authorities have been aware of this type of pollution, and everywhere they are insisting that open potable water reservoirs be covered. They are concerned for a number of reasons.

From the sky may come man-made radioactive fallout, the most common source being atomic or hydrogen bombs detonated above the ground surface. A number of these bombs have been set off in recent years. This activity seems to go in cycles, a bad year for bomb testing being followed by one with no testing at all. When these bombs are detonated, though, health authorities sound off in unison, and for good cause.

As the natural air currents move around the world, showers of radioactively charged particles settle on the land below. We know that cattle ingest some of these materials with the grasses they eat; and although the quantities may be small, they suffice to produce radioactive milk.

When radioactive particles fall into drinking water reservoirs, the path from bomb to consumer is more direct and considerably shorter. Some of the fallout is soluble in water and some is not, but in any event it takes very little imagination to see how these materials may end up in human beings. A particularly disturbing aspect is that very little is known concerning "safe" levels of radioactive contaminants. Even the experts disagree on where to place the invisible limiting threshold beyond which we should not venture. The Environmental Protection Agency has set maximum levels of *naturally* occurring radioactivity, specifically radium-226, at 4 picocuries per liter of water. Maximum dosages of man-made radiation, including fallout from nuclear weapons tests, medical and industrial uses, and effluents from nuclear power plants, are not to exceed 4.0 millirems per year. These regulations go into effect in June 1977 and will affect over 40,000 community water supply systems.[1] Perhaps the prudent thing to do is to follow the safest possible course as laid out in a typical state health department directive, one of which says, in part, "To maintain a finished water that is assuredly clean and protected from radioactive fallout, existing open storage reservoirs should be eliminated or covered as soon as possible."[2]

Despite the apprehension that radioactive pollution arouses in concerned, thinking people, there are many other things that can get into drinking water whose ramifications with respect to public health are equally sobering. In fact, these other sources actually represent more of a threat to public health than does radioactive fallout.

When potable water reservoirs are not roofed, birds of all types may use them as watering stations, bird baths, or resting places. Water fowl and other birds often deposit their excrement in reservoirs. Seagulls and ducks are common polluters of any open body of water, and their wastes are especially toxic. Of course, if the chlorinators are working well, the harmful, toxic components of bird feces are rendered safe. If something happens in the treatment plant, however (and this is always a possibility), public health is most certainly jeopardized. It is even possible that death could result, especially among the very young or the aged populations.

Animals can fall into a reservoir if it is uncovered. Since most cannot negotiate the steep sides of the reservoir, they eventually drown. Again, we are putting all our faith in the continued, unfailing performance of the chlorination system. Dead animals, such as rats, snakes, cats, dogs, and birds, that sink to the bottom of the reservoir join the muck from leaves, silt, and other windblown trash. This mass, under the usual conditions, will become anaerobic. In short, the slimy muck develops into a sort of cesspool within the reservoir itself. Those who have experienced the draining and cleaning of open reservoirs are well aware of the disagreeable odors that are produced when this muck deposit is agitated as the workers walk through it during the draining and cleaning process.

Insects can also be a problem in open reservoirs. Because there are many thousands of different types and because so little is known about insects in general, health departments generally feel that insects should be excluded from drinking water supplies on general principles. Anyone who has visited open storage reservoirs is familiar with the many types of insects that inhabit the water, either running about on top of it or swimming within.

One insect that has caused trouble is the midge fly (family Chironomidae, order Diptera). The fact that it is a member of the fly family is enough to elicit concern from anyone responsible for water purity and safety. Not much is known about this insect. Although the species that exist in this country are not believed to be harmful to man, the midge fly has psychological implications that far outweigh its clean bill of health. The larvae of Chironomidae are sometimes red in color and are called bloodworms. Normal chlorination will not destroy these larvae, and they are not removed by the usual filter systems. Hence they may end up in the consumer's home. As a water superintendent once commented, "You have never heard a complaint until you hear from the housewife who opens the kitchen tap and draws a glass of water à la bloodworms."

The midge fly is a somewhat mysterious insect. A reservoir that has never had bloodworms before will suddenly become infected. No known chemical method has yet been found to eliminate them permanently. Draining the reservoir and cleaning with high concentrations of chlorinating chemicals appears temporarily to solve the problem, but before long the midge larvae are back. Strangely enough, other reservoirs in the vicinity of a contaminated one are not bothered. It appears that the midge fly (or at least one species of it) has some sort of homing instinct with respect to where the eggs were laid originally, although this does not explain how the reservoir became contaminated in the first place. No one seems to know the answer to that question.

Last but not least there is man himself, the largest contributor to the problems of contamination in all environments. In addition to the atomic testing he does, he has a good right arm and seems always to be tempted to hurl sticks, rocks, and related items into any open body of water. From a pollution standpoint these objects probably do no harm. But sometimes he gets carried away and hurls such items as detergents or other chemicals whose presence may not be in the best interests of the consuming public. There have been instances in which chemicals of a more or less poisonous nature have made news in connection with the pollution of drinking waters by this route.

In large reservoirs the dilution factor is in favor of the reservoir, because it is difficult to transport large quantities of chemicals to a reservoir site without being detected. But, as any chemist knows, certain chemicals can be toxic in relatively small amounts. Some are broken down by the water treatment chemicals themselves and are rendered harmless, but others are not. In any event to depend on this possibility is placing a relatively thin line of defense against a possible disaster. The safer method is to isolate the reservoir contents from possible contact with harmful materials of any type.

In light of the preceding discussions, it would be assumed that all potable water reservoirs should be covered. Although many of them are, there are still thousands that are inadequately covered or have no cover at all. Most are older structures, built in the days when pollution controls were furthest from the designer's mind. Today new potable water reservoirs are covered when they are built, usually because construction permits will not be approved by health departments unless the plans call for a roof. In addition, older structures are being considered for roofs, partly because roofing makes good sense, and partly because state and federal laws are being enacted that force this type of action.

One deterrent to the construction of reservoir roofs is the fact that a conventional good-quality roof for a cut-and-fill reservoir costs more than everything else combined. Thus, if the roof is not built, a consider-

FIGURE 8.3. A first attempt in the 1950s at utilizing an unsupported 20 mil PVC lining membrane as an inflatable pollution control roof on a potable water reservoir in Los Angeles.

able savings can be effected. Of course, when mandatory health regulations dictate that one must be built, an expense out of proportion to the cost of the existing facility is indicated.

In this situation specially modified reservoir linings come to the rescue by means of a flexible, continuous, floating membrane roof cover system called ROOFLOAT® (Figure 8.4). It can be installed for but a small fraction of the cost of a good conventional roof structure, and the system also has several other advantages over conventional roofs. Properly designed, this type of reservoir cover system prevents all types of contamination by man, animals, birds, and insects. It eliminates evaporation, loss of chlorine residual, and taste and odor problems due to algae growths. Although there are no supporting columns, beams, or cables, the roof supports all ice and snow loads, including nonuniform loads. Moreover, this type of roof may be installed while the reservoir is in service. No

FIGURE 8.4. The 45 mil double-reinforced Hypalon lining membrane used as a floating cover to control pollution in a potable water reservoir in South Carolina.

significant stresses are transmitted to the side walls, even during seismic disturbances, and the cover is not affected by wind. Colors are now available, and they are permanently "built in" so that no painting of the roof is ever necessary. In every installation owners report improved water quality. This is truly a new and significant development utilizing the principle of the continuous, flexible membrane systems heretofor reserved for lining projects.

The cover has another unique property: it can be converted from a floating roof to a lining. This can be done by draining the reservoir, removing the stabilizing float battens, and making the seals to inlet, outlet, and overflow structures. The transition is quick and relatively simple.

Successful flexible membrane floating covers have been made from reinforced membrane constructions utilizing butyl rubber, EPDM, and Hypalon, all of which have carried long-term guarantees against degradation in sun exposure. They are attached directly to the top of the wall or slope, using the bolt and batten strip principle. In addition to the above advantages, they are also much less expensive than conventional roof structures. Average savings have been on the order of 70% of the cost of a conventional roof, but in some cases the savings have amounted to well over 90%.

Thermal Pollution

The most recent addition to the nation's problems with pollution involves heat. Thermal pollution is somewhat analogous to liquid pollution. In the early days sewage was dumped into the rivers, lakes, and oceans. At that time, it did not seem to make much difference because there was so little sewage compared to the extent of the bodies of water. The same was true with respect to heat pollution. Even as late as the 1940s the amount of heated water poured into the rivers and streams was insignificant compared to the water already in them. Thus they did not gain much total heat. Today, however, the situation is quite different. Our nation now has more industrial processes to cool, and they come in bigger sizes than before.

Heat pollution has many sources, but the biggest quantities come from power generating plants. When electricity is generated by the conversion of heat energy, more than half of the heat produced must be dissipated as waste. There used to be three good solutions as to what to do with all of this waste heat: it could be dumped into natural bodies of water, run into cooling ponds, or put through evaporative cooling towers. All of these systems worked well.

As our need for power expanded, generating capacity increased at a very rapid rate. Atomic energy plants were built, and to be efficient these must generate tremendous quantities of electricity. The cooling water requirements for a single plant may exceed 1 billion gal/day, all of which may experience a rise in temperature of over 10° F. The release of this water into rivers and lakes can cause a temperature change that alters the ecology of these natural bodies. Usually, what this means is that there is a decrease in the number and types of beneficial fish and a corresponding increase in the number and types of less desirable forms of life. Since natural balances are quite delicate, all sorts of problems can develop when they are disturbed.

When large quantities of water are cooled by evaporation towers (evaporation is a cooling process), tremendous quantities of moisture are released into the air. In fact, the quantities are so great that the weather pattern of the surrounding area may be drastically changed. When a power plant is proposed, all of these effects, such as weather patterns and the heating of rivers, must be studied and ecological reports issued before the plant can be built. Many sites are denied because the ecological impact of the plant will be greater than the environment can stand. The same type of problem arises when cooling ponds are used.

Another problem also occurs. When cooling towers or ponds are used, the cooling is effected at the sacrifice of water itself (that which is lost because of evaporation). Except at the coastlines very few areas suitable for power plants are left that can replenish the large quantities of water lost in the evaporative cooling process.

Some years ago the government began a research program on the problems of thermal pollution. One phase of this program dealt with the use of cooling ponds whose evaporation was suppressed while attempts were made to magnify the other natural cooling processes, namely, radiation and convection. At the low temperature gradients normally associated with cooling ponds, convection is not a very effective method of cooling, but radiation can be most effective. In fact, initial research indicates that properly pigmented lining membranes may be used as floating, flexible-type roof systems that will cool almost as efficiently as open ponds, even though the roof cover eliminates all evaporative cooling.[3] The analysis of the problem was pursued far enough to establish the fact that the process could develop into one of commercial importance, in view of the scarcity of generating plant sites, the cooling water shortage, and membrane economics. Since floating flexible membrane roof technology is well established, this technique may some day play a part in alleviating the growing energy crisis.

References

1. Water Information Center, *Water Newsletter,* Vol. 18, No. 15, Aug. 9, 1976, p. 1.
2. Colorado Department of Health, "1976 Drinking Water Standards," Sec. 2.3, Denver.
3. Lawrence D. Winiarski and Kenneth V. Bryam, "Reflective Cooling Ponds," Unpublished Paper, National Pollution Research Program, U. S. Department of the Interior, Federal Water Quality Administration, Pacific Northwest Water Laboratory, Corvallis, Ore., 1970.

9

Special Lining Problems

A number of lining situations demand special attention above and beyond that called for in the day-to-day situations which the designer faces. These cases have unusual features or requirements that must be handled by special techniques.

The most demanding of these special cases is the lining system that must pass the *zero leakage* tolerance test. Since this type of lining demand is the most severe that the engineer will encounter, it will be discussed first. Some of the principles developed in the next section will be utilized also in the other special problem areas that follow.

The Critical Reservoir—Zero Leakage

Although each owner of a lined hydraulic facility wants one that is literally bottletight, most will settle for some small amount of leakage. This is the only way to keep the lining costs within the bounds of the budget, and therefore the average owner will buy a reservoir lining whose leakage rate falls within the generally accepted tolerances. Sometimes, however, there occurs a need for a reservoir lining system that will absolutely not leak. Moreover, this system must be one that will never leak. This sounds like an exceptionally stringent specification (and it is), but these situations do occur. A common example is a pool that sits atop a building or serves as part of the roof of a conventional potable water reservoir (see Figures 9.1 and 9.2).

Zero-leakage facilities demand more careful attention to design details, construction, and inspection, and it is imperative that they have at

FIGURE 9.1. The critical, zero-leakage reservoir shown here sits atop an office building in San Diego. Beneath this small pool is occupied office space. The reservoir cannot leak a drop.

least one well-designed backup system. The latter policy is well known in such diverse activities as dam construction, space shots, and parachute jumping.

When a small lake or pool is built on top of a building, the designer is faced with a zero-leakage type of lining system. It seldom matters what lies below the pool; the owner will definitely be unhappy if any water gets through. Even a few drops an hour will cause a problem in the ceiling of an office or apartment below.

In the case of these pools the most effective backup system is a 3 in. deep metal pan with an appropriate drain line to take away any leakage or condensation that may develop. This simple, inexpensive underdrain can save costly repairs and maintenance expense later on. It will even be effective if seismic motion or building settlement disrupts the continuity of the basic pool lining membrane.

When existing reservoirs are roofed, the adjacent homeowners may

FIGURE 9.2. An excellent example of a critical, zero-leakage reservoir is this artificial river built into a conventional roof structure. Beneath the roof is a potable water reservoir, so no leakage is permitted by the State Department of Health. The reservoir is near Oakland, California.

raise strong objections, particularly if their property overlooks an irregularly shaped basin. Some have actually forced owners to replace the "lake" with some similar body of water so that, even after a roof is installed, they may still look down upon water. Bodies of water like these artificial lakes, which sit atop a potable water reservoir, are certainly critical with respect to leakage, and pollution control agencies will consider only zero-leakage performance for these designs. Here, again, a backup underdrain pan or similar provision should be incorporated into the design. With the present state of the art in manufacturing and construction, the economic feasibility of tracking down a few drops per minute of leakage is questionable. Also, over the long pull the thin membrane is always susceptible to mechanical damage from any number of sources.

Some pollution control agencies are insisting on zero leakage from facilities that contain potential polluting materials. At the same time they allow the use of very thin membranes in these facilities. The owners want to utilize these cheaper materials with the end result that a misunderstanding develops among owner, pollution control authorities, designer, and contractor. This type of problem is generated every day, because so many are unaware of the problems involved in creating zero-leakage lining systems.

In summary, it may be said that zero-leakage facilities can be and have been built. Although they are expensive by any standard, they may be executed with safety and at reasonable cost, based on overall efficiency, if the designer will only make full utilization of the underdrain principle as the backup protection system.

It is well to point out again that, just because water is only a few inches in depth, a leakage problem is not minor. A leak is a leak. Increased water depth merely increases the rate of leakage. Decreasing the water depth decreases leakage but unfortunately does not eliminate it until the water depth drops below all fissures in the lining material. It is true that surface tension can help when water depths are minute, but this is of no practical help in even the most shallow reflection pools.

Canals

All of the various lining types have been used in canals to limit seepage rates from these conveyance structures. The most active organization specifying linings in this field has been the U. S. Bureau of Reclamation in Denver, Colorado. This governmental branch has accounted for over 95% of commercial canal lining use, exclusive of the agricultural-type feeder canals and the California Aqueduct projects.

Because of the rapidly developing need for improved efficiency in water conveying systems, the USBR inaugurated a Lower Cost Canal Lining (LCCL) program in June 1946. The primary goal of this program was to reduce the cost of seepage control in canals, although the technology developed would also benefit other hydraulic structures.

Considerable activity in the manufacture, fabrication, and installation of lining systems was generated by the program. Many manufacturers and fabricators were made aware of the tremendous potential markets in the irrigation field. The USBR alone installed over 500 million sq ft of various lining systems in the 17 year period beginning in 1946.[1] The study evaluated linings that ranged in cost from 4½¢ to $1.06/sq ft in place.

There were many benefits of the USBR's LCCL program. Public awareness was certainly brought into focus, and the program corrected many problems caused by seepage from canal sections throughout the country. There is no doubt but that the program generated a strong motivation among manufacturers to develop more and better linings at the lowest possible cost.

On the other hand, the program had some obvious weaknesses. For example, many of the linings were laid with USBR forces, and hence the reported costs were often distorted in a way that made the lining systems appear to be more economical than they really were. This fact caused many would-be users to abandon their lining programs for economic reasons after they had been initially stimulated into action by the low cost figures of the LCCL programs.

Since canal lining involves some special skills and know-how, it was usually not feasible for owners outside USBR program boundaries to do their own lining. Thus they received a bid cost that was two or more times higher than what the USBR had reported in its studies. This experience did teach one thing: in studying lining costs in the literature, the reader should ascertain just how they were developed.

Another disadvantage of the government studies was that they did not delve into the mechanics or basic technology associated with linings. Of course, no agency has done this as yet; and, as previously mentioned, these types of studies are sorely needed if the industry is to move forward in the development of the ultimate, impervious lining system.

Canals pose some very special problems that often limit the use of continuous, flexible, impermeable membrane lining systems. First, there is the problem of mechanical damage, and this comes in several forms, such as vandalism, rodents, wind, hail, and the other rigors of an outdoor environment. All of these things are extremely destructive to thin linings. In addition, yet another factor places great restrictions on the type of lining to be used. This has to do with the innocent sounding word "algae."

In many parts of the country algae pose quite an operating problem. They normally begin their growth in the spring months; as the weather grows warmer, their growth rate speeds up. Their presence impedes the flow of water in the canal until dewatering takes place in the fall months, which of course kills all growths. In many areas, particularly in the warmer areas of the Southwest, however, canals operate the year around. In these cases, and sometimes also in cooler climates, the algae growths must be removed without taking the canal out of service. This is done by dragging the canal with pieces of metal attached to chains. It is obvious that, if drag cleaning is to be used, the canal lining must be able to withstand this type of punishment. Such a maintenance program eliminates from consideration the exposed thin membrane linings, including even the prefabricated asphalt panel. The only systems that can be considered are the various types of concretes, and perhaps the deeply buried thin membranes. The latter are not good candidates, however, unless the canal has low water velocities or very stable side slope material with which to bury the lining. Even then, erosion could be a problem, particularly if the canal is large.

Canal lining poses other special problems, not the least of which is the velocity of the water. In reservoirs the water velocity is very low everywhere except at the inlet or outlet, where, as already described, energy dissipaters are utilized. In canals, on the other hand, the water is always in motion because of the slope gradient, and hence energy arresters cannot be used.

At low water velocities, say less than 1.5 ft/sec, there are generally no problems with any type of soil used. As the velocity increases, however, unlined channels are plagued by erosion, first the fine sands, then the sandy and silty loams at around 2.0 ft/sec. Colloidal, stiff clays will carry up to 4.0 ft/sec, with shales and hardpans reaching 6.0 ft/sec.[2] These values are only approximate and depend on the age and depth of the canal, as well as design parameters.

Although linings are usually thought of in connection with seepage control, they also serve to control erosion on canal side slopes as well. Higher velocities may be carried in a lined than in an unlined canal. Even the linings have limitations with respect to the velocities they can carry, as already pointed out in the "Impingement" section of Chapter 7.

Burrowing rodents have been a problem with thin membrane lining use in canals. Sometimes, as in the case of some PVC lining formulations, the attraction is the sweet-smelling plasticizer system, but butyl types and some nonplasticized systems have experienced difficulties too. Any lining that blocks the rodent's only path to food or water is in jeopardy of being breached, as only armor-plate steel is strong enough to deter the action of teeth designed to gnaw through hard soils. Many canal owners will attest to this fact. Since animal burrows may run for considerable distances they can serve as conveyance tunnels to transport leakage water from the canal. When this happens, piping will usually enlarge the tunnel and lead to serious failure of the embankment.

In contrast to reservoirs, canals extend for great distances and in most cases their major portions are in remote, infrequently patrolled areas. Noncovered thin linings are therefore potential targets for vandalism. This has led to the practice of covering these flexible linings, except asphalt panels (Figure 9.3), to protect their integrity. Even butyl linings have been covered for this reason, despite their long-term guarantees against outdoor degradation.

FIGURE 9.3. A prefabricated asphalt panel lining in a Washington canal carrying waste chemical products.

Large animals also contribute to the destruction of unburied thin membrane linings in canals. This is a serious matter since breaches in the lining are sure to cause instant trouble because of the flowing water. Fences are good insurance for preventing this kind of problem, but even fences will not keep out high-leaping members of the deer family.

Obviously, the canal lining designer is in a difficult spot. The flexible, impervious linings are too flimsy for the rugged conditions imposed in canal use. On the other hand, the various concretes are not flexible enough when nonreinforced and are too expensive if they are designed to be structurally strong enough to compensate for the normal soil movements.

An obvious answer to this dilemma is to combine the linings: an impervious one to stop the leakage, and a concrete type to provide a protective, workable surface (Figure 9.4). This has been done, although not on a large scale because of the higher cost. Again, quality performance is sometimes overlooked in favor of economics.

The cement concrete lining, underlaid with a flexible, impervious, continuous membrane, has demonstrated its effectiveness at controlling seepage. The membrane adds another, quite unexpected advantage.

FIGURE 9.4. A PVC membrane lining being installed before placement of a concrete lining in the South Bay Aqueduct near Livermore, California.

Concrete cast on an impervious membrane does not exhibit the hairline cracks associated with an excessively rapid loss of moisture during the curing cycle. This renders the concrete more servicable, particularly against spalling actions.

Slip-form concrete linings may be laid on PVC-type membranes, but care should be taken in the design. Long 1:1 side slopes of a canal are difficult to pour, but with careful control of slump value slopes of 1½ : 1 have been used.

For three reasons soil is seldom used for protecting thin membrane linings in canals. First, properly compacted soil is too difficult to place in the narrow confines of most canals. Second, there is still the possibility of damage to the earth cover by water erosion. In addition, the lining acts as a low-friction slip sheet for the moist earthen cover, and sluffing is a continual problem that does not respond to any of the antiskid treatments. These objections restrict the protective blanket to the various concrete products, and all have been used for this purpose.

In the cement concrete–impermeable membrane combination the PVC lining is chosen as the underlay because of its rugged nature and low cost, although all of the membranes, including asphalt panels, have been utilized for this purpose at one time or another. When PVC is used in this capacity, a minimum thickness of 15 mils should be chosen. Any of these combinations will produce a total lining system that is probably the best overall value for the money of any canal lining on the market, if the designer is truly interested in ultimate performance.

Tunnels

The common method for lining tunnels is to utilize the structural concrete itself as the lining. This type of approach does have merit because in tunnel construction other design factors dictate thick and heavily reinforced walls. Even so, concrete tunnels develop leakage problems.

The correction of leakage problems in tunnels poses some special considerations. Two of these are very important. First, water in tunnels is subject to a velocity component, which may be more important than the pressure component. Second, the tunnel lining is subjected to reverse hydrostatic pressures, particularly in the upper half, although these may also occur at the bottom of the tunnel section as well. Also important is the fact that, because later remedial action is expensive, tunnel lining work should be considered in the same category as the lining of a critical reservoir.

Tunnels are seldom, if ever, designed with an interior-applied, flexible, thin membrane lining. If, after they have been constructed or after they have been placed in service, the need for a lining becomes apparent, the following guidelines will assist the designer in determining whether a continuous lining is feasible. It is always difficult to lay down a series of concise rules for anything, but a careful analysis of experience records in this field has yielded eight rules:

1. To facilitate proper handling of the lining, the tunnel shall be of the following minimum sizes, although these dimensions are subject to some modification: 3 ft wide × 4 ft high if rectangular, or 6.5 ft in diameter if circular.
2. The walls shall be very smooth.
3. The water velocity shall not exceed 6 ft/sec.
4. The flexible membrane lining shall be fabric reinforced and 100% bonded to the wall.
5. All transverse joints shall be lapped in the direction of the flow.
6. All transverse and longitudinal seams shall be mechanically anchored with metal bars and caulked to feather edges. The bars shall be of a metal that is inert to the action of the liquid being transported.

7. In addition to seam bars, a longitudinal ceiling bar shall be installed at the top center of the tunnel. Usually, the layout will provide for a seam at this location.
8. All side openings shall be properly cut out, flange anchored, and caulked.

In addition to these suggestions the normal rules for good lining practice should be observed as previously explained. There should be no wrinkles anywhere in the lining. Patches should be avoided if at all possible. All blisters should be removed before placing the tunnel in service.

Lining choice does not depend on outdoor aging, but the best adhesive bond to concrete system should be used. Successful installations have utilized a PVC-type wearing surface bonded to a gum butyl or neoprene backing (Figure 9.5). The backing adheres tenaciously to a

FIGURE 9.5. An example of a bonded tunnel lining in a sewage treatment plant influent line. The tunnel is approximately 3 x 4 ft.

special adhesive, applied over a properly primed concrete wall. Other systems have involved the use of special butyl rubber reinforced membranes, and a new one recently introduced features special on-the-job heat welded seams and a high-tenacity concrete adhesive backing on a reinforced PVC substrate.

The length of sheet segments that can be installed as one piece will vary with the shape and size of the tunnel and the width of the sheet lining material being used. Normally, a 48 or 54 in. sheet can be handled in the larger tunnels. Individual pieces may be 30 ft or more in length.

Lined tunnels should not be dewatered unless absolutely necessary and then only in dry weather, for they are subject to reverse hydrostatic problems unless protected by a good underdrain system or built in open and free draining materials. The use of air pressure in the tunnel during shutdown to counteract these reverse pressures is not practicable because of the many cross connections in most tunnels.

Reflection Pools

This type of facility may be classified into two important categories. In some instances the reflection pool may be ranked with zero-leakage critical installations. In other cases normal reservoir leakage tolerances may be specified. Although it is possible that a reflection pool could be built where no one cared whether it leaked or not, this would not be a common case.

How reflection pools are classified depends on several factors. The structures in the adjacent area, the way in which leakage would affect the total project, the maintenance procedures to be used, and the general type of neighborhood in which the pool will be located are four important considerations. Other factors that also have a bearing on the pool classification are its overall size and shape and its location, that is, within a building or exposed to the elements and weather conditions. What lies beneath the pool area is, of course, also extremely important. If the pool sits over areas that house people, products, equipment, or a parking lot, it should be considered a critical, zero-leakage type of facility.

Many times the reflection pool is adjacent to building foundations, support columns, or other load carrying structures. Most designers are particularly sensitive about the intrusion of water into foundation areas of any kind. Even though foundation engineering studies have defined bearing strengths for wet soils and the building has been designed accordingly, there is always apprehension by engineers and owners when water from any source makes its way into these critical areas. Those responsible for design and construction are never quite sure of the consequences, because there are always a few unknowns in connection with any project. Despite careful soils analysis this is still an area that is not as much of a science as we would like it to be.

When pools are adjacent to building foundations or are pierced by columns or other support members, they should be treated as critical, zero-leakage installations. This may be surprising because reflection pools are very shallow; most range in depth from a few inches to no more than 2 ft. But in a critical installation the only difference between a leakage of 1 pint per minute and 10 drops per hour is time. The latter leakage rate, fed to a soil with undesirable characteristics when wet, can cause many of the same serious problems over the long range as are caused by faster leakage rates in shorter time spans.

Except in special low-risk situations, reflection pool linings should be vandalproof. This is not a potential outlet for the unprotected thin membrane systems. Very special attention should be given to design and material selection, and a good backup system should be used if conditions warrant. If the lining is to be covered, care should be given also to this aspect. The more expensive and elaborate the cover system (tile, terazzo, etc.), the more important it is that the lining does not malfunction. Since the buried lining defies leak detection, corrective action on the lining itself is impossible. This type of action must be limited to the cover material. If the cover is relatively smooth concrete, corrective action is possible by relining or covering it. With extremely low leakage rates through the pool, direct water treatments may help. With an expensive, decorative tile cover, relining is to be avoided because the beauty of the exposed cover will be destroyed. The alternative of removing this material in order to find the leak source is very expensive and can seldom be justified.

If the pool is to be truly reflective in nature, the lining must be black or another dark color. But there may be some complications here. First, in many metropolitan areas air pollution will deposit residues in the pool that slowly change the black to a grayish or brownish color. Blowing dust and sand will also change the color effect. These contaminants of the pool are difficult to remove with normal circulation methods. Deposits will be swept clear in areas of noticeable water velocity, but it takes a very special design and surprisingly high water velocity to sweep clear all of the pool area. One way to handle this problem is by dye-masking, as explained below.

Provision should be made for treatment of the water against algae formation. In outdoor pools conditions for the development of algae are ideal: shallow depth and plenty of sunlight. Algae growth, in addition to being unsightly, also makes the pool surfaces extremely slick, a distinct safety hazard. Chlorination may be used to control algae, but because of evaporation this method is somewhat expensive. Other chemical treatments and algicides have been developed. For use in a reflection pool they should be nontoxic, because children may get into the pool and ingest some of the water. If the chemicals themselves add color to the water (and some of them do), the effect may be objectionable to the owner or architect. The problem may be handled by masking the color with a black, water-soluble dye, although this is not an inexpensive solution, particularly if the pool is large. Unfortunately, too, water-soluble dyes are not generally color fast, and replacement dye additions may be necessary from time to time. On the positive side, the use of black dyes enhances the effectiveness of the pool as a light reflecting medium.

The time-honored method of handling the finely divided debris that daily sifts into the pool from the air is hand cleaning. Swimming pool suction hose techniques may be used, or the pool may be drained from time to time and the contamination flushed down a drain. Alternatively, a combination method may be used. Each pool should have a mud drain or sump to facilitate cleaning or emergency dewatering. The cleaning operation can put the lining to test at the seam areas, particularly if high-pressure hoses are used. If this cleaning method is to be utilized, the lining system should be designed accordingly, keeping in mind the fact that fire hose lug connections are heavy and destructive to thin membrane linings. Sediment in the pool may be camouflaged by adding rock. This extends the time between cleaning operations but makes the job more difficult when it becomes unavoidable. Painting the pool bottom with a camouflaged design has also been used but is not as effective as the rock treatment.

Another problem in reflection pools involves the inlet nozzles. These have generally been adopted from swimming pool technology and involve dished profiles and other designs not compatible with good lining practice. The reason for this is the development trends in swimming pool marketing. Until recently, the usual practice in this field was to place the membrane behind the Gunite; hence nozzle compatibility with the lining was not a problem.

The inlet structures tend to introduce water at relatively high velocities. The lining to structure seals at these points are critical; otherwise velocity or vibration can loosen them and place water behind the lining. Eventual lining failure is not far distant if this condition is allowed to exist. All seals at these points, as well as all other places where pipes or other items pierce the lining, must be made using the proper primer, adhesives, mastics, and mechanical anchor system—the same technique as is employed on all critical facilities.

Reflection pools suffer from another problem, hydrostatic back pressure. This is caused primarily by irrigation of adjacent plants. Since the pool is recessed, irrigation water behind the pool walls may easily build up behind the lining to a head higher than that within the pool. The lining may be forced away from the wall, a situation that eventually manifests itself in leakage problems at the wall–floor intersections. This may occur even with water in the pool but will most certainly happen if the pool is dewatered quickly after a period of irrigation (or rainfall). The only satisfactory way to handle the problem caused by water behind the lining is to provide a drainage course behind the wall and a perforated drain pipe at the wall foundation area. Although drainage is the obvious solution, reflection pools have seldom been built in this manner, largely because designers believe that a shallow pool will not be subjected to the problems just discussed. This is a serious misconception. Naturally, the problem is not as critical as in a large reservoir, as back pressures are less, but everything is relative. A failure in a reflection pool can cause problems despite the low water heads involved.

The latest technology in lining reflection pools is to adhere the lining to the pool walls to prevent sag or drape. These conditions are very undesirable from the appearance standpoint. The thin membranes (except PVC and PE) are used for this purpose. If a two-ply asphalt panel vandalproof lining is used, a thin membrane lining is bonded to the vertical walls and a special transition toe seal is effected as shown in Figure 9.6. The bottom lining is not adhered to the substrate, except at seals to columns, pipes, and structures. Asphalt panels should not be used to line vertical wall surfaces below grade because of sag problems over the long period; also, if reverse hydrostatic forces ever develop, remedial action is difficult and costly. The cost of pinning the panels to the walls by mechanical fasteners to prevent the sag problem and to guard against reverse pressure problems is prohibitive in most cases.

Large Lakes

Lakes have become increasingly popular in recent years as a means of promoting housing sales in land development ventures. In many cases the lakes must be man-made as there are not enough natural ones to go around. Water must be pumped or otherwise purchased to fill them, unless runoff from rains feeds them. The water level must be main-

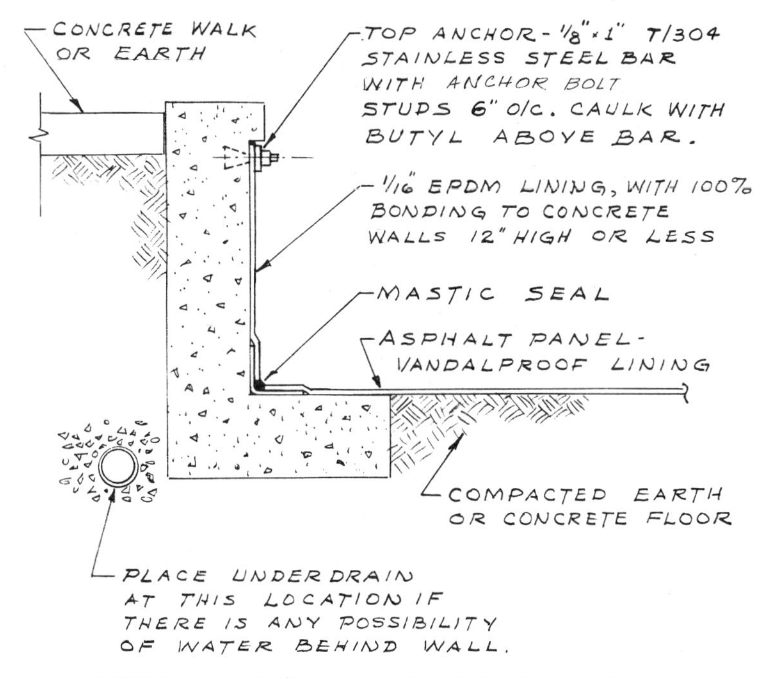

FIGURE 9.6. Reflection pool wall detail: typical section.

tained against losses from evaporation and seepage. Since no practical way has been found as yet for controlling evaporation from decorative lakes, for the replacement of this type of loss the owner must rely on rains and make up the difference by introducing additional water. The other loss, seepage, may have to be controlled in cases where replacement water cost is high.

Lakes generally do not present any serious liability problems due to seepage. No structures are liable to be undermined, and usually embankments do not exist, as the lake is merely a depression in the ground. Water depths are generally shallow, particularly if the lake is man-made.

Lakes have been lined with about every material discussed in this book, including all combinations of them. The most popular linings have been PE and PVC, because of their lower cost. The tendency is to bury these linings in the soil to protect them from boats, fishhooks, sticks, or other debris and from the degenerative effectives of sun aging. Sometimes this protection is given by utilizing a vandalproof and sun-resistant lining on the normally exposed side slopes. Concrete, asphalt concrete, asphalt panels (Figure 9.7), and riprap have all been used for this purpose. For PE, of course, burial is recommended because this lining, in

FIGURE 9.7. A prime example of the need for a vandalproof lining is this lined lake at Scherer Park, Long Beach, California, completed in 1957.

addition to having poor aging characteristics when exposed to sunlight, has a specific gravity of less than 1.0. Water behind a PE lining with no earthen cover on top will cause it to float to the surface with susequent destruction by man or weather.

General lining technology will apply for lakes as for any other hydraulic facility, with some possible relaxation of high standards if circumstances warrant. Of course, this depends on the cost of water. If PE is used, the surface texture must be very fine, as has already been discussed. Seals to structures and top anchor details should be the same as for reservoirs. There are, however, many variations in detail at the top due to the dictates of the landscape designer. At water inlets the design should protect the lining against "scour," and this should be the case also at any other locations where increased velocity or turbulence prevail. Although lakes seldom have utilized aerators, if they are present this is another area of turbulence. It can be extreme under certain conditions, and protection of the lining beneath this equipment is a necessity.

One of the very important things to remember is that defects or holes in a buried lining cannot be detected—even if they could, the cost of repairs is prohibitive. A soil sealant may be tried as a remedial technique, but if it fails, relining may be the only course. This point is very important, and the rather relaxed feeling that the designer and contractor may have with respect to low water depths and almost no contingent liability is in great part offset by this important consideration. Even with the least expensive PE lining, the cost with an earth cover will probably exceed 10¢ or 12¢/sq ft. If the lake is very large, even this small unit value multiplies into a rather sizable investment; for example, a 4 acre lake lining could amount to over $20,000. If it does not work, someone, somewhere, is going to be unhappy.

Soil sealants, such as SS-13 or Chevron Soil Sealer, have been used in large quantities as a leakage suppressant in lakes. Although they do not stop all seepage, they are easy to apply and under certain conditions can be effective enough to warrant their use. Another advantage of these materials is that they can be applied with the water in the lake. Generally, this type of material works best on the tighter soils with silt or clay content and are less effective on open soils such as those with a high content of sand, gravel, or broken rock. On open soils they may not be effective at all, unless these soils are underlaid with a more dense composite. The longevity of this type of treatment is variable. Sometimes it maintains reasonably good effectiveness with age and sometimes efficacy decreases as time passes. Soil complexities and their relationships with the treating materials and the water makeup are not well enough understood to allow precise predictions of their effectiveness.

Bentonite clays have been used as lake linings. Although there is a slurry process (liquid application without dewatering), these clays are usually much less effective in such cases than when they are properly worked into the soils mechanically and well compacted before the lake water is introduced. Bentonite linings may lose much of their effectiveness if the lake bottom dries up at any location. Soil cracks that develop do not reseal upon rewetting.

Since lakes do not have formal underdrain systems for the measurement of seepage, this is done by noting changes in the lake level with time. Naturally, inflows and outflows must be controlled or accounted for during the analysis for water loss. The evaporation of water from the surface can become quite a factor, depending on the time of year and the wind velocity. There is no way to control it, so correction factors

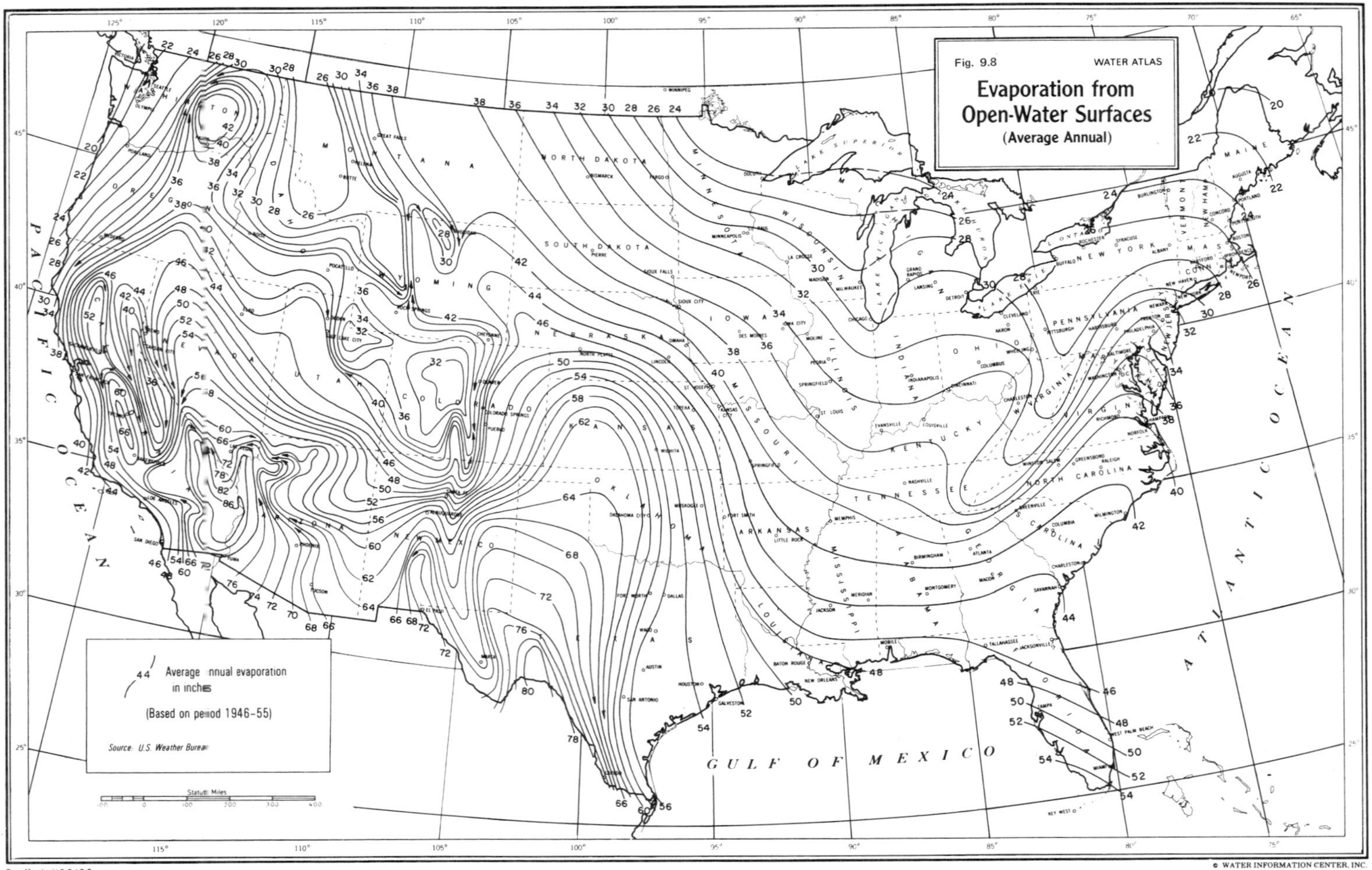
Fig. 9.8
WATER ATLAS
Evaporation from Open-Water Surfaces
(Average Annual)
44 Average annual evaporation in inches
(Based on period 1946–55)
Source: U.S. Weather Bureau
Statute Miles
PACIFIC OCEAN
ATLANTIC OCEAN
GULF OF MEXICO
Base Map by U.S.C.&G.S.

must be applied. These factors may be developed by placing a floating pan of water in the lake for a control. Rainwater must be excluded from the facility during the testing period; this can be done by means of appropriate adjustments based on water level data from the pan. Another method is to utilize evaporation charts such as the one shown in Figure 9.8.[3]

References

1. U. S. Bureau of Reclamation, *Linings for Irrigation Canals,* 1st edition, U. S. Government Printing Office, Washington, D. C., 1963, p. 2.
2. Ven Te Chow, *Open-Channel Hydraulics,* McGraw-Hill Book Co., New York, 1959, p. 165.
3. Miller, David W., Geraghty, James S., and Collins, Robert S., *Water Atlas of the United States,* Water Information Center, Port Washington, N.Y., 1962, Plate 12.

10

Modern Effective Design Practice

After a review of early design criteria, failure mechanisms, and current psychology, it will be well to take a look at up-to-date thinking on the subject of linings and, if possible, to establish some "rules of order" for a field that has had its share of confusion.

Although the choice of the lining should not be made before anything else, it should be considered throughout the design steps. It is as much a part of the total design as the underdrain system or the soils. All of these elements must work together as a team if the reservoir is to be effective in the use for which it is designed.

The first (and previously emphasized as the most important) point is that the lining be placed in a stable structure. Not only must the facility be stable when dry, but it must remain stable when wet. Although this rule is really aimed at the popular cut-and-fill reservoir, it obviously applies to steel and concrete tanks as well. No membrane lining material can hold a reservoir together if it is not stable in its own right.

This point is important when a request is made to line an existing facility to eliminate leakage. The first thing to determine is the current leakage rate in gallons per minute. Unless the leakage has liability overtones, very low leakage rates may not dictate the need for a lining. Extremely high rates, on the other hand, may indicate a serious problem, perhaps structural instability. If investigation confirms this fact, that problem needs to be resolved before the lining discussion gets very far.

In any type of project there exists a certain risk or probability of failure. With facilities that hold water or other liquids, a force is right at hand to take advantage of any careless mistakes in design or construction. Obviously, the job of minimizing risks takes on special significance in the case of cut-and-fill reservoirs.

It is, therefore, common sense to exercise every precaution. Facility design and inspection should be placed in the hands of professional experts with a heavy background in this field, and an experienced soils engineer must also be on the team if good performance is to be assured. The lining work itself should be left to experts in that field. Latest specifications state that the lining membrane and all ancillary items, such as adhesives, tapes, and caulks, be manufactured by a firm with at least 2 million sq ft of production experience in producing quality linings of the type specified. The lining firm, too, should be prequalified by having installed a minimum of 2 million sq ft of lining material meeting the specifications. Job lists detailing this work, together with owners' names and telephone numbers, are also being requested at bid time. Bids from nonqualifying parties are being rejected in favor of proven firms with good track records.

In cut-and-fill reservoir design the current trend is to use 3 : 1 slopes, particularly if the reservoir is a critical one. A continuous underdrain design is favored, and the drain is built either of small crushed rock with a membrane or filter cloth as a buffer to protect the lining, or of "popcorn" asphalt concrete with a pervious smoothing course for lining protection.

The underdrain is designed to operate at atmospheric pressure. On large installations the collector pipes are often segregated in order to pinpoint the general area of leakage. A leakage tolerance is included in the specifications and is based on the formula developed by the East Bay Water Company of Oakland, California, as described in the section of Chapter 7 entitled "The Lining." The denominator of the original formula has been slowly increasing as a result of improved lining materials and better adhesives, along with the development of high-tenacity field and factory joints that are stronger than the parent material. On facilities with large lining areas it is not unusual for the denominator constant to run from 100 to 200. The establishment of the final value depends on reservoir usage, design and budget considerations, and an appraisal of the critical nature of the facility. A leakage tolerance of *absolute zero* is to be avoided unless there is no alternative in design or reservoir location, as it drastically increases the final cost.

For new reservoirs the lining choice will depend on expected performance, operating conditions, maintenance procedures, and economics. For existing facilities the access and surface texture are additional points of concern. In combining all of these factors, it is necessary to choose the lining that is most compatible with them in a way that utilizes the good features of the lining to best advantage. At the same time the undesirable features of the lining system should not be detrimental to the final design and ultimate performance expected.

Linings of the continuous, thin, impermeable type should be laid on a smooth texture, whether it be concrete, earth, Gunite, or asphalt concrete. In the case of earth a rolled surface is best, but a sand cover is to be avoided unless it is part of the underdrain system. The linings should not be laid on a subbase of mud, ice, or other type of unstable material.

Except for asphalt panels all field joints should be made perpendicular to the toe of the slope. Factory joints should follow the same rule for all cured lining types (EPT, butyl, EPDM), together with PVC, 3110, and CPE. High-tenacity-type joints, such as can be produced with some Hypalon formulations and 3110 materials, can run in any direction. Nevertheless, joints are usually made perpendicular to the toe of the slope for ease of construction.

The anchor at the top of the slope may be formal or informal. A formal anchor system consists of ½ in. anchor bolts on 6 to 12 in. centers with an anchor bar ¼ × 2 in. cross section. Usually, the bar is a No. 6063T5 alloy of aluminum, but galvanized steel and stainless steel are also used. Wood is not recommended regardless of the type.

In new concrete, L-type galvanized anchor bolts are sometimes used, but the trend is to cadmium plated steel or, more recently, stainless steel. Nuts and washers are stainless in any event. No galvanic corrosion problems have occurred from combinations of aluminum, stainless steel, or cadmium plated steel hardware in these applications. Generally, no special precautions are taken, although occasionally fiber washers have been used beneath the stainless steel washer, insulating it from the aluminum anchor bar.

The bolts are set into existing concrete at the top of the slope (Figures 10.1 and 10.2); or, if none exists, a flush-mounted concrete beam is poured (Figure 10.3). It is normally poured in a trench without wooden forms to a width of about 9 in., depending on the trench digging equipment available. Its width should be no less than 6 in., and its depth consistent with the weather and frost heave conditions prevailing at the site. A single reinforcing bar is recommended to help guard against cracking.

At the top anchor point a chafer strip is placed between the lining and the concrete at the junction between the top of the slope and the berm. It is 12 in. in width in most cases, or as wide as is necessary to cover the

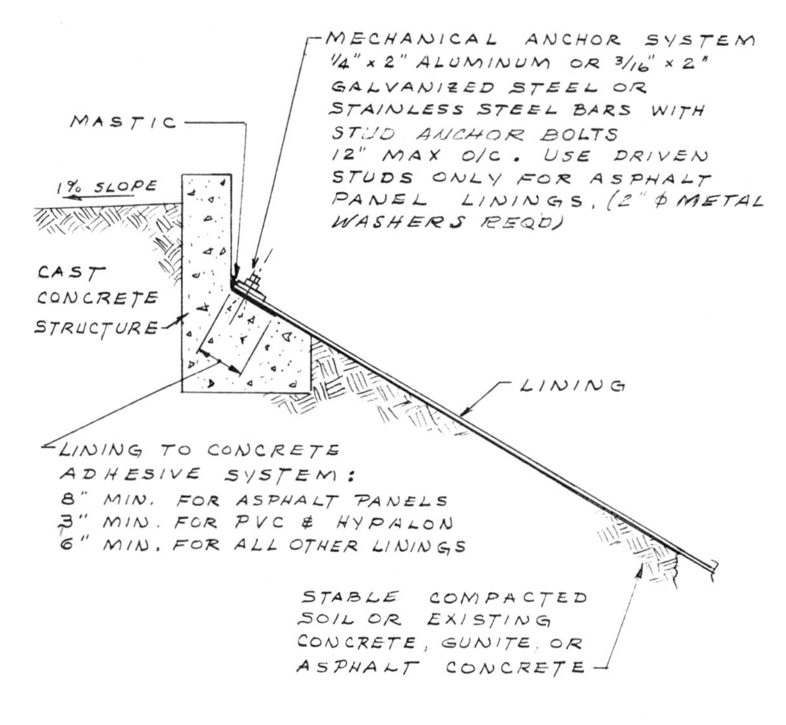

NOTE

1. TOP OF CONCRETE SHOULD BE SMOOTH AND FREE OF ALL CURING COMPOUNDS.
2. USE MIN. 1/32" x 2" GASKET (MAT'L COMPATIBLE WITH LINING) BETWEEN BAR AND LINING, EXCEPT NO GASKET REQUIRED FOR ASPHALT PANELS OR OTHER LININGS THICKER THAN .040".

FIGURE 10.1. Top anchor detail—alternative 1, all linings.

transition point. The strip protects the lining from being cut by sharp edges in the substrate toe–berm transition point or by any roughness of the concrete beam itself.

The informal anchor requires the cutting of a trench adjacent to the top of the slope and in the flat berm area. The trench is cut with a trenching machine, a backhoe, or, more commonly, a bulldozer or motor grader with the blade tilted to produce a V cross section. The depth of the trench is usually held in the range of 12 to 16 in., which is

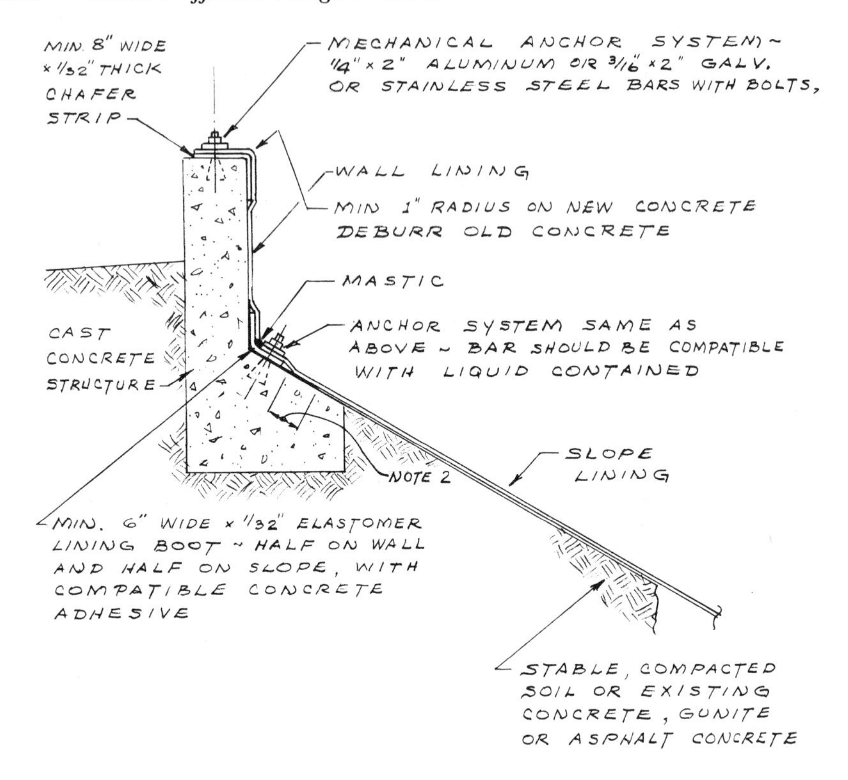

NOTE

1. ALL CONCRETE AT SEALS SHALL BE SMOOTH AND FREE OF ALL CURING COMPOUNDS.
2. USE COMPATIBLE ADHESIVE BETWEEN SLOPE LINING AND ELASTOMER BOOT, AND 3" MIN. WIDTH OF COMPATIBLE ADHESIVE BETWEEN SLOPE LINING AND CONCRETE.

FIGURE 10.2. Top anchor detail—alternative 2, all linings.

sufficient to contain any of the lining systems (Figure 10.4). In the case of asphalt panel linings the trench must have a V cross section rounded on the reservoir side, as the lining will not take bending into a vertical wall trench (Figure 10.5). The equipment weight is sufficient for compacting the backfilled trench after the lining is installed.

At inlet–outlet structures and other locations where the lining is pierced, seals are made in two accepted ways. In the first method the seal is made in the plane of the lining (Figures 10.6 and 10.7). With asphalt

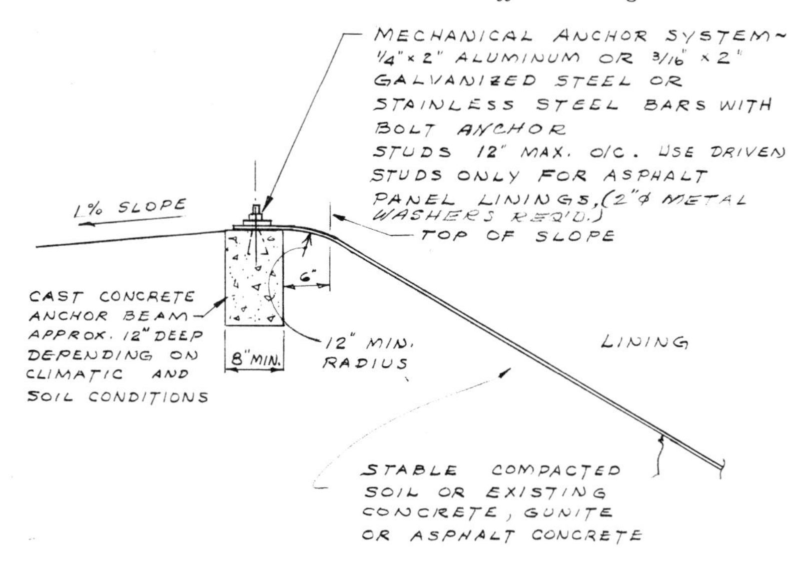

NOTE

TOP OF CONCRETE SHOULD BE SMOOTH AND FREE OF ALL CURING COMPOUNDS.

USE MIN. 1/32" x 2" GASKET (MAT'L COMPATIBLE WITH LINING) BETWEEN BAR & LINING, EXCEPT NO GASKET REQUIRED FOR ASPHALT PANELS OR OTHER LININGS THICKER THAN .040".

FIGURE 10.3. Top anchor detail—alternative 3, all linings.

panels the concrete is primed, the lining laid in a mastic–adhesive bed, and the area between the termination of the lining and the concrete caulked with asphalt mastic (Figure 10.7). It is not customary to use mechanical anchors unless the seals are being made on slopes. For the other thin lining systems, mechanical anchors and leveling gaskets are used in addition to the appropriate adhesives, mastics, and caulks to complete the seal.

The second method of making seals where pipes pierce the lining utilizes a pipe boot or shroud (Figure 10.8). The leg of the sleeve is the same as the outer diameter of the pipe. Fastened to this leg by factory seaming techniques is a flange. In the field this flange is adhered to the

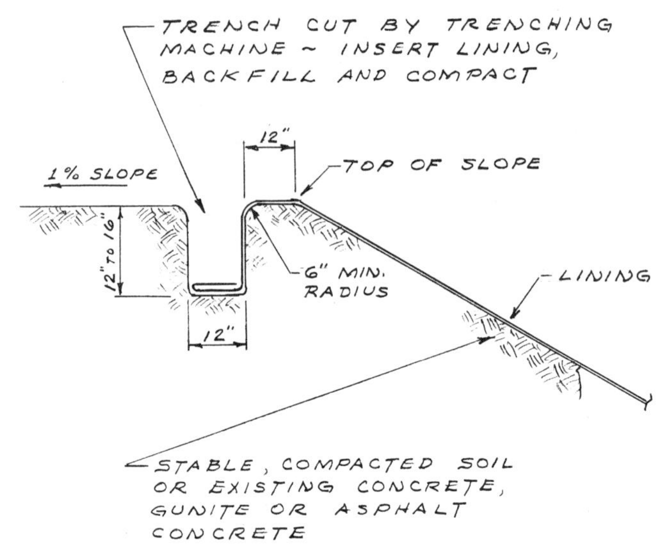

FIGURE 10.4. Top anchor detail—alternative 4, all linings except asphalt panels.

base lining material at the point where the pipe pierces the lining. Near the point where the boot terminates on the pipe, adhesive, followed by a stainless clamping band, is used. In an emergency the pipe boot may be made by an experienced lining technician in the field.

Inlet and/or outlet pipes should really be introduced into the reservoir through a structure (Figure 10.9). The seal between lining and structure is effected on top of the structure wall, again in the plane of the lining. The structure itself is not lined since it is a monolithic pour with thick wall sections. A mud drain located near the outlet structure connects directly to the sewer or other waste facility and serves for flushing any silt or debris through it during cleaning operations (Figure 10.10).

Large reservoirs above 20 or 30 million gal should be equipped with a quick-release device for rapid dumping of the water in case of emergency. This is usually a structure and may even be the same as the outlet, with proper valving to control its use. Lining seals are made to it as described above.

An existing reservoir that requires a lining must be smooth enough to receive the lining without damaging it. Rocks, small, sharp protrusions, and Gunite rebound are the greatest hazards and must be appropriately handled. Buffer sheets of lining or heavy synthetic filter cloth materials

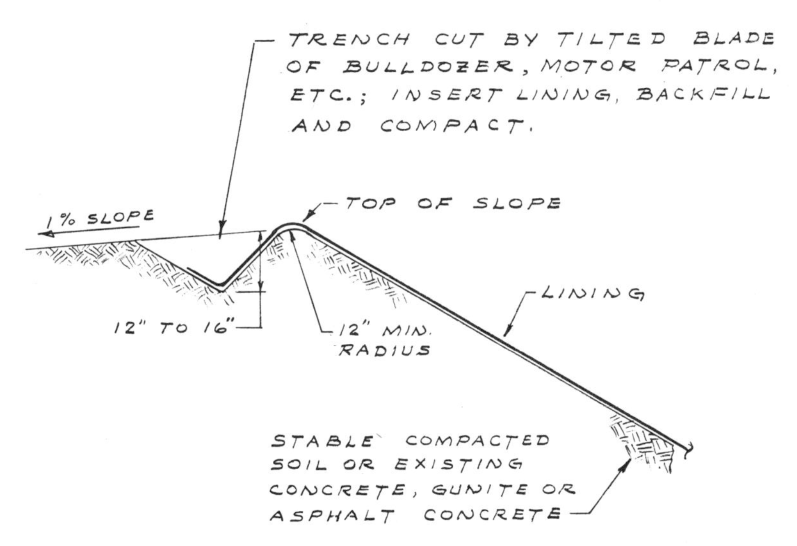

FIGURE 10.5. Top anchor detail—alternative 5, all linings.

may be used as cushions. If rough spots are local in character, they may be removed by grinding or smoothed by flood coating with cement or chemical grout materials. Sand cushions should be avoided if possible.

Cracks in existing reservoirs may be treated in a number of ways, as shown in Figures 10.11 and 10.12. Hairline cracks are not usually given any special treatment unless underwater photographs show that they widen enough under head to be detrimental to the lining system being used.

Exposed thin membrane lining systems, it will be recalled, may have problems with wind on the leeward berms. The most economical, time-tested method of handling this problem is by the use of vents built into the lining. They are installed just below the berm, usually on 50 ft. centers. They consist of a rigid tube, 3 in. in diameter, adhered to a

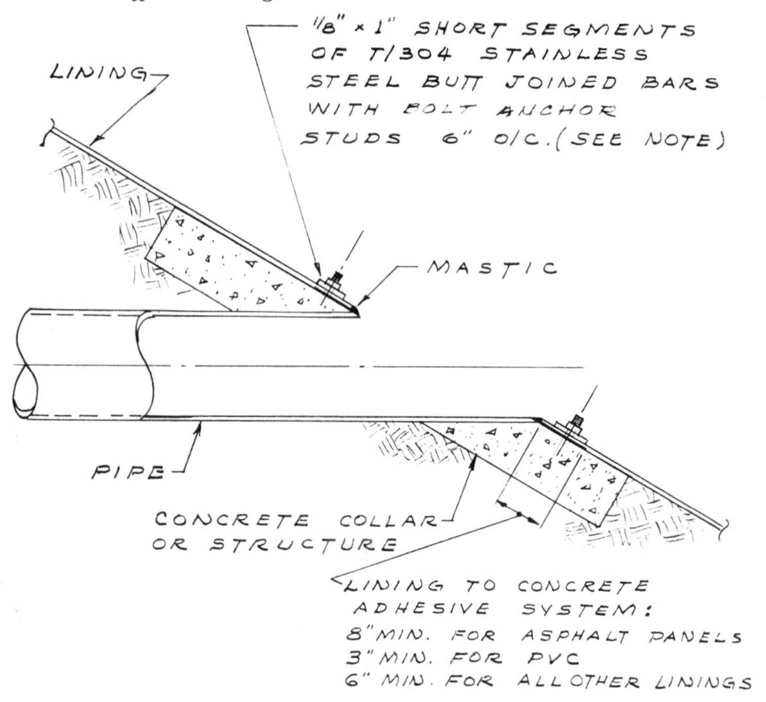

NOTE

FOR ASPHALT PANEL LININGS, PERCUSSION DRIVEN STUDS THRU 2" MIN. DIA. x 1/16" THICK GALVANIZED METAL DISCS AT 6" O/C ENCASED IN MASTIC MAY BE SUBSTITUTED FOR ANCHOR SHOWN.

FIGURE 10.6. Seal at pipes through slope, all linings.

flange in such a manner that the tube is in a vertical position when the flange is cemented to the lining on the slope. The top of the vent is protected with a conical roof and stainless steel mesh. These vents do an excellent job of equalizing the pressures above and below the lining so that no ballooning will occur in even the highest winds.

The lining should be laid in a smooth but relaxed attitude. The practice of inserting folds into the lining to compensate for future settlement is almost entirely ineffective, plus the fact that the wrinkles complicate the field seaming problems. Only if the settlement occurs *exactly* where

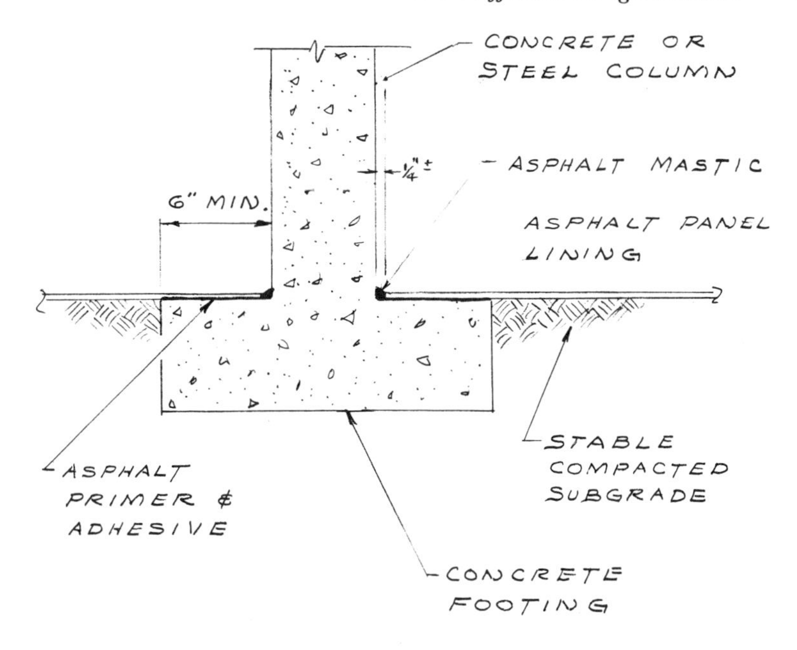

FIGURE 10.7. Seal at floor columns, asphalt panels.

the excess material is placed is the practice effective, and the odds do not favor this circumstance.

Regardless of the lining, great care should be taken to see that no bridging of the membrane occurs. The greatest trouble spots occur at 90 degree intersections between the floor and walls of concrete tanks, although structural intersections with the floor are other likely spots. The problem is minimized if factory and field seams are stronger than the parent material itself, and is greatly magnified if lining joints are weak. Circular tanks with vertical walls are particularly troublesome in this regard. With the lower strength joint materials, use of a factory-prefabricated boot with a joint whose angle matches that of the structure has been common practice in EPDM and butyl systems. Bridging can occur also in other parts of the reservoir, in areas of cracks, construction joints, or uneven concrete slabs. Obviously, all stake holes or other areas where rapid changes of elevation occur should be eliminated. If boots or pipe shrouds are used, the installing contractor should check them be-

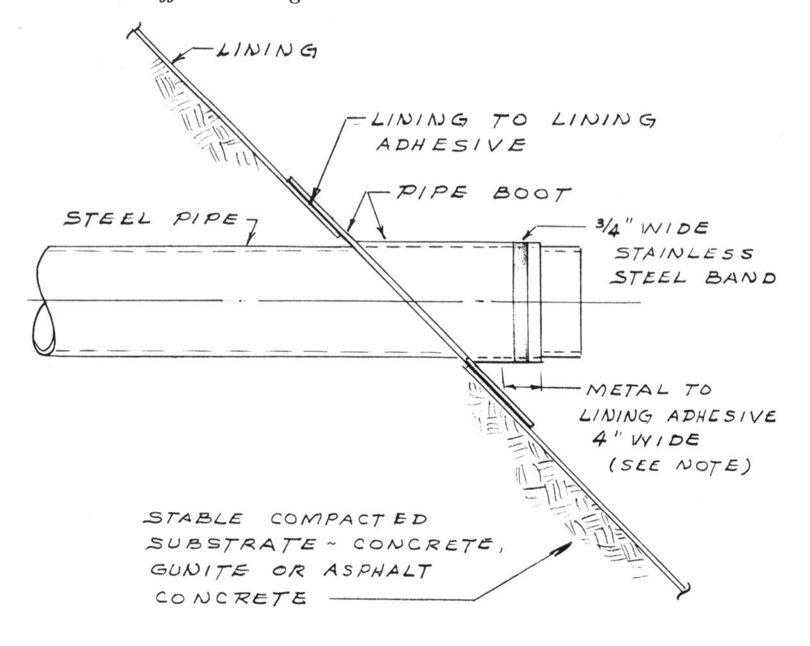

FIGURE 10.8. Pipe boot detail, all linings except asphalt panels.

fore installing to be sure that their intersections between sleeve and flange are at the same angle as the structure on which they are being used.

Much could be written on the subject of lining guarantees, but perhaps the following summary will suffice to get the main point across:

A guarantee is written by the manufacturer for the protection of the manufacturer and not for the protection of the buyer.

Although this statement covers all guarantees, few people really understand the basic concept until they have been through the claim process. In most cases, especially if a large sum of money is involved, all parties tend to relegate the responsibility for malfunction to someone else, with the end result that no one ever steps forward to resolve the problem.

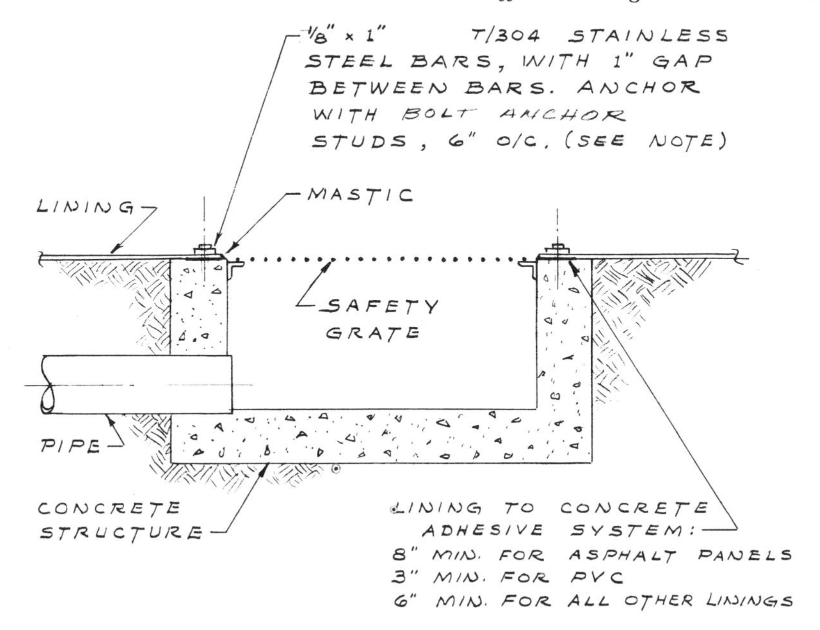

NOTE

WITH ASPHALT PANEL LININGS, PERCUSSION DRIVEN STUDS THRU 1" MIN. DIA. x 1/16" THICK GALVANIZED METAL DISCS AT 6" O/C, ENCASED IN MASTIC MAY BE SUBSTITUTED FOR ANCHOR SHOWN.

FIGURE 10.9. Seal at inlet–outlet structure, all linings.

This circumstance alone is certainly justification for seeking trustworthy manufacturers who have demonstrated their integrity in a number of ways and to place considerable reliance on their recommendations as to a reputable lining contractor with proven experience. On this point in particular, it pays to make the specification "bottletight."

Facilities that hold any liquid should be protected with adequate fencing, at least 6 ft in height. The fence should be placed on the outside berm slope and be designed to prevent entrance by undercutting. The top elevation of the fence should be below the berm elevation, and not, as usual, on top of the berm.

For linings in concrete and steel tanks the rules are somewhat different, although most of the basic concepts stated above still apply. The big

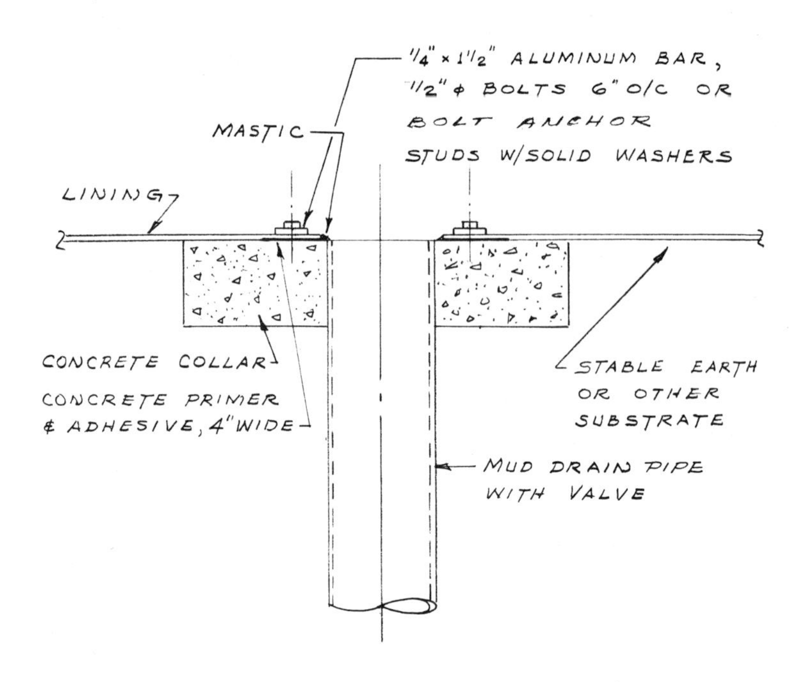

FIGURE 10.10. Mud drain detail, all linings.

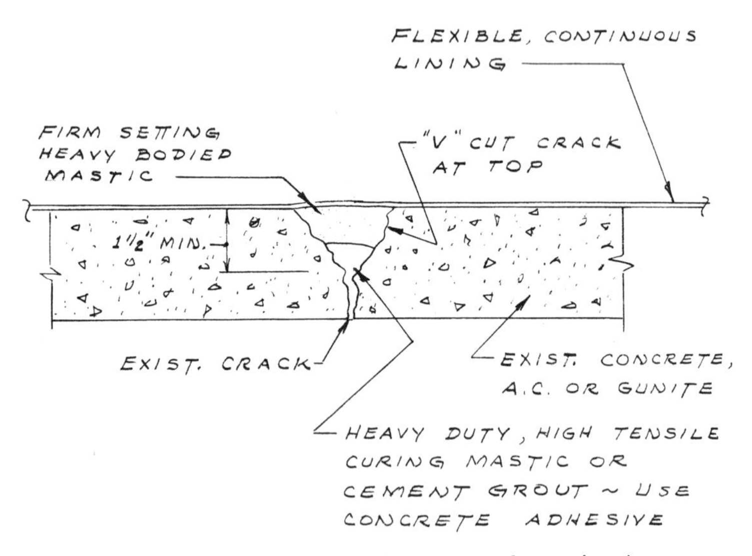

FIGURE 10.11. Crack treatment—alternative A.

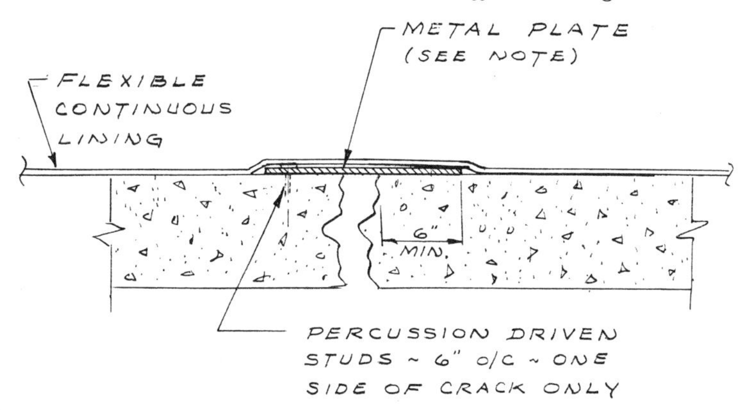

NOTE:

METAL PLATE MUST BE ABLE TO SPAN CRACK WITHOUT BUCKLING FROM WEIGHT OF WATER BRIDGING THE CRACK. COPPER & STAINLESS STEEL ARE MOST COMMON CHOICES.

FIGURE 10.12. Crack treatment—alternative B.

exception is adhering the lining to the substrate. This is never done in cut-and-fill reservoirs and is seldom done in concrete tanks, but in steel tanks the lining should be adequately bonded to the steel. Less is known about lining technology for steel tanks, with the result that sprayed systems are in more general use than are continuous sheet applied systems. As has already been emphasized, surface preparation of the steel tank surface before any bonded application is a most important key to successful performance of the total system. Again, an excellent supplier–contractor relationship is mandatory, and both must have top experience.

Wooden tanks are usually lined by means of the so-called drop-in lining. This has usually been fabricated of 20 mil nonreinforced PVC material in a factory and is sized to exactly fit the facility, plus a small allowance for shrinkage. Since wooden tanks are normally roofed, the plastic material is protected from sunlight degradation. Water entry should be at the bottom.

Top anchor systems for tanks are similar to those used in cut-and-fill reservoirs utilizing the bolt and bar principle. In steel tanks anchor bolts (such as Nelson studs) are welded to the tank, and the bar is of a metal that resists the liquid and vapor within the tank. Top anchor lining systems in wooden tanks utilize ¼ × 3 in. Redwood or cypress strips fastened to the inside of the wall with corrosion-resistant serrated nails.

In sewage treatment plants linings of butyl and EPDM have been bonded 100% to flat vertical concrete walls with good success. In launder* or conduit areas the bonded linings must be mechanically reinforced at their sheet edges if velocities of the liquid exceed 4 or 5 ft/sec or if lower velocities are accompanied by much turbulence. If these safeguards are followed, the lining system will enjoy good success for prolonged periods of time. Tank owners are eagerly eyeing the Hypalon and CPE materials for possible use in steel and concrete tanks. At this writing, however, practical adhesive systems for bonding large sheets of these materials to concrete substrates have not yet been developed.

Thin membrane lining materials have had widespread use in covering slopes for wave erosion control. When this is done, the lining must extend from the top of the slope completely to the toe, where it should terminate in a backfilled trench on the reservoir bottom, 30 in. or more in depth. In these types of applications the slopes that lie beneath the lining must be stable when wet, and there are no exceptions to this rule.

*The launder is the trough, usually circular, at the outer periphery of settling basins and tanks, into which the upper decanted water spills. Launders are common in water and sewage treatment plants.

11

Operations

There are certain well-established procedures for good cut-and-fill reservoir operation that have been worked out by large users of these facilities. Although procedures differ from owner to owner, they are very similar in their major points. Smaller facilities, such as shallow holding ponds, reflection pools, farm ponds, and noncritical installations, do not require the complex procedures associated with large reservoirs. With obvious modifications, though, what is said here can be applied to these facilities. The procedures for lined concrete and steel tanks are somewhat different and will be discussed in the latter part of the chapter.

As the earthen facility is constructed, careful and continual inspection of each step is made by the project inspector to assure conformity with the specifications. Any deviation necessary because of special site conditions or errors in the original specifications are recorded. The same on-the-job inspection procedure should be carried out with respect to the lining operation. This is usually one of the last steps in the construction schedule, so that upon its completion the reservoir itself should be ready for use. There is then conducted a final inspection of the lining installation itself, along with the other visible features of the structure as a whole.

Although provisions for protection of the lining against the energies associated with the inlet structure should be a part of the original design, these are sometimes overlooked. At the time of the final inspection, therefore, this area must be examined carefully to see whether any precautionary measures are in order before the filling operations. If water is going to cascade over or onto a portion of the lining during any step of the filling operation, some sort of lining protection should be considered for these areas. This protection can take the form of riprap, heavy rubber sheeting or conveyor belting, or lagging properly held down against flotation. Steel plates have also been used. Structures that inlet at the bottom of the reservoir utilize the reservoir water itself as an effective energy dissipater. This is the best method, but mechanical energy dissipaters are also quite common. What should be guarded against is no energy dissipater at all.

The procedure that has been established for the initial filling is to place in the reservoir no more than 2 or 3 vertical ft./day of water. This gives the reservoir lining ample time to adjust to the minor settlements always associated with the first filling. From the owner's standpoint this rate is usually no detriment, as the facility can probably not be filled any faster unless it is very small. The filling should be watched carefully, with frequent observations of the underdrain monitor system. Amount, time, and water depth should actually be recorded, as these values can serve as valuable data in determining the possible cause and location of any leakage that is noted.

A graph is most useful for plotting the data mentioned above; a typical one is shown in Figure 11.1. In addition to indicating the leakage rate at each underdrain, the chart should also show rainfall data, water depth, and pumping on–off cycles, if they exist. The chart will usually pick up the effects of rainfall (Feb. 8–10) if the runoff is intercepted by the underdrain (Feb. 10–11). A broken water main ¼ mile away from the reservoir was indicated on one chart. The graph will also spot structural malfunctions in appurtenances or in the embankment if they cause a lining rupture (Feb. 28). When repairs are made and the reservoir is placed back in service, comparison of the new graph line with the previous one will indicate the effectiveness of the remedial actions. If there is a structural problem involving the inlet structure or pipe, the chart indicates the problem when pumping operations are in progress (Feb. 18 and Feb. 23). This happens since the leakage rate from the system is noticeably higher when the pumps are running. Note that the underdrain actual response lags the predicted response, particularly on the down cycle with high leakage rates.

When water is first introduced into the reservoir, it is not uncommon for the drainage system to show some leakage. This is particularly true if the initial filling takes place during wet weather, and even more likely if the underdrain is running water during the construction work. The latter condition indicates that groundwater is being intercepted by the drains. As the weight of the water increases in the reservoir, the ground is compressed and water can actually be squeezed out of it and forced

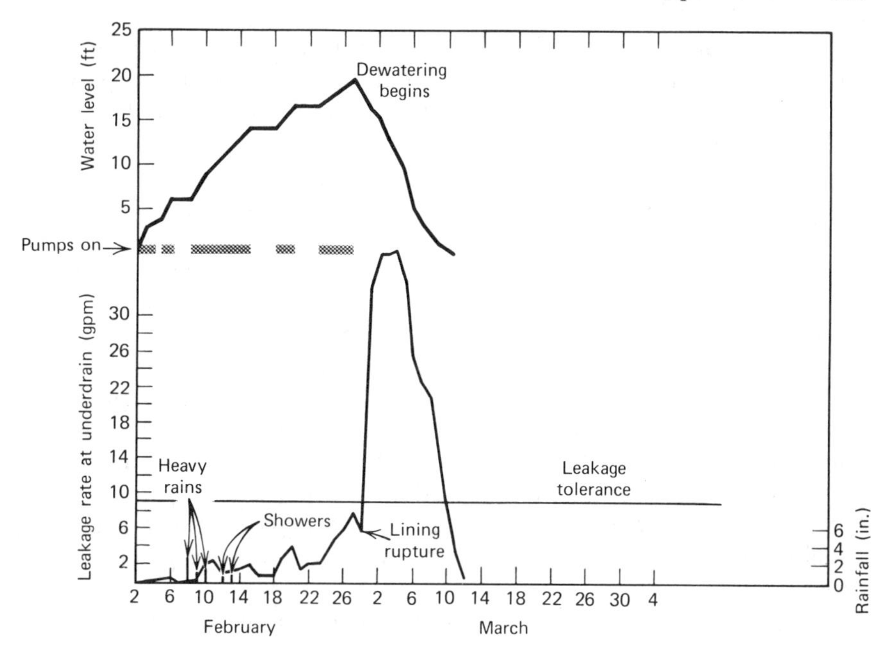

FIGURE 11.1 Reservoir leakage chart.

into the underdrain pipes. As more weight is loaded onto the reservoir substrate, more water may be squeezed out and this condition will continue until equilibrium is reached. Equilibrium may be a value of very low or zero leakage due to outside water; or if groundwater flows are involved, equilibrium will be a point of stabilized flow rate for a particular head in the reservoir.

It should be pointed out that this stabilized flow rate is not necessarily the same at identical heads on different days. This is so because water flow in the ground is variable and also because the loading of the reservoir can drastically change the groundwater flow characteristics. The weight of the water can decrease or increase these flow conditions, or, by a sort of damming analogy, it can decrease the flow rate and then allow it to increase later on. The situation gets rather complex, because the filling operation is continuous and the weight of the water in the reservoir is continually changing. In addition, any defects in the lining system are also contributing to the drainage from the pipes. For these reasons it was thought that, by comparing the water quality in the underdrain before filling with that during filling it might be possible to assess an approximate ratio of groundwater–reservoir water in the underdrain leakage and arrive at a rough gallon per minute figure for each. Past attempts to do this type of comparison, however, have not been very successful. Competent water chemists indicate that there is more to this analysis than meets the eye because water quality continually changes as it flows through the underground.

After the reservoir is filled, it is necessary to wait for the leakage rate to stabilize. If groundwater is present, this may take several days or even weeks, particularly if the reservoir is large. When drainage values have stabilized, the leakage in the underdrain, if any, just before filling is subtracted from the value determined at full depth. The resulting figure is said to be the net leakage associated with the facility itself. It is this figure that is compared to the leakage tolerance figure of the specifications to determine whether or not the reservoir is acceptable. It should be pointed out that this method does not necessarily give the true value of leakage from the reservoir, but it is the only known method whereby a logical attempt may be made to compensate for groundwater flows. Reference to standard leakage charts kept for some time will usually indicate the extent of the flows associated with groundwater movement.

If the leakage rate is too great in comparison to the leakage tolerance allowed by the specifications, the contractor must locate and repair the defects. Two methods may be used: one is underwater inspection, and the other is complete drawdown followed by visual inspection. If the leakage rate is very low, underwater inspection is probably the best method, even though it may be quite slow. The choice depends on the cost and inconvenience of emptying.

The first points to check, in any event, are the areas around all inlet–outlet or other drainage structures. If nothing is found there, then other structures are examined, followed by the columns and steps. A careful examination should also be made at areas where pipes pierce the lining. It is well to examine, too, the insides of the inlet–outlet pipes and structures and all other areas that are not lined initially. Leak detection underwater is aided if the divers make use of any silt on the floor of the facility. Since this is the first filling, there may not be enough silt available, and water-soluble dyes or fine fibrous materials may then be used to locate the leak. In either case, silt, dye, or other additives will "go for the leak." They must be placed close to the defects in order to speed up the detection process. If leakage is quite low, it is very easy for divers to miss it. This harks back to the relationships existing between quality before inspection and that afterwards. Inspection does not assure zero-defect quality levels.

Since the diver is working under less than optimum conditions, he may fail to spot leak areas. The leaks associated with the lining are usually the result of water flowing through a joint in the lining or through pinholes. Since the latter are almost invisible to the naked eye and the former are essentially close tolerance slits, it is possible to look right at a leak and never see it because the volume of water issuing from each particular defect is extremely small.

If large leaks are involved, say over 10 or 15 gpm, their locations can usually be spotted relatively quickly, either by diver or by visually inspecting the empty reservoir. Either system is effective here, and the choice depends on the circumstances.

If dewatering is to be the method of detecting leakage, there are one or two tricks that can be used to pinpoint leakage problems, even of the pinhole variety. As water is withdrawn from the facility, the lining surfaces should be carefully watched, for the visual phenomenon involved is somewhat elusive, particularly at the higher elevations in the reservoir. At any defect in the lining one of two things will happen as water level is reduced: (1) water will backflow through the defect, or (2) areas immediately adjacent to the defective point will not dry quickly. Fortunately, either of these phenomena will be of a duration exceeding the normal time for drying of the lining. Thus the lining will naturally begin to dry out as the water level decreases, *except* for the defective areas. To spot defects by this technique the inspector should be there as the water is lowered, as the tendency to remain damp at a defect decreases with time.

In regard to the above technique a word of caution should be added. All damp spots are not necessarily leaks. Such a spot may indicate a partially open seam that has trapped water by capillary action. These seams will show wet spots for a time. The rule may then be amended by stating that, *usually,* the longer a spot stays damp or runs water, the greater is the likelihood that it is a hole or defect in the lining. On the bottom of the reservoir the same rule applies, except that, since the bottom will have some degree of unevenness, puddles or wet spots may be merely accumulations of water in slight depressions. Continual examination, though, will usually serve to separate the defective areas from the good ones. For reservoirs with effective underdrain systems, this lingering dampness still applies, although the damp areas tend to dry faster than if no underdrain were present.

After the reservoir has been inspected, all defects have been corrected, and the facility has been put back into service, a comprehensive surveillance and inspection program should be instigated. This should include careful monitoring of leakage on a regular basis. The frequency of these checks will depend on the nature of the reservoir, including the depth of the water, and the consequences of a failure. Although almost all leakage monitoring today is done manually by an operator, this type of checking is easily automated and can be integrated into various types of alarm systems, or even systems that will initiate emergency action automatically. In large facilities these would seem to be most desirable in light of the disastrous consequences of large reservoir or dam embankment failures. For low water depth facilities, or those where a potential failure would not involve loss of property or life, surveillance may be of a much more limited nature. If an expensive plant operation depends on the reservoir, the inspections should be more frequent and undoubtedly more comprehensive.

In addition to reservoir leakage, such things as cracks in the embankments and in the area immediately surrounding the reservoir should be examined and recorded. Earth movements such as these, although not necessarily the result of water impounding, may have an effect on the integrity of the reservoir. No chances should be taken; and if anything looks unusual, the reservoir should be lowered for further inspection and checking. If only minor problems occur, this can be done in a normal manner by shutting off the inlet lines and allowing the demands of the system to reduce the water level.

When it is necessary to lower the reservoir level, there are two precautions of utmost importance. The first is that, except in an emergency, the level should be lowered slowly to prevent problems due to reverse hydrostatic heads. As has been discussed earlier, this is a situation with which impervious linings cannot cope, and even the more porous Gunite and concrete linings may buckle and crack under this type of loading. Second, and for the same reason as before, the reservoirs should not be lowered in wet weather; again, of course, emergency situations are excepted.

During operation the areas around the outside toe of the slopes should be watched for evidence of leakage. This may appear as actual streams of water, softness of the banks, or prolific growths of shrubbery, trees, or grass. When any of these things are noted, the causes leading to them should be thoroughly checked, especially if deep, critical reservoirs are involved.

Once a year, or once every 2 years, the reservoir should be emptied slowly for internal evaluations. If at all possible, this should not be done in periods of predominately wet weather, unless the reservoir is protected with a good underdrain system. Even then, most operators still lower water levels slowly as an extra precautionary measure. The lining should be thoroughly inspected for cracks, back flow of water, and any situation that might spell difficulty below the lining. All pipe seals column seals, and structure seals should be scrutinized and necessary repairs made. This inspection usually means the removal of silt and mud accumulations; and if the lining is of thin membrane type, extreme care should be exercised to prevent mechanical damage.

The reservoir itself can contribute to instability of the surrounding area because of saturation and other factors. Therefore surveillance should include the monitoring of soil movement of various kinds in the immediate vicinity of the site. Analysis should be done by competent people experienced in this type of work. Big reservoir and dam failures of recent times (Baldwin Hills Reservoir and St. Francis Dam in California, Malpasset Dam in France, and Vajont Dam in Italy) seem to stem from the fact that man failed to evaluate and properly analyze unstable and changing geological conditions. Although in the Italian disaster dam failure was not the cause of the problem, geological factors did play the major role that sent a 300 ft. wall of water down a canyon to wipe out the town of Longarone and 2000 people. Continual checks on elevations of embankments or large concrete structures should be made, and data analyzed. Rapid changes in elevation or slow, continued settlement should be cause for more detailed investigations.

In addition to a rapid increase in reservoir leakage through the underdrains, recent evidence indicates that a slow and steady rise in such leakage, even though spread out over a period of several months or longer, should signal investigative action.[1] One thing is clear about the cut-and-fill reservoir: no failure has ever occurred that has not been preceded by clear and unmistakable warning signs. Although the warning signs for steel and concrete tanks are often harder to detect, the same rule also applies to them. It has taken a number of failures of the past to point out the importance of these signs. Although their nature is much better understood today, there is still much knowledge to be gained in this analytical area.

The state of California, among others, has established through Dam Safety Division design and surveillance criteria for all cut-and-fill reservoirs and dams within the state subject to the following rules:

1. All structures whose maximum depth from stream bed to spillway does not exceed 6 ft are exempt from state jurisdiction.
2. Facilities whose depth exceeds 6 ft are exempt if they impound less than 50 acre-ft (16.3 million gal).
3. Facilities whose depth exceeds 25 ft are exempt if they impound less than 15 acre-ft (4.9 million gal).
4. The rules for depth are applied to each embankment separately.

Note that the height of the facility is measured from the streambed. Thus a reservoir 40 ft. deep dug 15 ft. into solid ground with the excavated earth used to extend the embankments 25 ft. above the original ground level would, for purposes of the law, be only 25 ft. in depth, because that portion below "streambed" is not counted as part of the height.

In cases where the state has jurisdiction, it must review and approve all phases of the plans. Until it issues a certificate of approval, construction work should not proceed. If work has proceeded, the facility may not be filled until the certificate has been issued. Reservoirs in operation that fall below the state's minimum standards are taken out of service by the state until they meet the requirements. The state's analysis includes such items as structural stability, underdrain system, emergency rapid drawdown procedures, and surveillance, to name some of the more important ones.

For concrete tanks surveillance is easier in some respects but more difficult in others. They are not generally built with underdrain systems that will monitor leakage. If the tank is all above grade, leakage through the walls and at the tow (junction of wall and floor members) will be easily detected from a qualitative standpoint. Leakage through the bottom will not show up, however, unless it attacks foundation stability in some way. The latter situation will be exposed if elevation readings are taken at key points on the structure. This is an important part of surveillance for the concrete tank, particularly if the floor of the tank is also concrete. Any adverse settlement should trigger an investigation to uncover the cause.

Since the concrete tank is reinforced, the concrete lining and the substrate are one and the same. Therefore leaks through the lining do not have the same effect on the substrate as they do when the substrate is the soil of the cut-and-fill reservoir. In the latter case the degradative action of water can be relatively quick. This does not mean that the action of water on concrete is passive; there is no doubt that the effect is there, even though the process proceeds slowly. Since the concrete tank sits on earth, seepage can affect its bearing ability, and it will be well to monitor this condition as the situation warrants.

As water leaches through the concrete past the reinforcing bars, the steel is corroded. The rate depends on the rate of water movement, the acidity (pH) of the water, the amount of dissolved oxygen, and the type of metal used for reinforcement.

Not much thought was originally given to this problem. It has received much more attention in recent years because of some tank failures in various parts of the country. These events led reinforced concrete tank owners to investigate the conditions of their tanks. These studies revealed many cases where corrosion had been at work over the years. Owners, engineers, and tank designers turned their interest toward an impervious lining system. For new prestressed tanks some designers began utilizing steel plate curtainwalls. On existing tanks the flexible membranes were called upon to do the job, as has already been discussed.

The brief historical background given above indicates the need for increased vigilance on the part of owners of concrete tanks. It is true that inspection may not have to be done with the frequency required for large ground reservoirs and dammed up canyons. But the record of experience does expose the myth that this type of construction is immune to attack from water within.

Surveillance procedures are better developed for steel than for concrete tanks. The steel is exposed directly to a relatively thin protective coating. Breaches in these coatings are not rare, and most tank owners are very familiar with the corrosion problems that can follow coating malfunctions. One problem is that the corrosion behind the coating may proceed undetected, hidden by the coating itself. A rust spot will often appear and be visible during inspections, but this is not a mandatory reaction in all cases. With the newer techniques now available, the thickness of the coating may be checked in ways nondestructive to both steel and coating, a boon so far denied to owners of cut-and-fill reservoirs.

Careful checking of the roof sections of steel tanks is required, especially where these members join the steel tank wall. At these points, corrosion is harder to check and coatings and surface preparation are much more difficult to control.

Even though a coating system is present, current practice requires cathodic protection as well. This is the backup system for the lining to render the steel passive at places where there are pinholes or other defects in the coating. It should be checked on a regular basis to ensure that it is performing in a satisfactory manner.

References

1. State of California, *Investigation of the Failure of Baldwin Hills Reservoir,* Resources Agency, Department of Water Resources, April 1964, p. 56.

12

Epilogue

The preceding pages have attempted to explain some of the more important points in the technology of linings for hydraulic facilities. The contents have been based on the writer's experience in the engineering and construction of lining systems throughout the world. This work has utilized all of the impermeable thin membranes (less than ¾ in. in thickness) in common use today. The text reflects also some of the experiences of other engineers and contractors working in this specialty field.

Lining technology is improving almost every day, so in no way does the author imply that this is the last word on the subject. What is said, however, should bring the reader up to date and give him a better understanding of the basic factors involved in lining work of all types.

We will never know all there is to know about this complex subject. The attaining of ultimate knowledge is a journey, not a destination, but it is hoped that what has been learned and recorded herein will make the road less rocky for those who will walk it in the future.

All cost estimates have been based on 1976 prices and will tend to be more or less comparable with each other, even though time will cause them to be obsolete insofar as their absolute values are concerned. Prices are based on a "clean"* facility, with good access and no unusual factors that would grossly influence normal construction costs. It is also assumed that that the work was done during the normal construction season, thus eliminating snow and ice conditions.

*"Clean," in this sense, means a facility with no unusual structures or design features. It is assumed that the reservoir has normally designed inlet–outlet structures, a mud drain and overflow, and slopes near 2½ : 1. If the facility is roofed, it is assumed that column spacing is on 20 × 20 ft centers or greater and that all penetrations through the lining are in accord with good engineering practice insofar as the lining is concerned.

As time goes by, prices will increase. But it should be emphasized again that lining prices are but a part of the overall picture. Any lining system is expensive, regardless of which material is used. The aim is to get the one installed that has the lowest total cost based on the life expectancy and performance characteristics best suited for the project. If this type of economic analysis is kept in mind, and strict attention is given to proper design, inspection, and installation procedures, together with good operation practices, everyone involved in the project will reap the greatest possible benefit from it.

What is down the road? It would appear that the answer is, "More membranes." Every 2 or 3 years a new one is introduced amid much advertising fanfare to the effect that it is greatly superior to anything else on the market. Actually, no really great advances have been made in this field for over 20 years. All that has happened is that we have seen a parade of membranes, and, even at that, they have not varied much from each other, as we learned from discussing their properties. We still have the cured membranes and the plastic membranes.

It would appear to be time to give some serious thought to a lining system that is both more efficient and more economical at the same time. Maybe there is a way to eliminate the costly resins, the expensive processing, the factory fabrication, the high freight charges, and the time-consuming field installation procedures. Then, too, there is the problem that all linings have—malfunctions. This is not to say that they are no good, but merely to point out that this business is complex and that lots of things can happen to these thin membranes. To make them stronger is a good step from the technical standpoint, but a poor one from the standpoint of economics. The user is already appalled by the high cost of linings.

What about a natural process? Is it possible to utilize some of the chemicals (hardness) of the water to build a lining *in situ?* Iron and calcium deposits can be very hard, dense, and impervious; and although these attributes do not necessarily make a lining, they are something to think about. Maybe they can be modified. If so, we could effect substantial savings for lining users everywhere. Chemicals to bring about changes that would modify these elements within the water to give us a workable lining system might be within our capability to produce. Maybe we already have them! Perhaps all we really need is a catalyst.

There are other approaches to the problem, of course, and it would be interesting to apply the "think tank" approach to linings. It is certainly a good time to think in this direction, for we are now fighting a very serious battle to control the pollution within our environment, and also to keep what good water we have in storage from getting away from us.

APPENDIX I

Engineering Data

Useful Information

Area Conversions

Square Inches	Square Feet	Square Meters	Acres
1	0.0833	0.0078	0.00000016
12	1	0.0931	0.0000229
1547.1	10.744	1	0.000247
6,272,640	43,560	4047	1

Volume Conversions

Cubic Feet	Gallons	Cubic Meters	Acre-Feet
43,560	325,829	1233.51	1
35.314	264.18	1	0.00081
0.1337	1	0.003785	0.000003
1	7.481	0.0283	0.000023

Rate Conversions

Cubic Feet per Second	Gallons per Minute	Gallons per Hour	Gallons per Day	Acre-Feet per Day	Acre-Feet per Month	Acre-Feet per Year
1	450	26,932	646,360	1.983	59.5	714
0.00222	1	60	1440	0.00442	0.1326	1.609
0.000037	0.0167	1	24	0.000074	0.0022	0.026
0.0000015	0.00069	0.0417	1	0.000003	0.000092	0.0011
0.50	226.3	13,576	325,829	1	30	260
0.125	7.53	452	10,861	0.0333	1	12
0.0105	0.63	37.7	905	0.0028	0.0833	1

Miscellaneous Data

1 million gal	=	3.07 acre-feet
1 million gal/day	=	1122 acre-feet/year
1000 gpm	=	2.23 cu ft/sec
	=	4.42 acre-feet/day
	=	132.6 acre-feet/month
	=	1592 acre-feet/year

The Lining Area of a Reservoir

To calculate the lining area of a reservoir in which 4 ft. of the lining is anchored into a trench at the top of the slope, use the following formulas:

$$A = 4P_t + A_s + A_b$$

$$A_s = \left(\frac{P_t + P_b}{2} \right) SL$$

$$A_b = L_b W_b$$

where; A = lining area (sq ft),
P_t = perimenter at the top (ft),
P_b = perimeter at the bottom (ft),
SL = slope length (ft),
S = slope, that is, ratio between horizontal and vertical distance,
A_s = area of the slope (sq ft),
A_b = area of the bottom (sq ft),
L_b = length of the bottom (ft),
W_b = width of the bottom (ft).

Example: Find the lining area of a reservoir measuring 200 × 200 ft at the top with a depth of 20 ft and side slope ratio of 2 : 1.
Solution: The reservoir is diagrammed on the facing page.

Calculations: In the accompanying "Slope Factors" tables, the 2 : 1 slope length factor is given as 2.24. Therefore slope length is (2.24)(20) = 44.8 ft and

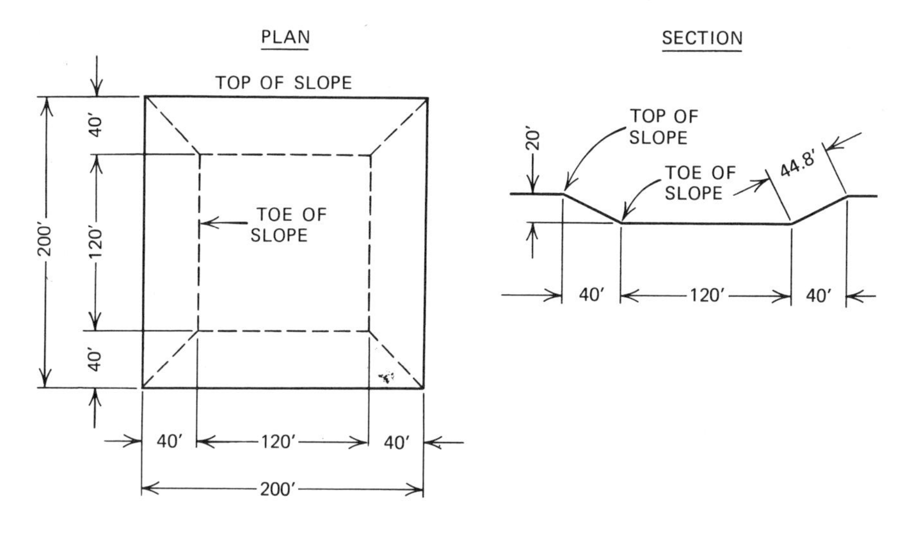

$$P_t = 4(200) = 800 \text{ ft}$$
$$P_b = 4(120) = 480 \text{ ft}$$
$$A_s = \left(\frac{800 + 480}{2}\right)44.8 = 28{,}672 \text{ sq ft}$$
$$A_b = (120)(120) = 14{,}400$$
$$4P_t = (4)(800) = 3{,}200$$

$$46{,}272 \text{ sq ft}$$

This value checks with the one shown in the lining area section of the Standard Reservoir Tables (Appendix III).

The example above was worked out for a square reservoir, but the same formula applies for a rectangular reservoir. It can also be used for any other shape, except that, if the bottom is not a true rectangle, the area must be calculated accordingly. If the slope ratio varies with each slope, the area of each slope must be calculated separately.

In some reservoirs the lining will be anchored at the top of the slope or atop a flush-mounted concrete anchor beam. In these cases the lining material expended in the anchoring process will usually be less than 4 ft. The lining area calculation is amended to compensate for this difference.

SLOPE FACTORS

Slope	Factor
1/6:1	1.01
1/5:1	1.02
1/4:1	1.03
1/3:1	1.05
1/2:1	1.12
2/3:1	1.20
3/4:1	1.25
1:1	1.41
1-1/4:1	1.60
1-1/2:1	1.80
1-3/4:1	2.02
2:1	2.24
2-1/4:1	2.46
2-1/2:1	2.69
2-3/4:1	2.93
3:1	3.16
3-1/4:1	3.40
3-1/2:1	3.64
3-3/4:1	3.88
4:1	4.12
4-1/4:1	4.37
4-1/2:1	4.61
4-3/4:1	4.85
5:1	5.10
5-1/4:1	5.35
5-1/2:1	5.60
5-3/4:1	5.84
6:1	6.08
6-1/4:1	6.32
6-1/2:1	6.58
6-3/4:1	6.82
7:1	7.07
8:1	8.06
9:1	9.06
10:1	10.05

Example: Find the slope length of a reservoir 18 feet deep with side slopes 2-3/4:1?

Answer: 2.93 X 18 = 52.74 ft.

The Volume of a Reservoir

To calculate the total volume of a reservoir, use the following formula:

$$V = (A_t + A_b + 4A_m)\ \frac{H}{6}\ (7.48)$$

where; V = total volume (gal),
A_t = area at the top (sq ft),
A_b = area at the bottom (sq ft),
A_m = area at the midpoint (sq ft),
H = total height (ft).

Example: Find the volume of a reservoir measuring 200 × 200 ft at the top with a depth of 20 ft and a side slope ratio of 2 : 1.
Solution: The drawing of the reservoir is the same as the one shown in the preceding example.

Calculations:

$$\begin{aligned} A_t &= (200)(200) &&= 40{,}000 \\ A_b &= (120)(120) &&= 14{,}400 \\ 4A_m &= (4)(160)(160) &&= 102{,}400 \\ & && \overline{156{,}800} \times 20/6 = 522{,}144.0 \times 7.48 \\ & && = 3{,}905{,}637 \text{ gal} \end{aligned}$$

Adjustments for the volume lost because of the freeboard are approximations, based on the volume of water in the top area multiplied by the freeboard depth. In the above case, if the freeboard is 18 in, the approximate volume of water will be (1.5) (40,000) (7.48) or 448,800 gal. The net gallons in the reservoir will then be 3,905,637 less 448,800 or 3,456,837. Note that the total capacity, 3.906 million gal, checks perfectly with the value shown in the volume section of the Standard Reservoir Tables.

APPENDIX II

Trade Names and Sources of Common Lining Materials

Trade Names

Trade Name	Product Description	Manufacturer
Aqua Sav	Butyl rubber	Plymouth Rubber Canton, Mass.
Armor last	Reinforced neoprene and Hypalon	Cooley, Inc. Pawtucket, R. I.
Armorshell	PVC–nylon laminates	Cooley, Inc. Pawtucket, R. I.
Armortite	PVC coated fabrics	Cooley, Inc. Pawtucket, R. I.
Arrowhead	Bentonite	Dresser Minerals Houston, Tex.
Biostate Liner	Biologically stable PVC	Goodyear Tire & Rubber Co. Akron, Ohio
Careymat	Prefabricated asphalt panels	Phillip Carey Co. Cincinnati, Ohio
CPE (resin)	Chlorinated PE resin	Dow Chemical Co. Midland, Mich.
Coverlight	Reinforced butyl and Hypalon	Reeves Brothers, Inc. New York, N. Y.
Driliner	Butyl rubber	Goodyear Tire & Rubber Co. Akron, Ohio
EPDM (resin)	Ethylene propylene diene monomer resins	U. S. Rubber Co. New York, N. Y.
Flexseal	Hypalon and Reinforced Hypalon	B. F. Goodrich Co. Akron, Ohio

Trade Name	Product Description	Manufacturer
Geon (resin)	PVC resin	B. F. Goodrich Co. Akron, Ohio
Griffolyn 45	Reinforced Hypalon	Griffolyn Co., Inc. Houston, Tex.
Griffolyn E	Reinforced PVC	Griffolyn Co., Inc. Houston, Tex.
Griffolyn V	Reinforced PVC, oil resistant	Griffolyn Co., Inc. Houston, Tex.
Hydroliner	Butyl rubber	Goodyear Tire & Rubber Co. Akron, Ohio
Hydromat	Prefabricated asphalt panels	W. R. Meadows, Inc. Elgin, Ill.
Hypalon (resin)	Chlorosulfonated PE resin	E. I. Du Pont Co. Wilmington, Del.
Ibex	Bentonite	Chas. Pfizer & Co. New York, N. Y.
Koroseal	PVC films	B. F. Goodrich Co. Akron, Ohio
Kreene	PVC films	Union Carbide & Chemical Co. New York, N. Y.
Meadowmat	Prefabricated asphalt panels with PVC Core	W. R. Meadows, Inc. Elgin, Ill.
National Baroid	Bentonite	National Lead Co. Houston, Tex.
Nordel (resin)	Ethylene propylene diene monomer resin	E. I. Du Pont Co. Wilmington, Del.
Panelcraft	Prefabricated asphalt panels	Envoy-APOC Long Beach, Calif.
Paraqual	EPDM and butyl	Aldan Rubber Co. Philadelphia, Pa.
Petromat	Polypropylene woven fabric (Base fabric-spray linings)	Phillips Petroleum Co. Bartlesville, Okla.
Pliobond	PVC adhesive	Goodyear Tire & Rubber Co. Akron, Ohio

Trade Name	Product Description	Manufacturer
Polyliner	PVC–CPE alloy film	Goodyear Tire & Rubber Co. Akron, Ohio
Red Top	Bentonite	Wilbur Ellis Co. Fresno, Calif.
Royal Seal	EPDM and butyl	U. S. Rubber Co. Mishawaka, Ind.
SS-13	Waterborne dispersion	Lauratan Corp. Anaheim, Calif.
Sure Seal	Butyl, EPDM, neoprene, and Hypalon, plain and reinforced	Carlisle Corp. Carlisle, Pa.
Vinaliner	PVC	Goodyear Tire & Rubber Co. Akron, Ohio
Vinyl Clad	PVC, reinforced	Sun Chemical Co. Paterson, N. J.
Visqueen	PE resin	Ethyl Corp. Baton Rouge, La.
Volclay	Bentonite	American Colloid Co. Skokie, Ill.
Water Seal	Bentonite	Wyo-Ben Products Billings, Mont.

Sources

Materials	Manufacturers	Locations
Bentonite	American Colloid Co.	Skokie, Ill.
	Archer-Daniels-Midland	Minneapolis, Minn.
	Ashland Chemical Co.	Cleveland, Ohio
	Chas. Pfizer & Co.	New York, N. Y.
	Dresser Minerals	Houston, Tex.
	National Lead Co.	Houston, Tex.
	Wilbur Ellis Co.	Fresno, Calif.
	Wyo-Ben Products, Inc.	Billings, Mont.
Butyl and EPDM	Carlisle Corp.	Carlisle, Pa.
	Goodyear Tire & Rubber Co.	Akron, Ohio

Materials	Manufacturers	Locations
Butyl and EPDM, reinforced	Aldan Rubber Co.	Philadelphia, Pa.
	Carlisle Corp.	Carlisle, Pa.
	Plymouth Rubber Co.	Canton, Mass.
	Reeves Brothers, Inc.	New York, N. Y.
CPE, reinforced	Goodyear Tire & Rubber Co.	Akron, Ohio
Hypalon	Burke Rubber Co.	San Jose, Calif.
	B. F. Goodrich Co.	Akron, Ohio
Hypalon, reinforced	Burke Rubber Co.	San Jose, Calif.
	Carlisle Corp.	Carlisle, Pa.
	B. F. Goodrich Co.	Akron, Ohio
	Plymouth Rubber Co.	Canton, Mass.
	J. P. Stevens Co.	New York, N.Y.
EPDM	See "Butyl and EPDM"	
EPDM, reinforced	See "Butyl and EPDM, reinforced"	
Neoprene	Carlisle Corp.	Carlisle, Pa.
	Firestone Tire & Rubber Co.	Akron, Ohio
	B. F. Goodrich Co.	Akron, Ohio
	Goodyear Tire & Rubber Co.	Akron, Ohio
Neoprene, reinforced	Carlisle Corp.	Carlisle, Pa.
	B. F. Goodrich Co.	Akron, Ohio
	Firestone Tire & Rubber Co.	Akron, Ohio
	Plymouth Rubber Co.	Canton, Mass.
	Reeves Brothers, Inc.	New York, N. Y.
PE	Monsanto Chemical Co.	St. Louis, Mo.
	Union Carbide, Inc.	New York, N. Y.
	Ethyl Corp.	Baton Rouge, La.
PE, reinforced	Griffolyn Co., Inc.	Houston, Tex.
PVC	Firestone Tire & Rubber Co.	Akron, Ohio
	B. F. Goodrich Co.	Akron, Ohio

Materials	Manufacturers	Locations
	Goodyear Tire & Rubber Co.	Akron, Ohio
	Pantasote Co.	New York, N. Y.
	Stauffer Chemical Co.	New York, N. Y.
	Union Carbide, Inc.	New York, N. Y.
PVC, reinforced	Firestone Tire & Rubber Co.	Akron, Ohio
	B. F. Goodrich Co.	Akron, Ohio
	Goodyear Tire & Rubber Co.	Akron, Ohio
	Reeves Brothers, Inc.	New York, N. Y.
	Cooley, Inc.	Pawtucket, R. I.
	Sun Chemical Co.	Paterson, N. J.
Prefabricated asphalt panels	Envoy-APOC	Long Beach, Calif.
	Gulf Seal, Inc.	Houston, Tex.
	W. R. Meadows, Inc.	Elgin, Ill.
	Phillip Carey Co.	Cincinnati, Ohio
3110	E. I. Du Pont Co.	Louisville, Ky.

APPENDIX III

Standard Reservoir Tables

The following tables are a collection of data on the volumes and lining areas of circular, square, and rectangular flat-bottom, vertical-wall (slope = 0:1) tanks and cut-and-fill reservoirs with side slopes of 1:1, 2:1, and 3:1 in depths from 2 to 50 ft at 2 ft increments. Blank spaces in the tables indicate that the design is not feasible in that no flat-bottom area exists.

The initial series of tables (Tables 1 through 31) shows the volumes and lining areas of circular reservoirs from 50 to 500 ft in diameter. The value of the perimeter at the top in feet is indicated to the right of the table titles after the letter *P*. Below this is listed the values for 2*P*, 3*P*, and 4*P*.

The lining area, sometimes called the customer area, is the sum of the area of the flat bottom plus the area of the slope or wall plus the anchor area. The anchor value is standardized at 4 times the perimeter (4*P*), on the assumption that 4 ft of lining is required for burial in the top anchor trench. For linings that terminate at the top of the slope, the value of 4*P* must be subtracted from the lining area given in the table. For each additional foot of anchor material required beyond 4 ft, a value equal to *P* is added.

As an example, in Table 23 the volume contained in a 340 ft diameter reservoir 30 ft deep with 2:1 side slopes is given as 14.03 million gal (MG). The area to line in this facility, assuming a 4 ft width in the anchor trench, is 101,398 sq ft. If, however, the lining is anchored at the top of the slope instead of in the conventional trench, the value of 4*P* (4272 sq ft) must be subtracted to give a lining area of 97,126 sq ft.

If the reservoir is designed to operate with a 2 ft freeboard, the value of 1.325 MG must be subtracted from the 14.03 MG capacity figure to yield a net capacity of 12.71 MG.

Except for vertical-wall reservoirs (slope = 0:1) extrapolation of intermediate depth values within the tables is not accurate, although it will yield an approximation that may be close enough for most estimating work. Extrapolation between sheets to obtain values for intermediate diameters or between columns of any sheet to obtain values for intermediate slope ratios is not valid either, although again it will yield values close enough for most estimating work.

The next section of tables (Tables 32 to 93) shows the volumes of square and rectangular reservoirs, based on the general arrangement used for the circular systems just discussed. For each slope there are three columns of data. The first column gives the volume for a square reservoir whose side dimension is indicated at the top of the column and also in the table title. The second column is the volume addition gained for each additional increment of width added in one direction, the increment being indicated at the top of the column. The third column is the volume addition gained for each additional 100 ft increment of length added in one direction.

As an example, in Table 76, the volume of a square reservoir whose top dimension is 340 x 340 ft, with a depth of 30 ft and side slopes of 2:1, is given as 17.86 MG. If the shape is changed to a rectangle 340 x 680 ft, the volume addition will be 21.36 MG and the rectangular facility will have a total volume of 39.22 MG (17.86 + 21.36). With a 2 ft freeboard, 1.687 MG is deducted from the total capacity of the square reservoir, or 3.396 MG (1.687 + 1.709) from the capacity of the rectangular variation, to obtain net capacity.

In the same table it will be seen that each additional 100 ft of length added to a 340 ft wide reservoir will produce 6.283 MG of added total capacity. Thus a reservoir measuring 340 x 440 ft at the top, 30 ft deep with 2:1 slopes, will have a total volume of 24.143 MG (17.86 + 6.283), the effect of adding 100 ft to one dimension. If only 80 ft is added to one dimension, only 80% of the 100 ft increment (0.8 × 6.283) = 5.026 MG) will be added to the 340 x 340 ft square reservoir capacity value to obtain the capacity of a 340 x 420 ft facility, and so on.

Within the bounds of these tables, then, it is possible to obtain the volume of any square reservoir up to 500 x 500 ft or of any rectangular reservoir up to 500 ft in width. As in the first section of tables, the value of the perimeter at the top is indicated after the letter *P*, followed by the values for 2*P*, 3*P*, and 4*P*.

Following through on the example just given, Table 77 shows a reservoir 340 x 340 x 30 ft deep with 2:1 side slopes having a lining area of 129,104 sq ft. To obtain the lining area for a 340 x 680 ft rectangular reservoir of the same depth and slope, add 123,216 sq ft to obtain a value of 252,320 sq ft. For every 100 ft of length added to a basic 340 x 340 x 30 ft deep 2:1 slope reservoir, add 36,240 sq ft of lining area to make a total lining area of 165,344 sq ft for the 340 x 440 ft facility.

Metric System Conversion. To convert a square foot reading in the table to square meters, multiply by 0.093. To convert millions of gallons in the table to cubic meters, multiply by 3789. For example, 1000 m³ = 0.264 MG.

TABLE 1. Volume and Lining Area for 50 Ft Diameter Circular Reservoirs from 2 to 50 Ft Deep, with Slopes from 0:1 to 3:1

P = 157
2P = 314
3P = 471
4P = 628

Depth (Ft.)	Volume, Millions of Gallons				Area, Square Feet			
	0:1	1:1	2:1	3:1	0:1	1:1	2:1	3:1
2	0.029	0.027	0.025	0.023	2,435	2,715	2,567	2,636
4	0.059	0.049	0.042	0.035	2,749	2,829	2,718	2,668
6	0.088	0.068	0.053	0.040	3,063	2,932	2,764	2,688
8	0.118	0.084	0.058	0.041	3,377	3,025	2,797	2,696
10	0.147	0.096	0.060		3,691	3,107	2,818	
12	0.176	0.105	0.061		4,006	3,179	2,827	
14	0.206	0.112			4,320	3,241		
16	0.235	0.117			4,634	3,292		
18	0.265	0.119			4,948	3,334		
20	0.294	0.122			5,262	3,365		
22	0.323	0.123			5,576	3,385		
24	0.353	0.123			5,891	3,395		
26	0.382				6,205			
28	0.412				6,519			
30	0.441				6,833			
32	0.470				7,147			
34	0.500				7,461			
36	0.529				7,775			
38	0.558				8,090			
40	0.587				8,404			
42	0.617				8,718			
44	0.646				9,032			
46	0.675				9,346			
48	0.705				9,660			
50	0.734				9,975			

TABLE 2. Volume and Lining Area for 60 Ft Diameter Circular Reservoirs from 2 to 50 Ft Deep, with Slopes from 0:1 to 3:1

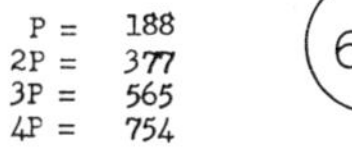

P = 188
2P = 377
3P = 565
4P = 754

60

Depth (Ft.)	Volume, Millions of Gallons				Area, Square Feet			
	0:1	1:1	2:1	3:1	0:1	1:1	2:1	3:1
2	0.042	0.039	0.037	0.034	3,393	3,731	3,666	3,636
4	0.085	0.074	0.064	0.056	3,770	3,870	3,739	3,678
6	0.127	0.103	0.083	0.066	4,147	3,998	3,798	3,708
8	0.169	0.128	0.095	0.070	4,524	4,117	3,847	3,726
10	0.211	0.149	0.102		4,901	4,225	3,883	
12	0.254	0.166	0.105		5,278	4,324	3,907	
14	0.296	0.179	0.105		5,655	4,411	3,919	
16	0.339	0.190			6,032	4,488		
18	0.380	0.198			6,409	4,555		
20	0.423	0.203			6,786	4,612		
22	0.465	0.207			7,163	4,658		
24	0.507	0.210			7,540	4,694		
26	0.550	0.210			7,917	4,720		
28	0.592	0.211			8,294	4,735		
30	0.634				8,671			
32	0.677				9,048			
34	0.719				9,425			
36	0.761				9,802			
38	0.803				10,179			
40	0.846				10,556			
42	0.888				10,933			
44	0.931				11,310			
46	0.973				11,687			
48	1.016				12,064			
50	1.057				12,441			

TABLE 3. Volume and Lining Area for 70 Ft Diameter Circular Reservoirs from 2 to 50 Ft Deep, with Slopes from 0:1 to 3:1

P = 220
2P = 440
3P = 660
4P = 880

70

Depth (Ft.)	Volume, Millions of Gallons 0:1	1:1	2:1	3:1	Area, Square Feet 0:1	1:1	2:1	3:1
2	0.057	0.054	0.051	0.048	4,508	4,903	4,828	4,793
4	0.115	0.103	0.091	0.080	4,948	5,068	4,915	4,844
6	0.173	0.145	0.120	0.099	5,388	5,223	4,990	4,885
8	0.230	0.181	0.141	0.108	5,828	5,367	5,054	4,913
10	0.287	0.214	0.155	0.112	6,267	5,501	5,105	4,929
12	0.348	0.240	0.163		6,707	5,624	5,144	
14	0.403	0.263	0.167		7,147	5,738	5,171	
16	0.460	0.282	0.168		7,587	6,077	5,187	
18	0.518	0.298			8,027	5,934		
20	0.576	0.309			8.467	6,016		
22	0.633	0.319			8,906	6,088		
24	0.691	0.325			9,346	6,150		
26	0.748	0.330			9,786	6,202		
28	0.806	0.333			10,226	6,243		
30	0.864	0.335			10,666	6,274		
32	0.921	0.335			11,106	6,294		
34	0.979	0.336			11,545	6,304		
36	1.036				11,985			
38	1.094				12,424			
40	1.151				12,865			
42	1.209				13,305			
44	1.267				13,745			
46	1.324				14,184			
48	1.382				14,624			
50	1.440				15,064			

TABLE 4. Volume and Lining Area for 80 Ft Diameter Circular Reservoirs from 2 to 50 Ft Deep, with Slopes from 0:1 to 3:1

P = 251
2P = 503
3P = 754
4P = 1005

80

Depth (Ft.)	Volume, Millions of Gallons 0:1	1:1	2:1	3:1	Area, Square Feet 0:1	1:1	2:1	3:1
2	0.075	0.071	0.068	0.064	5,781	6,233	6,147	6,106
4	0.150	0.137	0.123	0.110	6,283	6,424	6,249	6,169
6	0.225	0.193	0.165	0.139	6,786	6,604	6,340	6,219
8	0.300	0.244	0.196	0.156	7,289	6,774	6,418	6,257
10	0.375	0.291	0.220	0.165	7,791	6,934	6,484	6,283
12	0.451	0.329	0.235	0.167	8,294	7,083	6,538	6,297
14	0.525	0.363	0.243		8,796	7,222	6,581	
16	0.601	0.393	0.249		9,299	7,351	6,611	
18	0.676	0.418	0.251		9,802	7,469	6,629	
20	0.751	0.438			10,304	7,578		
22	0.826	0.456			10,807	7,676		
24	0.901	0.469			11,310	7,763		
26	0.976	0.479			11,812	7,841		
28	1.052	0.489			12,315	7,907		
30	1.126	0.493			12,818	7,964		
32	1.202	0.497			13,320	8,010		
34	1.276	0.500			13,823	8,046		
36	1.352	0.501			14,326	8,072		
38	1.427	0.501			14,828	8,087		
40	1.502				15,331			
42	1.577				15,834			
44	1.652				16,336			
46	1.727				16,839			
48	1.802				17,342			
50	1.877				17,844			

TABLE 5. Volume and Lining Area for 90 Ft Diameter Circular Reservoirs from 2 to 50 Ft Deep, with Slopes from 0:1 to 3:1

P = 283
2P = 565
3P = 848
4P = 1131

90

Depth (Ft.)	Volume, Millions of Gallons				Area, Square Feet			
	0:1	1:1	2:1	3:1	0:1	1:1	2:1	3:1
2	0.095	0.091	0.087	0.083	7,210	7,720	7,622	7,578
4	0.190	0.174	0.159	0.145	7,775	7,936	7,740	7,650
6	0.286	0.249	0.216	0.187	8,341	8,142	7,845	7,709
8	0.381	0.317	0.262	0.214	8,906	8,338	7,939	7,758
10	0.476	0.378	0.296	0.229	9,472	8,523	8,021	7,794
12	0.571	0.432	0.320	0.236	10,037	8,698	8,090	7,819
14	0.666	0.481	0.338	0.238	10,603	8,863	8,147	7,830
16	0.761	0.523	0.349		11,168	9,018	8,193	
18	0.857	0.560	0.354		11,734	9,162	8,225	
20	0.952	0.591	0.357		12,299	9,296	8,247	
22	1.047	0.619	0.357		12,865	9,419	8,256	
24	1.142	0.641			13,430	9,533		
26	1.237	0.660			13,996	9,636		
28	1.332	0.675			14,561	9,729		
30	1.428	0.687			15,127	9,811		
32	1.523	0.697			15,692	9,883		
34	1.618	0.704			16,258	9,946		
36	1.713	0.708			16,823	9,997		
38	1.808	0.711			17,389	10,038		
40	1.904	0.713			17,954	10,069		
42	1.999	0.714			18,520	10,089		
44	2.093	0.714			19,085	10,099		
46	2.189				19,651			
48	2.284				20,216			
50	2.379				20,782			

TABLE 6. Volume and Lining Area for 100 Ft Diameter Circular Reservoirs from 2 to 50 Ft Deep, with Slopes from 0:1 to 3:1

P = 314
2P = 628
3P = 942
4P = 1257

Depth (Ft.)	Volume, Millions of Gallons				Area, Square Feet			
	0:1	1:1	2:1	3:1	0:1	1:1	2:1	3:1
2	0.118	0.113	0.108	0.103	8,796	9,363	9,255	9,205
4	0.235	0.217	0.202	0.184	9,425	9,605	9,387	9,287
6	0.353	0.314	0.275	0.241	10,053	9,837	9,509	9,358
8	0.470	0.399	0.335	0.280	10,681	10,059	9,617	9,416
10	0.587	0.478	0.385	0.306	11,310	10,270	9,714	9,462
12	0.705	0.549	0.421	0.319	11,938	10,471	9,799	9,497
14	0.822	0.613	0.447	0.324	12,566	10,662	9,871	9,519
16	0.940	0.672	0.467	0.327	13,195	10,842	9,931	9,528
18	1.057	0.723	0.479		13,823	11,012	9,979	
20	1.175	0.767	0.485		14,451	11,172	10,015	
22	1.293	0.807	0.489		15,080	11,321	10,040	
24	1.410	0.841	0.489		15,708	11,460	10,052	
26	1.528	0.870			16,336	11,589		
28	1.645	0.896			16,965	11,707		
30	1.762	0.917			17,593	11,816		
32	1.880	0.933			18,221	11,914		
34	1.997	0.948			18,850	12,001		
36	2.115	0.957			19,478	12,079		
38	2.232	0.965			20,106	12,145		
40	2.350	0.971			20,735	12,202		
42	2.468	0.975			21,363	12,248		
44	2.585	0.977			21,991	12,256		
46	2.703	0.979			22,620	12,310		
48	2.820	0.979			23,248	12,325		
50	2.937				23,876			

TABLE 7. Volume and Lining Area for 110 Ft Diameter Circular Reservoirs from 2 to 50 Ft Deep, with Slopes from 0:1 to 3:1

P = 346
2P = 691
3P = 1037
4P = 1382

(110)

Depth	Volume, Millions of Gallons				Area, Square Feet			
(Ft.)	0:1	1:1	2:1	3:1	0:1	1:1	2:1	3:1
2	0.143	0.137	0.132	0.127	10,540	11,164	11,040	10,990
4	0.284	0.264	0.245	0.227	11,231	11,431	11,194	11,083
6	0.426	0.382	0.340	0.302	11,922	11,689	11,329	11,163
8	0.569	0.490	0.419	0.357	12,614	11,937	11,453	11,321
10	0.711	0.589	0.484	0.393	13,305	12,174	11,564	11,288
12	0.853	0.680	0.535	0.416	13,996	12,401	11,664	11,332
14	0.995	0.763	0.575	0.428	14,687	12,617	11,751	11,364
16	1.137	0.839	0.604	0.434	15,378	12,823	11,827	11,384
18	1.279	0.906	0.624	0.434	16,069	13,019	11,890	11,392
20	1.422	0.968	0.639		16,760	13,204	11,941	
22	1.564	1.022	0.646		17,452	13,379	11,980	
24	1.707	1.070	0.650		18,143	13,544	12,007	
26	1.848	1.112	0.652		18,834	13,699	12,023	
28	1.990	1.149			19,525	13,843		
30	2.132	1.118			20,216	13,977		
32	2.275	1.208			20,907	14,100		
34	2.417	1.231			21,599	14,211		
36	2.559	1.250			22,290	14,317		
38	2.702	1.264			22,981	14,410		
40	2.843	1.277			23,672	14,492		
42	2.986	1.286			24,363	14,564		
44	3.127	1.293			25,054	14,627		
46	3.270	1.297			25,745	14,678		
48	3.412	1.301			26,389	14,719		
50	3.554	1.302			27,128	14,750		

TABLE 8. Volume and Lining Area for 120 Ft Diameter Circular Reservoirs from 2 to 50 Ft Deep, with Slopes from 0:1 to 3:1

P = 377
2P = 754
3P = 1131
4P = 1508

Depth	Volume, Millions of Gallons				Area, Square Feet			
(Ft.)	0:1	1:1	2:1	3:1	0:1	1:1	2:1	3:1
2	0.169	0.162	0.156	0.152	12,441	13,122	12,993	12,932
4	0.339	0.318	0.297	0.276	13,195	13,415	13,155	13,034
6	0.507	0.459	0.413	0.371	13,949	13,699	13,306	13,125
8	0.677	0.589	0.511	0.441	14,703	13,971	13,445	13,204
10	0.846	0.714	0.596	0.495	15,457	14,235	13,572	13,270
12	1.016	0.825	0.663	0.528	16,211	14,487	13,686	13,324
14	1.184	0.928	0.716	0.548	16,965	14,729	13,788	13,367
16	1.353	1.026	0.761	0.560	17,719	14,961	13,880	13,397
18	1.523	1.111	0.792	0.563	18,473	15,183	13,958	13,415
20	1.692	1.189	0.814		19,227	15,394	14,024	
22	1.861	1.263	0.831		19,981	15,595	14,078	
24	2.030	1.324	0.840		20,735	15,786	14,121	
26	2.200	1.383	0.844		21,489	15,966	14,151	
28	2.369	1.436	0.847		22,243	16,136	14,169	
30	2.538	1.480			22,977	16,295		
32	2.703	1.519			23,750	16,445		
34	2.876	1.555			24,504	16,584		
36	3.046	1.583			25,258	16,713		
38	3.215	1.608			26,012	16,831		
40	3.384	1.630			26,766	16,940		
42	3.553	1.646			27,520	17,038		
44	3.722	1.659			28,274	17,125		
46	3.892	1.671			29,028	17,203		
48	4.061	1.678			29,782	17,269		
50	4.230	1.683			30,536	17,326		

TABLE 9. Volume and Lining Area for 130 Ft Diameter Circular Reservoirs from 2 to 50 Ft Deep, with Slopes from 0:1 to 3:1

P = 408
2P = 817
3P = 1225
4P = 1634

130

Depth (Ft.)	Volume, Millions of Gallons 0:1	1:1	2:1	3:1	Area, Square Feet 0:1	1:1	2:1	3:1
2	0.199	0.192	0.187	0.181	14,514	15,237	15,097	15,032
4	0.397	0.373	0.350	0.328	15,331	15,556	15,274	15,144
6	0.595	0.542	0.492	0.446	16,148	15,865	15,441	15,245
8	0.794	0.701	0.615	0.537	16,965	16,164	15,595	15,333
10	0.992	0.848	0.719	0.605	17,781	16,453	15,736	15,410
12	1.191	0.985	0.806	0.653	18,598	16,731	15,866	15,474
14	1.390	1.112	01877	0.686	19,415	16,998	15,984	15,526
16	1.589	1.230	0.935	0.705	20,231	17,256	16,089	15,567
18	1.787	1.338	0.980	0.714	21,049	16,448	16,182	15,595
20	1.985	1.437	1.015	0.717	21,866	17,741	16,264	15,611
22	2.184	1.528	1.039		22,682	17,968	16,333	
24	2.383	1.612	1.056		23,499	18,184	16,391	
26	2.582	1.686	1.067		24,316	18,390	16,436	
28	2.780	1.755	1.073		25,133	18,586	16,469	
30	2.978	1.815	1.075		25,950	18,771	16,490	
32	3.177	1.869	1.075		26,766	18,946	16,500	
34	3.376	1.918			27,583	19,111		
36	3.574	1.960			28,400	19,266		
38	3.773	1.996			29,217	19,410		
40	3.971	2.029			30,034	19,544		
42	4.170	2.056			30,851	19,667		
44	4.368	2.078			31,667	19,781		
46	4.567	2.098			32,484	19,884		
48	4.766	2.113			33,301	19,977		
50	4.965	2.125			34,118	20,059		

TABLE 10. Volume and Lining Area for 140 Ft Diameter Circular Reservoirs from 2 to 50 Ft Deep, with Slopes from 0:1 to 3:1

P = 440
2P = 880
3P = 1319
4P = 1759

140

Depth (Ft.)	Volume, Millions of Gallons 0:1	1:1	2:1	3:1	Area, Square Feet 0:1	1:1	2:1	3:1
2	0.230	0.221	0.215	0.209	16,713	17,509	17,358	17,288
4	0.460	0.437	0.412	0.388	17,593	17,854	17,551	17,411
6	0.691	0.633	0.579	0.529	18,473	18,189	17,732	17,521
8	0.921	0.818	0.725	0.640	19,352	18,513	17,901	17,624
10	1.151	0.997	0.855	0.730	20,232	18,828	18,058	17,706
12	1.382	1.158	0.962	0.793	21,112	19,132	18,202	17,781
14	1.612	1.309	1.052	0.837	21,991	19,425	18,335	17,843
16	1.843	1.455	1.130	0.869	22,871	19,709	18,456	17,893
18	2.073	1.586	1.189	0.885	23,750	19,981	18,564	17,931
20	2.303	1.706	1.236	0.892	24,630	20,244	18,661	17,957
22	2.533	1.822	1.275	0.896	25,510	20,497	18,746	17,739
24	2.764	1.924	1.301		26,389	20,739	18,818	
26	2.994	2.018	1.319		27,269	20,971	18,878	
28	3.224	2.108	1.334		28,149	21,192	18,927	
30	3.454	2.186	1.339		29,028	21,404	18,963	
32	3.684	2.256	1.341		29,908	21,605	18,987	
34	3.915	2.322	1.344		30,788	21,796	18,999	
36	4.145	2.379			31,667	21,975		
38	4.375	2.428			32,547	22,146		
40	4.606	2.476			33,427	22,305		
42	4.836	2.515			34,306	22,455		
44	5.067	2.548			35,186	22,594		
46	5.297	2.579			36,066	22,722		
48	5.527	2.604			36,945	22,841		
50	5.757	2.623			37,825	22,949		

TABLE 11. Volume and Lining Area for 150 Ft Diameter Circular Reservoirs from 2 to 50 Ft Deep, with Slopes from 0:1 to 3:1

P = 471
2P = 942
3P = 1414
4P = 1885

150

Depth (Ft.)	Volume, Millions of Gallons 0:1	1:1	2:1	3:1	Area, Square Feet 0:1	1:1	2:1	3:1
2	0.265	0.258	0.251	0.243	19,085	19,937	19,776	19,701
4	0.529	0.502	0.474	0.448	20,028	20,309	19,985	19,834
6	0.793	0.731	0.673	0.618	20,970	20,980	20,181	19,955
8	1.057	0.949	0.848	0.755	21,913	21,020	20,365	20,063
10	1.322	1.153	1.001	0.864	22,855	21,360	20,537	20,160
12	1.587	1.346	1.133	0.946	23,798	21,690	20,697	20,244
14	1.850	1.527	1.246	1.008	24,740	22,009	20,845	20,317
16	2.115	1.696	1.341	1.050	25,683	22,318	20,980	20,376
18	2.379	1.854	1.420	1.078	26,625	22,617	21,104	20,425
20	2.644	2.001	1.484	1.092	27,568	22,905	21,215	20,461
22	2.908	2.139	1.535	1.100	28,510	23,183	21,315	20,486
24	3.172	2.266	1.576	1.101	29,453	23,451	21,402	20,497
26	3.437	2.383	1.605		30,395	23,709	21,478	
28	3.701	2.491	1.626		31,337	23,956	21,541	
30	3.965	2.591	1.639		32,280	24,193	21,592	
32	4.230	2.682	1.647		33,222	24,420	21,631	
34	4.494	2.765	1.651		34,165	24,636	21,658	
36	4.759	2.840	1.652		35,107	24,843	21,674	
38	5.023	2.908			36,050	25,039		
40	5.287	2.969			36,992	25,224		
42	5.552	3.023			37,994	25,399		
44	5.816	3.071			38,877	25,564		
46	6.081	3.113			39,820	25,719		
48	6.344	3.150			40,762	25,862		
50	6.609	3.182			41,705	25,997		

TABLE 12. Volume and Lining Area for 160 Ft Diameter Circular Reservoirs from 2 to 50 Ft Deep, with Slopes from 0:1 to 3:1

P = 503
2P = 1005
3P = 1508
4P = 2011

Depth (Ft.)	Volume, Millions of Gallons 0:1	1:1	2:1	3:1	Area, Square Feet 0:1	1:1	2:1	3:1
2	0.301	0.291	0.284	0.276	21,614	22,524	22,352	22,272
4	0.602	0.575	0.546	0.518	22,620	22,920	22,576	22,415
6	0.902	0.836	0.774	0.715	23,625	23,307	22,786	22,545
8	1.203	1.084	0.976	0.876	24,630	23,683	22,986	22,664
10	1.504	1.326	1.162	1.012	25,635	24,049	23,172	22,770
12	1.805	1.547	1.317	1.114	26,641	24,405	23,348	22,865
14	2.106	1.756	1.452	1.192	27,646	24,750	23,510	22,947
16	2.406	1.960	1.594	1.253	28,651	25,085	23,661	23,018
18	2.707	2.143	1.671	1.291	29,657	25,409	23,800	23,076
20	3.008	2.316	1.753	1.315	30,662	25,723	23,926	23,122
22	3.309	2.484	1.823	1.331	31,667	26,027	24,041	23,157
24	3.610	2.635	1.877	1.335	32,673	26,320	24,143	23,179
26	3.911	2.775	1.918	1.336	33,678	26,605	24,234	23,189
28	4.211	2.911	1.953		34,683	26,877	24,313	
30	4.512	3.032	1.974		35,689	27,140	24,378	
32	4.813	3.142	1.988		36,693	27,392	24,433	
34	5.115	3.250	2.000		37,699	27,635	24,475	
36	5.415	3.343	2.004		38,705	27,867	24,505	
38	5.715	3.428	2.004		39,710	28,088	24,523	
40	6.016	3.511			40,715	28,300		
42	6.316	3.581			41,720	28,501		
44	6.617	3.643			42,726	28,691		
46	6.918	3.704			43,731	28,871		
48	7.219	3.754			44,736	29,057		
50	7.519	3.797			45,742	29,201		

TABLE 13. Volume and Lining Area for 170 Ft Diameter Circular Reservoirs from 2 to 50 Ft Deep, with Slopes from 0:1 to 3:1

P = 534
2P = 1068
3P = 1602
4P = 2136

(170)

Depth	Volume, Millions of Gallons				Area, Square Feet			
(Ft.)	0:1	1:1	2:1	3:1	0:1	1:1	2:1	3:1
2	0.339	0.331	0.323	0.315	24,300	25,267	25,085	24,999
4	0.679	0.648	0.617	0.587	25,368	25,690	25,323	25,152
6	1.019	0.949	0.882	0.818	26,437	26,102	25,549	25,293
8	1.358	1.235	1.118	1.011	27,505	26,503	25,763	25,422
10	1.698	1.506	1.330	1.169	28,573	26,895	25,965	25,538
12	2.037	1.763	1.517	1.297	29,641	27,276	26,155	25,643
14	2.377	2.007	1.680	1.396	30,709	27,648	26,333	25,735
16	2.717	2.238	1.822	1.471	31,777	28,008	26,499	25,815
18	3.056	2.454	1.945	1.526	32,845	28,358	26,653	25,884
20	3.395	2.659	2.048	1.563	33,914	28,699	26,795	25,940
22	3.735	2.852	2.136	1.586	34,982	29,028	26,924	25,984
24	4.075	3.032	2.207	1.598	36,050	29,348	27,042	26,016
26	4.414	3.202	2.264	1.602	37,118	29,657	27,147	26,037
28	4.754	3.360	2.310	1.604	38,186	29,956	27,241	26,045
30	5.093	3.508	2.344		39,254	30,244	27,332	
32	5.433	3.644	2.369		40,322	30,522	27,392	
34	5.773	3.771	2.386		41,391	30,790	27,449	
36	6.112	3.889	2.396		42,459	31,048	27,494	
38	6.452	3.997	2.403		43,527	31,295	27,527	
40	6.791	4.097	2.405		44,595	31,532	27,549	
42	7.131	4.188	2.405		45,663	31,759	27,558	
44	7.471	4.270			46,731	31,975		
46	7.810	4.346			47,799	32,182		
48	8.152	4.414			48,868	32,377		
50	8.490	4.474			49,936	32,413		

TABLE 14. Volume and Lining Area for 180 Ft Diameter Circular Reservoirs from 2 to 50 Ft Deep, with Slopes from 0:1 to 3:1

P = 565
2P = 1131
3P = 1696
4P = 2262

(180)

Depth	Volume, Millions of Gallons				Area, Square Feet			
(Ft.)	0:1	1:1	2:1	3:1	0:1	1:1	2:1	3:1
2	0.380	0.368	0.360	0.353	27,143	28,168	27,974	27,884
4	0.761	0.732	0.699	0.668	28,274	28,616	28,227	28,047
6	1.142	1.067	0.997	0.929	29,405	29,054	28,469	28,197
8	1.523	1.388	1.265	1.150	30,536	29,419	28,699	28,336
10	1.904	1.704	1.515	1.342	31,667	29,889	28,915	28,463
12	2.284	1.993	1.729	1.492	32,798	30,305	29,120	28,578
14	2.665	2.268	1.919	1.612	33,929	30,702	29,313	28,680
16	3.046	2.539	2.094	1.712	35,060	31,088	29,494	28,771
18	3.426	2.787	2.238	1.781	36,191	31,465	29,663	28,849
20	3.807	3.021	2.363	1.831	37,322	31,831	29,820	28,915
22	4.188	3.250	2.476	1.869	38,453	32,186	29,965	28,969
24	4.568	3.458	2.565	1.888	39,584	32,531	30,097	29,012
26	4.949	3.654	2.638	1.898	40,715	32,867	30,218	29,042
28	5.330	3.846	2.703	1.905	41,846	33,191	30,327	29,060
30	5.711	4.018	2.750		42,977	33,509	30,423	
32	6.091	4.179	2.784		44,108	33,809	30,508	
34	6.472	4.337	2.815		45,239	34,103	30,580	
36	6.853	4.477	2.832		46,370	34,386	30,640	
38	7.233	4.606	2.843		47,501	34,659	30,689	
40	7.614	4.734	2.853		48,632	34,922	30,725	
42	7.995	4.844	2.854		49,763	35,174	30,749	
44	8.372	4.946	2.854		50,894	35,417	30,761	
46	8.757	5.045			52,025	35,649		
48	9.134	5.130			53,156	35,870		
50	9.519	5.207			54,287	36,081		

TABLE 15. Volume and Lining Area for 190 Ft Diameter Circular Reservoirs from 2 to 50 Ft Deep, with Slopes from 0:1 to 3:1

P = 597
2P = 1194
3P = 1791
4P = 2388

Depth	Volume, Millions of Gallons				Area, Square Feet			
(Ft.)	0:1	1:1	2:1	3:1	0:1	1:1	2:1	3:1
2	0.424	0.415	0.407	0.398	30,144	31,225	31,021	30,926
4	0.848	0.813	0.779	0.745	31,337	31,699	31,290	31,099
6	1.272	1.194	1.118	1.047	32,531	32,163	31,546	31,259
8	1.696	1.557	1.427	1.304	33,725	32,616	31,790	31,408
10	2.121	1.905	1.706	1.521	34,919	33,059	32,022	31,545
12	2.545	2.237	1.956	1.703	36,113	33,492	32,242	31,670
14	2.969	2.553	2.180	1.850	37,307	33,914	32,450	31,782
16	3.393	2.854	2.379	1.967	38,500	34,327	32,647	31,883
18	3.818	3.140	2.553	2.059	39,694	34,728	32,831	31,971
20	4.242	3.411	2.706	2.127	40,888	35,120	33,003	32,047
22	4.666	3.669	2.838	2.175	42,082	35,501	33,163	32,112
24	5.090	3.912	2.952	2.207	43,276	35,872	33,310	32,164
26	5.514	4.143	3.047	2.226	44,469	36,233	33,445	32,205
28	5.938	4.360	3.126	2.235	45.663	36,583	33,570	32,233
30	6.363	4.565	3.190	2.238	46,857	47,012	33,681	32,249
32	6.787	4.757	3.241		48,051	37,253	33,781	
34	7.211	4.938	3.281		49,245	37,573	33,868	
36	7.635	5.107	3.310		50,438	37,881	33,943	
38	8.058	5.265	3.331		51,632	38,181	34,007	
40	8.482	5.413	3.345		52,826	38,469	34,058	
42	8.906	5.550	3.353		54,020	38,747	34,097	
44	9.331	5.677	3.357		55,214	39,015	34,124	
46	9.755	5.795	3.358		56,407	39,272	34,140	
48	10.18	5.902			57,601	39,520		
50	10.60	6.002			58,795	39,757		

TABLE 16. Volume and Lining Area for 200 Ft Diameter Circular Reservoirs from 2 to 50 Ft Deep, with Slopes from 0:1 to 3:1

P = 628
2P = 1257
3P = 1885
4P = 2513

Depth	Volume, Millions of Gallons				Area, Square Feet			
(Ft.)	0:1	1:1	2:1	3:1	0:1	1:1	2:1	3:1
2	0.469	0.456	0.447	0.437	33,301	34,439	34,225	34,116
4	0.940	0.907	0.871	0.836	34,558	34,939	34,508	34,307
6	1.410	1.327	1.247	1.171	35,814	35,421	34,780	34,478
8	1.880	1.729	1.591	1.461	37,071	35,908	35,039	34,637
10	2.350	2.127	1.912	1.719	38,328	36,377	35,286	34,784
12	2.820	2.495	2.198	1.927	39,584	36,835	35,522	34,919
14	3.290	2.847	2.454	2.099	40,841	37,284	35,745	35,041
16	3.760	3.194	2.688	2.247	42,097	37,720	35,956	35,151
18	4.230	3.514	2.889	2.357	43,354	38,149	36,155	35,250
20	4.700	3.819	3.068	2.442	44,611	38,566	36,342	35,337
22	5.170	4.119	3.232	2.511	45,867	38,973	36,517	35,411
24	5.640	4.394	3.366	2.553	47,124	39,370	36,680	35,473
26	6.110	4.655	3.481	2.582	48,381	39,756	36,831	35,524
28	6.580	4.913	3.585	2.602	49,637	40,132	36,970	35,562
30	7.050	5.146	3.666	2.608	50,894	40,498	37,096	35,588
32	7.519	5.367	3.731	2.609	52,151	40,854	37,211	35,601
34	7.988	5.584	3.790		53,407	41,199	37,313	
36	8.459	5.780	3.830		54,664	41,534	37,404	
38	8.930	5.963	3.860		55,920	41,859	37,482	
40	9.401	6.144	3.887		57,177	42,173	37,548	
42	9.872	6.304	3.900		58,434	42,477	37,603	
44	10.34	6.454	3.908		59,690	42,771	37,645	
46	10.81	6.603	3.916		60,947	43,054	37,675	
48	11.28	6.732	3.916		62,204	43,327	37,693	
50	11.75	6.851			63,460	43,950		

TABLE 17. Volume and Lining Area for 220 Ft Diameter Circular Reservoirs from 2 to 50 Ft Deep, with Slopes from 0:1 to 3:1

P = 691
2P = 1382
3P = 2073
4P = 2765

220

Depth (Ft.)	Volume, Millions of Gallons 0:1	1:1	2:1	3:1	Area, Square Feet 0:1	1:1	2:1	3:1
2	0.568	0.553	0.543	0.533	40,087	41,340	41,104	40,993
4	1.137	1.102	1.062	1.023	41,469	41,891	41,417	41,196
6	1.706	1.615	1.527	1.436	42,851	42,423	41,719	41,387
8	2.275	2.108	1.955	1.810	44,234	42,962	42,009	41,567
10	2.843	2.598	2.362	2.143	45,616	43,483	42,286	41,733
12	3.412	3.054	2.721	2.417	46,998	43,993	42,551	41,888
14	3.980	3.490	3.049	2.651	48,381	44,493	42,804	42,031
16	4.550	3.925	3.358	2.856	49,763	44,982	43,046	42,161
18	5.118	4.326	3.626	3.017	51,145	45,461	43,276	42,280
20	5.687	4.711	3.866	3.146	52,528	45,930	43,492	42,386
22	6.256	5.093	4.090	3.255	53,910	46,389	43,697	42,481
24	6.824	5.444	4.280	3.332	55,292	46,837	43,891	42,563
26	7.393	5.779	4.445	3.387	56,674	47,275	44,071	42,634
28	7.964	6.111	4.599	3.432	58,057	47,703	44,240	42,692
30	8.529	6.415	4.723	3.454	59,439	48,120	44,397	42,738
32	9.103	6.704	4.829	3.466	60,821	48,527	44,538	42,773
34	9.668	6.992	4.925	3.474	62,204	48,923	44,674	42,795
36	10.23	7.252	4.998	3.475	63,586	49,310	44,795	42,804
38	10.81	7.498	5.056		64,968	49,686	44,904	
40	11.37	7.743	5.110		66,351	50,052	45,000	
42	11.95	7.964	5.144		67,733	50,408	45,085	
44	12.51	8.168	5.169		69,115	50,753	45,157	
46	13.08	8.380	5.192		70,498	51,088	45,217	
48	13.65	8.561	5.202		71,880	51,412	45,266	
50	14.22	8.734	5.207		73,262	51,726	45,302	

TABLE 18. Volume and Lining Area for 240 Ft Diameter Circular Reservoirs from 2 to 50 Ft Deep, with Slopes from 0:1 to 3:1

P = 754
2P = 1508
3P = 2262
4P = 3016

Depth (Ft.)	Volume, Millions of Gallons 0:1	1:1	2:1	3:1	Area, Square Feet 0:1	1:1	2:1	3:1
2	0.677	0.659	0.648	0.637	47,501	48,868	48,611	48,491
4	1.353	1.316	1.272	1.229	49,009	49,471	48,955	48,714
6	2.030	1.931	1.834	1.741	50,517	50,064	49,286	48,924
8	2.707	2.524	2.357	2.196	52,025	50,646	49,606	49,124
10	3.384	3.116	2.857	2.608	53,533	51,218	49,914	49,311
12	4.061	3.668	3.303	2.964	55,041	51,779	50,209	49,486
14	4.738	4.200	3.713	3.273	56,549	52,330	50,493	49,648
16	5.415	4.730	4.104	3.537	58,057	52,872	50,764	49,799
18	6.091	5.223	4.446	3.761	59,565	53,402	51,024	49,938
20	6.768	5.697	4.758	3.948	61,073	53,922	51,271	50,065
22	7.445	6.169	5.053	4.105	62,581	54,432	51,507	50,179
24	8.121	6.605	5.306	4.223	64,089	54,932	51,730	50,281
26	8.796	7.024	5.532	4.315	65,597	55,422	51,940	50,372
28	9.472	7.442	5.745	4.390	67,105	55,901	52,140	50,451
30	10.16	7.825	5.922	4.441	68,613	56,370	52,326	50,517
32	10.83	8.192	6.076	4.476	70,121	56,828	52,502	50,571
34	11.51	8.561	6.220	4.500	71,628	57,277	52,664	50,614
36	12.18	8.891	6.334	4.507	73,136	57,714	52,815	50,643
38	12.86	9.213	6.431	4.511	74,644	58,142	52,954	50,661
40	13.53	9.527	6.520		76,152	58,559	53,080	
42	14.22	9.818	6.585		77,660	58,966	53,195	
44	14.89	10.09	6.637		79,168	59,363	53,297	
46	15.57	10.37	6.682		80,676	59,749	53,388	
48	16.24	10.61	6.714		82,184	60,126	53,467	
50	16.92	10.85	6.736		83,692	60,492	53,533	

TABLE 19. Volume and Lining Area for 260 Ft Diameter Circular Reservoirs from 2 to 50 Ft Deep, with Slopes from 0:1 to 3:1

P = 817
2P = 1634
3P = 2450
4P = 3267

Depth (Ft.)	Volume, Millions of Gallons 0:1	1:1	2:1	3:1	Area, Square Feet 0:1	1:1	2:1	3:1
2	0.794	0.781	0.770	0.757	55,543	57,025	56,747	56,616
4	1.589	1.541	1.493	1.447	57,177	57,679	57,121	56,859
6	2.383	2.275	2.169	2.068	58,811	58,323	57,482	57,090
8	3.177	2.985	2.802	2.626	60,444	58,957	57,832	57,309
10	3.971	3.674	3.392	3.126	62,078	59,580	58,170	57,516
12	4.766	4.339	3.940	3.568	63,712	60,194	58,495	57,711
14	5.553	4.982	4.449	3.957	65,345	60,796	58,809	57,894
16	6.354	5.604	4.919	4.297	66,979	61,389	59,111	58,065
18	7.149	6.205	5.352	4.590	68,613	61,971	59,401	58,224
20	7.940	6.783	5.749	4.840	70,246	62,543	59,678	58,371
22	8.734	7.342	6.114	5.052	71,880	63,105	59,943	58,506
24	9.535	7.878	6.445	5.227	73,513	63,656	60,196	58,629
26	10.33	8.396	6.746	5.365	75,147	64,197	60,438	58,738
28	11.12	8.899	7.018	5.483	76,781	64,727	60,667	58,837
30	11.91	9.378	7.261	5.569	78,414	65,248	60,884	58,924
32	12.71	9.841	7.478	5.634	80,048	65,758	61,089	58,998
34	13.50	10.28	7.672	5.679	81,682	66,258	61,282	59,061
36	14.29	10.71	7.841	5.708	83,315	66,747	61,463	59,111
38	15.10	11.11	7.988	5.726	84,949	67,226	61,632	59,149
40	15.89	11.50	8.113	5.734	86,582	67,695	61,789	59,175
42	16.68	11.88	8.223	5.737	88,216	68,154	61,934	59,189
44	17.48	12.22	8.317		89,850	68,602	62,066	
46	18.27	12.57	8.388		91,483	69,040	62,187	
48	19.06	12.89	8.451		93,117	69,468	62,296	
50	19.85	13.20	8.498		94,751	69,885	62,392	

TABLE 20. Volume and Lining Area for 280 Ft Diameter Circular Reservoirs from 2 to 50 Ft Deep, with Slopes from 0:1 to 3:1

P = 880
2P = 1759
3P = 2639
4P = 3519

Depth (Ft.)	Volume, Millions of Gallons 0:1	1:1	2:1	3:1	Area, Square Feet 0:1	1:1	2:1	3:1
2	0.921	0.907	0.895	0.881	64,214	65,810	65,510	65,370
4	1.843	1.791	1.740	1.689	65,975	66,516	65,914	65,633
6	2.764	2.647	2.532	2.424	67,733	67,211	66,307	65,884
8	3.684	3.477	3.279	3.081	69,492	67,896	66,687	66,124
10	4.606	4.293	3.987	3.697	71,251	68,572	67,054	66,351
12	5.527	5.067	4.634	4.228	73,011	69,236	67,410	66,566
14	6.448	5.817	5.244	4.706	74,770	69,890	67,754	66,768
16	7.377	6.560	5.814	5.138	76,529	70,534	68,086	66,960
18	8.294	7.270	6.341	5.504	78,289	71,168	68,405	67,139
20	9.213	7.956	6.830	5.827	80,048	71,792	68,711	67,306
22	10.13	8.624	7.282	6.107	81,807	72,405	69,008	67,460
24	11.05	9.268	7.697	6.344	83,567	73,008	69,292	67,603
26	11.98	9.888	8.074	6.542	85,326	73,640	69,564	67,734
28	12.90	10.49	8.427	6.707	87,085	74,183	69,823	67,852
30	13.82	11.07	8.742	6.338	88,844	74,754	70,070	67,959
32	14.74	11.62	9.024	6.942	90.604	75,316	70,306	68,053
34	15.66	12.17	9.283	7.021	92,363	75,867	70,529	68,136
36	16.59	12.68	9.519	7.079	94,122	76,394	70,739	68,206
38	17.50	13.18	9.723	7.118	95,882	76,939	70,939	68,265
40	18.43	13.66	9.904	7.144	97,641	77,459	71,126	68,311
42	19.34	14.12	10.06	7.157	99,400	77,969	71,300	68,346
44	20.26	14.56	10.19	7.163	101,160	78,469	71,464	68,367
46	21.19	14.99	10.32	7.168	102,919	78,959	71,614	68,377
48	22.11	15.39	10.41		104,678	79,438	71,753	
50	23.03	15.79	10.49		106,437	79,907	71,880	

TABLE 21. Volume and Lining Area for 300 Ft Diameter Circular Reservoirs from 2 to 50 Ft Deep, with Slopes from 0:1 to 3:1

P = 942
2P = 1885
3P = 2827
4P = 3770

300

Depth (Ft.)	Volume, Millions of Gallons 0:1	1:1	2:1	3:1	Area, Square Feet 0:1	1:1	2:1	3:1
2	1.057	1.042	1.029	1.015	73,513	75,223	74,902	74,751
4	2.115	2.060	2.005	1.951	75,398	75,981	75,336	75,035
6	3.172	3.047	2.926	2.807	77,283	76,728	75,759	75,307
8	4.230	4.007	3.793	3.588	79,168	77,464	76,169	75,566
10	5.287	4.953	4.614	4.301	81,053	78,191	76,567	75,813
12	6.344	5.850	5.384	4.944	82,938	78,908	76,953	76,048
14	7.402	6.723	6.106	5.522	84,823	79,614	77,327	76,272
16	8.459	7.590	6.784	6.049	86,708	80,309	77,689	76,482
18	9.519	8.419	7.416	6.502	88,593	80,994	78,039	76,682
20	10.57	9.228	8.003	6.908	90.478	81,669	78,377	76,869
22	11.63	10.02	8.553	7.265	92,363	82,333	78,703	77,044
24	12.69	10.77	9.064	7.573	94,248	82,988	79,016	77,206
26	13.74	11.50	9.535	7.837	96,133	83,632	79,318	77,357
28	14.80	12.21	9.967	8.066	98,018	84,266	79,607	77,496
30	15.87	12.90	10.37	8.247	99,903	84,889	79,885	77,623
32	16.92	13.56	10.73	8.404	101,788	85,503	80,150	77,737
34	17.98	14.21	11.06	8.522	103,673	86,105	80,403	77,839
36	19.04	14.83	11.36	8.616	105,558	86,698	80,645	77,931
38	20.09	15.43	11.63	8.687	107,443	87,280	80,874	78,009
40	21.15	16.01	11.88	8.742	109,328	87,852	81,091	78,075
42	22.20	16.57	12.10	8.773	111,213	88,413	81,296	78,129
44	23.26	17.11	12.28	8.796	113,098	88,965	81,489	78,172
46	24.32	17.62	12.46	8.812	114,983	89,506	81,670	78,201
48	25.38	18.13	12.61	8.812	116,868	90,036	81,839	78,220
50	26.44	18.61	12.73		118,752	90,557	81,996	

TABLE 22. Volume and Lining Area for 320 Ft Diameter Circular Reservoirs from 2 to 50 Ft Deep, with Slopes from 0:1 to 3:1

P = 1005
2P = 2011
3P = 3016
4P = 4021

Depth (Ft.)	Volume, Millions of Gallons 0:1	1:1	2:1	3:1	Area, Square Feet 0:1	1:1	2:1	3:1
2	1.203	1.187	1.172	1.158	83,441	85,265	84,923	84,762
4	2.406	2.348	2.289	2.231	85,452	86,074	85,387	85,065
6	3.610	3.476	3.346	3.219	87,462	86,873	85,840	85,356
8	4.813	4.575	4.346	4.126	89,473	86,875	86,280	85,637
10	6.016	5.649	5.296	4.959	91,483	88,439	86,708	85,904
12	7.219	6.691	6.191	5.716	93,494	89,207	87,124	86,159
14	8.419	7.706	7.033	6.404	95,505	89,964	87,528	86,403
16	9.629	8.694	7.830	7.027	97,515	90,711	87,921	86,634
18	10.83	9.653	8.577	7.585	99,526	91,447	88,301	86,853
20	12.03	10.59	9.276	8.082	101,537	92,175	88,669	87,060
22	13.23	11.50	9.927	8.529	103,547	92,891	89,024	87,255
24	14.44	12.38	10.54	8.914	105,558	93,597	89,368	87,438
26	15.65	13.23	11.11	9.252	107,568	94,292	89,700	87,609
28	16.85	14.07	11.64	9.550	109,579	94,978	90,019	87,768
30	18.05	14.88	12.13	9.802	111,590	95,652	90,327	87,915
32	19.25	15.65	12.57	10.01	113,600	96,317	90,623	88,050
34	20.45	16.41	12.99	10.19	115,611	96,971	90,906	88,172
36	21.65	17.15	13.38	10.30	117,622	97,615	91,118	88,282
38	22.86	17.86	13.72	10.44	119,632	98,249	91,437	88,381
40	24.06	18.55	14.04	10.53	121,643	98,872	91,684	88,467
42	25.27	19.21	14.33	10.60	123,653	99,486	91,920	88,542
44	26.47	19.85	14.58	10.63	125,664	100,088	92,143	88,604
46	27.67	20.48	14.81	10.67	127,675	100,681	92,354	88,654
48	28.88	21.08	15.02	10.68	129,685	101,263	92,553	88,692
50	30.08	21.66	15.20	10.69	131,696	101,834	92,740	88,719

TABLE 23. Volume and Lining Area for 340 Ft Diameter Circular Reservoirs from 2 to 50 Ft Deep, with Slopes from 0:1 to 3:1

P = 1068
2P = 2136
3P = 3204
4P = 4272

(340)

Depth (Ft.)	Volume, Millions of Gallons 0:1	1:1	2:1	3:1	Area, Square Feet 0:1	1:1	2:1	3:1
2	1.358	1.341	1.325	1.309	93,997	95,936	95,571	95,401
4	2.717	2.655	2.592	2.531	96,133	96,796	96,066	95,725
6	4.075	3.932	3.794	3.658	98,269	97,646	96,548	96,036
8	5.433	5.180	4.936	4.701	100,406	98,486	97,019	96,335
10	6.791	6.401	6.025	5.664	102,542	99,315	97,478	96,623
12	8.152	7.588	7.053	6.546	104,678	100,135	97,924	96,899
14	9.511	8.742	8.027	7.310	106,814	100,944	98,357	97,130
16	10.87	9.880	8.954	8.090	108,951	101,744	98,781	97,382
18	12.22	10.98	9.818	8.749	111,087	102,530	99,191	97,653
20	13.58	12.05	10.63	9.354	113,223	103,308	99,589	97,880
22	14.94	13.09	11.41	9.896	115,360	104,076	99,975	98,095
24	16.30	14.11	12.13	10.37	117,496	104,834	100,349	98,298
26	17.66	15.10	12.81	10.79	119,632	105,581	100,710	98,489
28	19.01	16.05	13.44	11.17	121,768	106,316	101,360	98,668
30	20.37	16.99	14.03	11.49	123,905	107,044	101,398	98,835
32	21.73	17.90	14.58	11.77	126,041	107,760	101,724	98,989
34	23.09	18.78	15.09	12.01	128,177	108,466	102,038	99,132
36	24.45	19.64	15.56	12.21	130,314	109,161	102,339	99,263
38	25.81	20.47	15.99	12.37	132,450	109,705	102,629	99,381
40	27.17	21.28	16.39	12.50	134,586	110,521	102,906	99,488
42	28.53	22.06	16.75	12.61	136,722	111,186	103,172	99,582
44	29.88	22.82	17.08	12.68	138,859	111,840	103,425	99,665
46	31.24	23.55	17.38	12.74	140,995	112,484	103,667	99,736
48	32.60	24.26	17.66	12.79	143,131	113,118	103,896	99,794
50	33.95	24.94	17.90	12.81	145,268	113,742	104,113	99,840

TABLE 24. Volume and Lining Area for 360 Ft Diameter Circular Reservoirs from 2 to 50 Ft Deep, with Slopes from 0:1 to 3:1

P = 1131
2P = 2262
3P = 3393
4P = 4524

(360)

Depth (Ft.)	Volume, Millions of Gallons 0:1	1:1	2:1	3:1	Area, Square Feet 0:1	1:1	2:1	3:1
2	1.523	1.504	1.488	1.471	105,181	107,234	106,849	106,668
4	3.046	2.980	2.914	2.849	107,443	108,146	107,374	107,012
6	4.568	4.418	4.270	4.126	109,705	109,047	107,886	107,343
8	6.091	5.823	5.565	5.313	111,967	109,939	108,387	107,663
10	7.614	7.200	6.801	6.417	114,229	110,820	108,875	107,971
12	9.134	8.545	7.972	7.431	116,691	111,691	109,352	108,266
14	10.66	9.849	9.087	8.365	118,752	112,551	109,816	108,549
16	12.18	11.13	10.15	9.221	121,014	113,402	110,269	108,821
18	13.71	12.38	11.14	10.01	123.276	114,241	110,709	109,080
20	15.23	13.60	12.10	10.71	125,538	115,071	111,137	109,328
22	16.75	14.79	12.99	11.36	127,800	115,890	111,554	109,563
24	18.28	15.94	13.83	11.94	130,062	116,699	111,957	109,786
26	19.80	17.07	14.62	12.46	132,324	117,497	112,350	109,997
28	21.32	18.17	15.38	12.92	134,586	118,285	112,730	110,196
30	22.84	19.24	16.08	13.33	136,848	119,063	113,098	110,383
32	24.36	20.29	16.73	13.68	139,110	119,831	113,453	110,558
34	25.87	21.31	17.34	13.99	141,372	120,589	113,797	110,721
36	27.41	22.29	17.91	14.26	143,634	121,336	114,129	110,872
38	[illegible]	[illegible]	[illegible]	14.47	145,096	[illegible]	[illegible]	111,011
40	30.46	24.19	18.93	14.66	148,158	122,799	114,756	111,137
42	31.98	25.09	19.38	14.82	150,420	123,515	115,052	111,252
44	33.50	25.98	19.79	14.94	152,682	124,221	115,335	111,354
46	35.02	26.84	20.18	15.03	154,944	124,916	115,607	111,445
48	36.54	27.67	20.52	15.10	157,206	125,602	115,866	111,524
50	38.07	28.47	20.84	15.16	159,468	126,277	116,114	111,590

TABLE 25. Volume and Lining Area for 380 Ft Diameter Circular Reservoirs from 2 to 50 Ft Deep, with Slopes from 0:1 to 3:1

P = 1194
2P = 2388
3P = 3581
4P = 4775

380

Depth (Ft.)	Volume, Millions of Gallons 0:1	1:1	2:1	3:1	Area, Square Feet 0:1	1:1	2:1	3:1
2	1.696	1.677	1.660	1.642	116,993	119,161	118,754	118,563
4	3.393	3.324	3.254	3.185	119,381	120,125	119,309	118,927
6	5.090	4.931	4.775	4.623	121,768	121,077	119,852	119,279
8	6.787	6.503	6.230	5.964	124,156	122,021	120,383	119,619
10	8.482	8.050	7.623	7.215	126,544	122,953	120,901	119,946
12	10.18	9.550	8.946	8.372	128,931	123,875	121,407	120,262
14	11.88	11.02	10.21	9.441	131,319	124,787	121,903	120,565
16	13.57	12.46	11.42	10.44	133,706	125,688	122,385	120,857
18	15.27	13.87	12.56	11.34	136,094	126,580	122,855	121,137
20	16.96	15.24	13.64	12.17	138,482	127,461	123,314	121,404
22	18.66	16.59	14.68	12.94	140,869	128,332	123,760	121,659
24	20.36	17.90	15.65	13.62	143,257	129,192	124,202	121,903
26	22.05	19.17	16.57	14.24	145,645	130,043	124,617	122,134
28	23.75	20.43	17.44	14.80	148,032	130,882	125,027	122,353
30	25.45	21.65	18.26	15.30	150,420	131,712	125,425	122,560
32	27.14	22.83	19.03	15.74	152,807	132,531	125,812	122,755
34	28.84	23.99	19.75	16.13	155,195	133,340	126,186	122,938
36	30.54	25.12	20.43	16.47	157,583	134,138	126,547	123,109
38	32.23	26.22	21.06	16.76	159,970	134,926	126,897	123,268
40	33.93	27.29	21.65	17.01	162,358	135,705	127,235	123,415
42	35.63	28.34	22.20	17.22	164,746	136,472	127,561	123,549
44	37.33	29.35	22.71	17.40	167,133	137,231	127,874	123,672
46	39.03	30.34	23.18	17.55	169,521	137,977	128,176	123,782
48	40.72	31.30	23.61	17.66	171,908	138,713	128,466	123,881
50	42.42	32.23	24.01	17.74	174,296	139,440	128,743	123,968

TABLE 26. Volume and Lining Area for 400 Ft Diameter Circular Reservoirs from 2 to 50 Ft Deep, with Slopes from 0:1 to 3:1

P = 1257
2P = 2513
3P = 3770
4P = 5027

Depth (Ft.)	Volume, Millions of Gallons 0:1	1:1	2:1	3:1	Area, Square Feet 0:1	1:1	2:1	3:1
2	1.880	1.859	1.841	1.822	129,434	131,716	131,287	131,086
4	3.760	3.687	3.614	3.541	131,947	132,731	131,873	131,470
6	5.640	5.473	5.309	5.148	134,460	133,736	132,446	131,843
8	7.519	7.221	6.933	6.652	136,974	134,730	133,007	132,202
10	9.401	8.938	8.490	8.066	139,487	135,714	133,556	132,550
12	11.28	10.62	9.982	9.370	142,000	136,688	134,093	132,887
14	13.16	12.26	11.40	10.59	144,513	137,651	134,618	133,210
16	15.04	13.87	12.76	11.72	147,027	138,604	135,130	133,521
18	16.92	15.44	14.06	12.76	149,540	139,548	135,631	133,821
20	18.80	16.98	15.29	13.72	152,053	140,480	136,119	134,109
22	20.68	18.49	16.46	14.61	154,567	141,402	136,596	134,384
24	22.56	19.96	17.58	15.41	157,080	142,314	137,060	134,647
26	24.44	21.40	18.64	16.15	159,593	143,215	137,513	134,899
28	26.32	22.81	19.64	16.82	162,107	144,107	137,953	135,138
30	28.20	24.18	20.59	17.41	164,620	144,988	138,381	135,365
32	30.08	25.53	21.48	17.95	167,133	145,859	138,797	135,580
34	31.96	26.84	22.33	18.43	169,646	146,719	139,201	135,783
36	33.84	28.12	23.12	18.86	172,160	147,570	139,594	135,975
38	35.72	29.36	23.86	19.23	174,673	148,409	139,974	136,154
40	37.60	30.58	24.57	19.56	177,186	149,239	140,342	136,320
42	39.48	31.77	25.22	19.83	179,700	150,058	140,697	136,475
44	41.36	32.92	25.83	20.07	182,213	150,867	141,041	136,618
46	43.24	34.05	26.40	20.26	184,726	151,665	141,373	136,748
48	45.12	35.15	26.92	20.43	187,239	152,453	141,692	136,867
50	47.00	36.23	27.42	20.56	189,753	153,232	142,000	136,974

TABLE 27. Volume and Lining Area for 420 Ft Diameter Circular Reservoirs from 2 to 50 Ft Deep, with Slopes from 0:1 to 3:1

P = 1319
2P = 2639
3P = 3958
4P = 5278

Depth (Ft.)	Volume, Millions of Gallons				Area, Square Feet			
	0:1	1:1	2:1	3:1	0:1	1:1	2:1	3:1
2	2.073	2.053	2.033	2.014	142,503	144,899	144,450	144,239
4	4.145	4.067	3.989	3.913	145,142	145,966	145,065	144,642
6	6.218	6.042	5.869	5.700	147,781	147,022	145,668	145,035
8	8.294	7.980	7.674	7.379	150,420	148,068	146,259	145,415
10	10.36	9.880	9.409	8.954	153,059	149,103	146,838	145,783
12	12.43	11.74	11.07	10.42	155,698	150,128	147,405	146,139
14	14.51	13.56	12.66	11.80	158,337	151,144	147,960	146,483
16	16.58	15.35	14.18	13.08	160,976	152,148	148,503	146,814
18	18.65	17.10	15.64	14.27	163,615	153,143	149,034	147,134
20	20.73	18.82	17.03	15.37	166,253	154,127	149,553	147,442
22	22.80	20.49	18.35	16.38	168,892	155,100	150,059	147,737
24	24.87	22.14	19.62	17.37	171,531	156,064	150,554	148,020
26	26.95	23.74	20.82	18.17	174,170	157,017	151,036	148,292
28	29.02	25.32	21.97	18.95	176,809	157,960	151,507	148,551
30	31.09	26.86	23.05	19.67	179,448	158,893	151,965	148,799
32	33.16	28.37	24.08	20.31	182,087	159,815	152,412	149,034
34	35.23	29.84	25.05	20.89	184,726	160,727	152,846	149,257
36	37.31	31.27	25.98	21.41	187,365	161,628	153,268	149,468
38	39.38	32.68	26.84	21.87	190,004	162,520	153,678	149,667
40	41.45	34.05	27.67	22.27	192,643	163,401	154,077	149,854
42	43.53	35.40	28.44	22.64	195,282	164,272	154,463	150,029
44	45.60	36.71	29.16	22.94	197,921	165,132	154,837	150,192
46	47.67	37.99	29.84	23.21	200,560	165,983	155,198	150,343
48	49.74	39.24	30.47	23.43	203,199	166,821	155,548	150,482
50	51.81	40.46	31.05	23.57	205,838	167,651	155,886	150,608

TABLE 28. Volume and Lining Area for 440 Ft Diameter Circular Reservoirs from 2 to 50 Ft Deep, with Slopes from 0:1 to 3:1

P = 1382
2P = 2765
3P = 4147
4P = 5529

Depth (Ft.)	Volume, Millions of Gallons				Area, Square Feet			
	0:1	1:1	2:1	3:1	0:1	1:1	2:1	3:1
2	2.275	2.254	2.234	2.213	156,200	158,711	158,240	158,019
4	4.550	4.467	4.386	4.306	158,965	159,829	158,886	158,443
6	6.824	6.640	6.458	6.281	161,730	160,937	159,519	158,855
8	9.103	8.773	8.451	8.145	164,494	162,034	160,140	159,256
10	11.37	10.86	10.37	9.896	167,259	163,121	160,749	159,644
12	13.65	12.92	12.21	11.54	170,023	164,198	161,198	160,020
14	15.92	14.93	13.98	13.08	172,788	165,265	161,931	160,383
16	18.20	16.91	15.68	14.51	175,553	166,321	162,505	160,735
18	20.48	18.84	17.30	15.86	178,317	167,366	163,066	161,075
20	22.75	20.74	18.87	17.09	181,082	168,402	163,615	161,403
22	25.02	22.60	20.35	18.27	183,846	169,427	164,152	161,719
24	27.30	24.43	21.77	19.34	186,611	170,443	164,676	162,022
26	29.57	26.22	23.13	20.33	189,376	171,447	165,188	162,314
28	31.85	27.97	24.43	21.24	192,140	172,442	165,690	162,594
30	34.12	29.68	25.66	22.06	194,905	173,426	166,178	162,861
32	36.40	31.36	26.84	22.82	197,669	174,399	166,655	163,116
34	38.67	33.00	27.95	23.51	200.434	175,376	167,119	163,359
36	40.94	34.61	29.00	24.14	203,199	176,316	167,571	163,590
38	43.22	36.18	30.01	24.69	205,963	177,258	168,012	163,809
40	45.50	37.72	30.95	25.19	208,728	178,192	168,440	164,017
42	47.77	39.23	31.85	25.64	211,493	179,114	168,856	164,211
44	50.05	40.70	32.70	26.02	214,257	180,025	169,260	164,394
46	52.32	42.14	33.49	26.37	217,022	180,927	169,653	164,566
48	54.59	43.55	34.24	26.66	219,786	181,819	170,032	164,724
50	56.87	44.92	34.93	26.91	222,551	182,700	170,400	164,871

TABLE 29. Volume and Lining Area for 460 Ft Diameter Circular Reservoirs from 2 to 50 Ft Deep, with Slopes from 0:1 to 3:1

P = 1445
2P = 2890
3P = 4335
4P = 5781

460

Depth (Ft.)	Volume, Millions of Gallons 0:1	1:1	2:1	3:1	Area, Square Feet 0:1	1:1	2:1	3:1
2	2.487	2.465	2.443	2.422	170,526	173,526	172,659	172,428
4	4.972	4.887	4.801	4.717	173,416	174,320	173,335	172,872
6	7.459	7.266	7.076	6.890	176,307	175,480	173,998	173,304
8	9.943	9.605	9.268	8.946	179,197	176,629	174,649	173,724
10	12.43	11.90	11.38	10.88	182,087	177,767	175,289	174,133
12	14.91	14.15	13.41	12.71	184,977	178,896	175,916	174,528
14	17.40	16.37	15.37	14.42	187,868	180,014	176,531	174,913
16	19.89	18.54	17.25	16.03	190,758	181,122	177,134	175,285
18	22.38	20.67	19.05	17.53	193,648	182,219	177,726	175,644
20	24.87	22.76	20.79	18.94	196,538	183,306	178,305	175,992
22	27.35	24.82	22.45	20.25	199,429	184,383	178,872	176,329
24	29.84	26.83	24.04	21.47	202,319	185,449	179,426	176,652
26	32.32	28.81	25.56	22.60	205,209	186,506	179,970	176,963
28	34.81	30.74	27.02	23.64	208,100	187,551	180,501	177,263
30	37.29	32.64	28.41	24.61	210,990	188,587	181,019	177,551
32	39.78	34.50	29.74	25.49	213,880	189,612	181,526	177,826
34	42.26	36.32	31.00	26.30	216,770	190,628	182,020	178,089
36	44.75	38.12	32.20	27.03	219,661	191,632	182,503	178,341
38	47.24	39.87	33.35	27.69	222,551	192,626	182,973	178,580
40	49.72	41.58	34.43	28.29	225,441	193,611	183,432	178,807
42	52.21	43.26	35.46	28.83	228,331	194,584	183,878	179,023
44	54.70	44.90	36.43	29.31	231,222	195,548	184,312	179,225
46	57.18	46.50	37.36	29.74	234,112	196,501	184,735	179,417
48	59.67	48.08	38.23	30.10	237,002	197,444	185,145	179,596
50	62.16	49.62	39.05	30.43	239,893	198,376	185,543	179,762

TABLE 30. Volume and Lining Area for 480 Ft Diameter Circular Reservoirs from 2 to 50 Ft Deep, with Slopes from 0:1 to 3:1

P = 1508
2P = 3016
3P = 4524
4P = 6032

Depth (Ft.)	Volume, Millions of Gallons 0:1	1:1	2:1	3:1	Area, Square Feet 0:1	1:1	2:1	3:1
2	2.707	2.684	2.663	2.640	185,480	188,219	187,706	187,465
4	5.415	5.324	5.236	5.148	188,496	189,441	188,411	187,929
6	8.121	7.917	7.722	7.527	191,512	190,512	189,105	188,381
8	10.83	10.47	10.12	9.778	194,528	191,852	189,787	188,821
10	13.53	12.98	12.44	11.91	197,544	193,042	190,456	189,250
12	16.24	15.44	14.67	13.93	200,560	194,222	191,114	189,666
14	18.95	17.87	16.82	15.83	203,576	195,391	191,759	190,070
16	21.65	20.25	18,90	17,62	206,592	196,550	192,392	190,463
18	24.36	22.58	20.89	19.29	209,608	197,699	193,014	190,856
20	27.07	24.88	22.81	20.87	212,623	198,838	193,623	191,210
22	29.77	27.13	24.65	22.34	215,639	199,967	194,220	191,566
24	32.48	29.34	26.42	23.71	218,655	201,084	194,805	191,910
26	35.19	31.52	28.12	24.99	221,671	202,193	195,378	192,242
28	37.90	33.65	29.74	26.19	224,687	203,290	195,939	192,561
30	40.61	35.74	31.30	27.28	227,703	204,377	196,488	192,869
32	43.31	37.79	32.79	28.30	230,719	205,454	197,025	193,164
34	46.02	39.81	34.21	29.23	233,735	206,520	197,550	193,448
36	48.73	41.78	35.57	30.09	236,751	207,577	198,062	193,720
38	51.44	43.72	36.87	30.87	239,767	208,622	198,563	193,979
40	54.15	45.61	38.10	31.58	242,783	209,783	199,052	194,226
42	56.85	47.48	39.27	32.22	245,799	210,683	199,529	194,462
44	59.56	49.31	40.39	32.81	248,815	211,698	199,993	194,685
46	62.27	51.09	41.45	33.32	251,831	212,703	200,445	194,895
48	64.97	52.84	42.45	33.79	254,847	213,697	200,885	195,095
50	67.68	54.55	43.39	34.19	257,863	214,681	201,314	195,282

TABLE 31. Volume and Lining Area for 500 Ft Diameter Circular Reservoirs from 2 to 50 Ft Deep, with Slopes from 0:1 to 3:1

P = 1571
2P = 3142
3P = 4712
4P = 6283

Depth (Ft.)	Volume, Millions of Gallons 0:1	1:1	2:1	3:1	Area, Square Feet 0:1	1:1	2:1	3:1
2	2.937	2.914	2.890	2.867	201,062	203,381	203,381	203,130
4	5.875	5.781	5.689	5.598	204,204	205,189	204,117	203,614
6	8.812	8.600	8.396	8.192	207,346	206,451	204,841	204,087
8	11.75	11.38	11.01	10.66	210,487	207,703	205,553	204,547
10	14.69	14.11	13.54	13.00	213,629	208,945	206,252	204,996
12	17.62	16.79	15.98	15.21	216,770	210,176	206,940	205,432
14	20.56	19.43	18.35	17.30	219,912	211,397	207,616	205,856
16	23.50	22.03	20.62	19.27	223,054	212,608	208,279	206,268
18	26.44	24.59	22.82	21.14	226,195	213,809	209,931	206,669
20	29.37	27.09	24.93	22.89	229,337	214,999	209,570	207,057
22	32.31	29.55	26.95	24.53	232,478	216,178	210,197	207,433
24	35.25	31.97	28.91	26.08	235.620	217,348	210,812	207,796
26	38.19	34.35	30.79	27.50	238,762	218,507	211,416	208,148
28	41.12	36.69	32.60	28.86	241,903	219,656	212,007	208,488
30	44.06	38.99	34.33	30.10	245,045	220,795	212,586	208,816
32	47.00	41.24	35.99	31.26	248,186	221,923	213,153	209,132
34	49.94	43.36	37.58	32.33	251,328	223,041	213,707	209,435
36	52.87	45.62	39.11	33.32	254,470	224,149	214,251	209,727
38	55.81	47.76	40.57	34.23	257,611	225,246	214,782	210,007
40	58.75	49.85	41.96	35.14	260,753	226,333	215,300	210,274
42	61.69	51.90	43.28	35.82	263,894	227,410	215,807	210,529
44	64.62	53.92	44.55	36.51	267,036	228,477	216,302	210,772
46	67.56	55.89	45.75	37.13	270,178	229,533	216,784	211,003
50	70.50	57.83	46.89	37.68	273,319	230,579	217,254	211,222

TABLE 32. Volume in Millions of Gallons for 50 Ft Square and 50 Ft Wide Rectangular Reservoirs from 2 to 50 Ft Deep, with Slopes from 0:1 to 3:1

50

Depth (Ft.)	0:1			1:1			2:1			3:1		
	50	+50	+100	50	+50	+100	50	+50	+100	50	+50	+100
2	0.037	0.037	0.075	0.034	0.036	0.072	0.032	0.034	0.069	0.029	0.033	0.066
4	0.075	0.075	0.150	0.063	0.069	0.138	0.053	0.063	0.126	0.045	0.057	0.114
6	0.112	0.112	0.224	0.087	0.099	0.197	0.067	0.085	0.171	0.051	0.072	0.144
8	0.150	0.150	0.299	0.107	0.126	0.251	0.074	0.102	0.204	0.052	0.078	0.156
10	0.187	0.187	0.374	0.122	0.150	0.299	0.077	0.112	0.224			
12	0.224	0.224	0.448	0.134	0.171	0.341	0.078	0.117	0.233			
14	0.262	0.262	0.524	0.143	0.188	0.377						
16	0.299	0.299	0.598	0.149	0.204	0.407						
18	0.337	0.337	0.673	0.152	0.210	0.421						
20	0.374	0.374	0.748	0.155	0.224	0.449						
22	0.411	0.411	0.823	0.156	0.230	0.461						
24	0.449	0.449	0.898	0.156	0.233	0.467						
26	0.486	0.486	0.972									
28	0.524	0.524	1.047									
30	0.561	0.561	1.122									
32	0.598	0.598	1.197									
34	0.636	0.636	1.272									
36	0.673	0.673	1.346									
38	0.711	0.711	1.421									
40	0.748	0.748	1.496									
42	0.785	0.785	1.570									
44	0.823	0.823	1.646									
46	0.860	0.860	1.720									
48	0.898	0.898	1.795									
50	0.935	0.935	1.870									

TABLE 33. Lining Area in Square Feet for 50 Ft Square and 50 Ft Wide Rectangular Reservoirs from 2 to 50 Ft Deep, with Slopes from 0:1 to 3:1

P = 200
2P = 400
3P = 600
4P = 800

50

Depth (Ft.)	0:1			1:1			2:1			3:1		
	50	+50	+100	50	+50	+100	50	+50	+100	50	+50	+100
2	3,100	2,800	5,600	3,457	2,982	5,964	3,388	2,948	5,896	3,356	2,932	5,864
4	3,500	3,000	6,000	3,602	3,064	6,128	3,461	2,996	5,992	3,397	2,964	5,928
6	3,900	3,200	6,400	3,733	3,146	6,292	3,519	3,044	6,088	3,423	2,996	5,992
8	4,300	3,400	6,800	3,851	3,228	6,456	3,561	3,092	6,184	3,433	3,028	6,056
10	4,700	3,600	7,200	3,956	3,310	6,620	3,588	3,140	6,280			
12	5,100	3,800	7,600	4,048	3,392	6,784	3,600	3,188	6,376			
14	5,500	4,000	8,000	4,127	3,474	6,948						
16	5,900	4,200	8,400	4,192	3,556	7,112						
18	6,300	4,400	8,800	4,245	3,638	7,276						
20	6,700	4,600	9,200	4,284	3,720	7,440						
22	7,100	4,800	9,600	4,310	3,802	7,604						
24	7,500	5,000	10,000	4,323	3,884	7,768						
26	7,900	5,200	10,400									
28	8,300	5,400	10,800									
30	8,700	5,600	11,200									
32	9,100	5,800	11,600									
34	9,500	6,000	12,000									
36	9,900	6,200	12,400									
38	10,300	6,400	12,800									
40	10,700	6,600	13,200									
42	11,100	6,800	13,600									
44	11,500	7,000	14,000									
46	11,900	7,200	14,400									
48	12,300	7,400	14,800									
50	12,700	7,600	15,200									

TABLE 34. Volume in Millions of Gallons for 60 Ft Square and 60 Ft Wide Rectangular Reservoirs from 2 to 50 Ft Deep, with Slopes from 0:1 to 3:1

60

Depth (Ft.)	0:1			1:1			2:1			3:1		
	60	+60	+100	60	+60	+100	60	+60	+100	60	+60	+100
2	0.054	0.054	0.090	0.050	0.052	0.087	0.047	0.050	0.084	0.043	0.048	0.081
4	0.108	0.108	0.180	0.094	0.101	0.168	0.082	0.093	0.156	0.071	0.086	0.144
6	0.162	0.162	0.269	0.131	0.145	0.243	0.106	0.129	0.215	0.084	0.113	0.188
8	0.215	0.215	0.359	0.163	0.187	0.311	0.121	0.158	0.263	0.089	0.129	0.215
10	0.269	0.269	0.449	0.190	0.224	0.374	0.130	0.179	0.299			
12	0.323	0.323	0.539	0.211	0.259	0.431	0.134	0.199	0.323			
14	0.377	0.377	0.628	0.228	0.289	0.481	0.134	0.201	0.335			
16	0.431	0.431	0.718	0.242	0.316	0.527						
18	0.484	0.484	0.808	0.252	0.339	0.565						
20	0.538	0.538	0.898	0.259	0.359	0.598						
22	0.592	0.592	0.987	0.264	0.375	0.625						
24	0.646	0.646	1.077	0.267	0.388	0.646						
26	0.700	0.700	1.167	0.268	0.397	0.661						
28	0.754	0.754	1.257	0.269	0.402	0.670						
30	0.807	0.807	1.346									
32	0.862	0.862	1.436									
34	0.916	0.916	1.525									
36	0.969	0.969	1.616									
38	1.023	1.023	1.705									
40	1.077	1.077	1.795									
42	1.131	1.131	1.885									
44	1.185	1.185	1.975									
46	1.239	1.239	2.064									
48	1.293	1.293	2.154									
50	1.346	1.346	2.244									

TABLE 35. Lining Area in Square Feet for 60 Ft Square and 60 Ft Wide Rectangular Reservoirs from 2 to 50 Ft Deep, with Slopes from 0:1 to 3:1

P = 240
2P = 480
3P = 720
4P = 960

60

Depth (Ft.)	0:1			1:1			2:1			3:1		
	60	+60	+100	60	+60	+100	60	+60	+100	60	+60	+100
2	4,320	3,960	6,600	4,750	4,178	6,964	4,668	4,138	6,896	4,629	4,118	6,864
4	4,800	4,200	7,000	4,927	4,277	7,128	4,760	4,195	6,992	4,683	4,157	6,928
6	5,280	4,440	7,400	5,091	4,375	7,292	4,836	4,253	7,088	4,721	4,195	6,992
8	5,760	4,680	7,800	5,242	4,474	7,456	4,898	4,310	7,184	4,744	4,234	7,056
10	6,240	4,920	8,200	5,380	4,572	7,620	4,944	4,368	7,280			
12	6,720	5,160	9,600	5,505	4,670	7,784	4,975	4,426	7,376			
14	7,200	5,400	9,000	5,616	4,769	7,948	4,990	4,483	7,472			
16	7,680	5,640	9,400	5,714	4,867	8,112						
18	8,160	5,880	9,800	5,800	4,966	8,276						
20	8,640	6,120	10,200	5,872	5,064	8,440						
22	9,120	6,360	10,600	5,931	5,162	8,604						
24	9,600	6,600	11,000	5,977	5,261	8,768						
26	10,080	6,840	11,400	6,010	5,359	8,932						
28	10,560	7,080	11,800	6,029	5,458	9,096						
30	11,040	7,320	12,200									
32	11,520	7,560	12,600									
34	12,000	7,800	13,000									
36	12,480	8,040	13,400									
38	12,960	8,280	13,800									
40	13,440	8,520	14,200									
42	13,920	8,760	14,600									
44	14,400	9,000	15,000									
46	14,880	9,240	15,400									
48	15,360	9,480	15,800									
50	15,840	9,720	16,200									

TABLE 36. Volume in Millions of Gallons for 70 Ft Square and 70 Ft Wide Rectangular Reservoirs from 2 to 50 Ft Deep, with Slopes from 0:1 to 3:1

70

Depth	0:1			1:1			2:1			3:1		
(Ft.)	70	+70	+100	70	+70	+100	70	+70	+100	70	+70	+100
2	0.073	0.073	0.105	0.069	0.071	0.102	0.065	0.069	0.099	0.061	0.067	0.096
4	0.146	0.146	0.209	0.131	0.138	0.197	0.116	0.130	0.186	0.102	0.121	0.174
6	0.220	0.220	0.314	0.184	0.201	0.287	0.153	0.182	0.260	0.126	0.163	0.233
8	0.293	0.293	0.419	0.231	0.260	0.371	0.180	0.226	0.323	0.138	0.193	0.275
10	0.366	0.366	0.524	0.272	0.314	0.449	0.197	0.261	0.374	0.142	0.209	0.299
12	0.440	0.440	0.628	0.306	0.364	0.521	0.207	0.289	0.413			
14	0.513	0.513	0.733	0.335	0.411	0.586	0.212	0.308	0.440			
16	0.586	0.586	0.838	0.359	0.452	0.646	0.214	0.318	0.455			
18	0.660	0.660	0.942	0.379	0.490	0.700						
20	0.733	0.733	1.047	0.394	0.524	0.748						
22	0.806	0.806	1.152	0.406	0.553	0.790						
24	0.880	0.880	1.257	0.414	0.578	0.826						
26	0.953	0.953	1.361	0.420	0.599	0.856						
28	1.026	1.026	1.466	0.424	0.616	0.880						
30	1.100	1.100	1.571	0.426	0.628	0.898						
32	1.173	1.173	1.678	0.427	0.637	0.910						
34	1.246	1.246	1.780	0.428	0.641	0.916						
36	1.319	1.319	1.885									
38	1.393	1.393	1.990									
40	1.466	1.466	2.094									
42	1.539	1.539	2.199									
44	1.613	1.613	2.304									
46	1.686	1.686	2.409									
48	1.759	1.759	2.513									
50	1.833	1.833	2.618									

TABLE 37. Lining Area in Square Feet for 70 Ft Square and 70 Ft Wide Rectangular Reservoirs from 2 to 50 Ft Deep, with Slopes from 0:1 to 3:1

P = 280
2P = 560
3P = 840
4P = 1120

70

Depth	0:1			1:1			2:1			3:1		
(Ft.)	70	+70	+100	70	+70	+100	70	+70	+100	70	+70	+100
2	5,740	5,320	7,600	6,243	5,575	7,964	6,147	5,527	7,896	6,102	5,505	7,864
4	6,300	5,600	8,000	6,453	5,690	8,128	6,258	5,594	7,992	6,168	5,550	7,928
6	6,860	5,880	8,400	6,650	5,804	8,292	6,354	5,662	8,088	6,220	5,594	7,992
8	7,420	6,160	8,800	6,833	5,919	8,456	6,435	5,729	8,184	6,256	5,639	8,056
10	7,980	6,440	9,200	7,004	6,034	8,620	6,500	5,796	8,280	6,276	5,690	8,120
12	8,540	6,720	9,600	7,161	6,149	8,784	6,550	5,863	8,376			
14	9,100	7,000	10,000	7,306	6,264	8,948	6,584	5,930	8,472			
16	9,660	7,280	10,400	7,737	6,378	9,112	6,604	5,998	8,568			
18	10,220	7,560	10,800	7,555	6,493	9,276						
20	10,780	7,840	11,200	7,660	6,608	9,440						
22	11,340	8,120	11,600	7,752	6,723	9,604						
24	11,900	8,400	12,000	7,831	6,838	99,768						
26	12,460	8,680	12,400	7,896	6,952	9,932						
28	13,020	8,960	12,800	7,949	7,067	10,096						
30	13,580	9,240	13,200	7,988	7,182	10,260						
32	14,140	9,520	13,600	8,014	7,297	10,424						
34	14,700	9,800	14,000	8,027	7,412	10,588						
36	15,260	10,080	14,400									
38	15,820	10,360	14,800									
40	16,380	10,640	15,200									
42	16,940	10,920	15,600									
44	17,500	11,200	16,000									
46	18,060	11,480	16,400									
48	18,620	11,760	16,800									
50	19,180	12,040	17,200									

TABLE 38. Volume in Millions of Gallons for 80 Ft Square and 80 Ft Wide Rectangular Reservoirs from 2 to 50 Ft Deep, with Slopes from 0:1 to 3:1

Depth	0:1			1:1			2:1			3:1		
(Ft.)	80	+80	+100	80	+80	+100	80	+80	+100	80	+80	+100
2	0.096	0.096	0.120	0.090	0.094	0.117	0.086	0.091	0.113	0.081	0.087	0.111
4	0.191	0.191	0.239	0.174	0.182	0.227	0.157	0.172	0.215	0.140	0.163	0.203
6	0.287	0.287	0.359	0.246	0.266	0.332	0.210	0.244	0.305	0.177	0.223	0.278
8	0.382	0.382	0.479	0.311	0.345	0.431	0.250	0.306	0.383	0.199	0.268	0.335
10	0.478	0.478	0.598	0.370	0.419	0.524	0.280	0.359	0.449	0.210	0.292	0.374
12	0.574	0.574	0.718	0.419	0.488	0.610	0.299	0.402	0.503	0.213	0.315	0.395
14	0.669	0.669	0.838	0.462	0.553	0.691	0.310	0.436	0.545			
16	0.765	0.765	0.957	0.501	0.613	0.766	0.317	0.460	0.574			
18	0.861	0.861	1.077	0.532	0.668	0.835	0.319	0.474	0.592			
20	0.956	0.956	1.197	0.558	0.718	0.898						
22	1.052	1.052	1.316	0.581	0.764	0.954						
24	1.147	1.147	1.436	0.597	0.804	1.005						
26	1.243	1.243	1.556	0.610	0.840	1.050						
28	1.339	1.339	1.676	0.622	0.871	1.089						
30	1.434	1.434	1.795	0.628	0.898	1.122						
32	1.530	1.530	1.915	0.633	0.919	1.149						
34	1.625	1.625	2.035	0.637	0.936	1.170						
36	1.721	1.721	2.154	0.638	0.948	1.185						
38	1.817	1.817	2.274	0.638	0.955	1.194						
40	1.912	1.912	2.394									
42	2.008	2.008	2.513									
44	2.104	2.104	2.632									
46	2.199	2.199	2.753									
48	2.295	2.295	2.872									
50	2.390	2.390	2.992									

TABLE 39. Lining Area in Square Feet for 80 Ft Square and 80 Ft Wide Rectangular Reservoirs from 2 to 50 Ft Deep, with Slopes from 0:1 to 3:1

P = 320
2P = 640
3P = 960
4P = 1280

Depth	0:1			1:1			2:1			3:1		
(Ft.)	80	+80	+100	80	+80	+100	80	+80	+100	80	+80	+100
2	7,360	6,880	8,600	7,936	7,171	8,964	7,826	7,117	8,896	7,775	7,091	8,864
4	8,000	7,200	9,000	8,179	7,302	9,128	7,956	7,146	8,932	7,854	7,142	8,928
6	8,640	7,520	9,400	8,408	7,434	9,292	8,072	7,270	9,088	7,918	7,194	8,992
8	9,280	7,840	9,800	8,625	7,565	9,456	8,172	7,347	9,184	7,967	7,245	9,056
10	9,920	8,160	10,200	8,828	7,696	9,620	8,256	7,424	9,280	8,000	7,296	9,120
12	10,560	8,480	10,600	9,018	7,827	9,784	8,325	7,501	9,376	8,018	7,347	9,184
14	11,200	8,800	11,000	9,195	7,958	9,948	8,379	7,578	9,472			
16	11,840	9,120	11,400	9,359	8,090	10,112	8,417	7,654	9,568			
18	12,480	9,440	11,800	9,510	8,221	10,276	8,440	7,731	9,664			
20	13,120	9,760	12,200	9,648	8,352	10,440						
22	13,760	10,080	12,600	9,773	8,483	10,604						
24	14,400	10,400	13,000	9,884	8,614	10,768						
26	15,040	10,720	13,400	9,983	8,746	10,932						
28	15,680	11,040	13,800	10,068	8,877	11,096						
30	16,320	11,360	14,200	10,140	9,008	11,260						
32	16,960	11,680	14,600	10,199	9,139	11,424						
34	17,600	12,000	15,000	10,245	9,270	11,588						
36	18,240	12,320	15,400	10,278	9,402	11,752						
38	18,880	12,640	15,800	10,297	9,532	11,916						
40	19,520	12,960	16,200									
42	20,160	13,280	16,600									
44	20,800	13,600	17,000									
46	21,440	13,920	17,400									
48	22,080	14,240	17,800									
50	22,720	14,560	18,200									

TABLE 40. Volume in Millions of Gallons for 90 Ft Square and 90 Ft Wide Rectangular Reservoirs from 2 to 50 Ft Deep, with Slopes from 0:1 to 3:1

90

Depth	0:1			1:1			2:1			3:1		
(Ft.)	90	+90	+100	90	+90	+100	90	+90	+100	90	+90	+100
2	0.121	0.121	0.135	0.116	0.118	0.132	0.111	0.116	0.129	0.106	0.113	0.126
4	0.242	0.242	0.269	0.222	0.232	0.257	0.202	0.221	0.245	0.184	0.210	0.233
6	0.364	0.364	0.404	0.317	0.339	0.377	0.275	0.315	0.350	0.238	0.291	0.323
8	0.485	0.485	0.539	0.404	0.442	0.491	0.333	0.399	0.443	0.272	0.355	0.395
10	0.606	0.606	0.673	0.481	0.539	0.598	0.377	0.471	0.524	0.292	0.404	0.449
12	0.727	0.727	0.808	0.550	0.630	0.700	0.408	0.533	0.592	0.301	0.436	0.485
14	0.848	0.848	0.942	0.612	0.716	0.796	0.430	0.584	0.649	0.303	0.452	0.503
16	0.969	0.969	1.077	0.666	0.797	0.886	0.444	0.625	0.694			
18	1.091	1.091	1.212	0.713	0.873	0.969	0.451	0.654	0.727			
20	1.212	1.212	1.346	0.753	0.943	1.047	0.454	0.673	0.748			
22	1.333	1.333	1.481	0.788	1.007	1.119	0.454	0.681	0.757			
24	1.454	1.454	1.616	0.816	1.066	1.185						
26	1.575	1.575	1.750	0.840	1.122	1.245						
28	1.696	1.696	1.885	0.860	1.169	1.299						
30	1.818	1.818	2.020	0.875	1.212	1.346						
32	1.939	1.939	2.154	0.887	1.250	1.388						
34	2.060	2.060	2.289	0.896	1.282	1.424						
36	2.181	2.181	2.424	0.902	1.309	1.454						
38	2.302	2.302	2.559	0.905	1.330	1.478						
40	2.424	2.424	2.693	0.908	1.346	1.496						
42	2.545	2.545	2.827	0.909	1.357	1.508						
44	2.666	2.666	2.962	0.909	1.362	1.514						
46	2.787	2.787	3.097									
48	2.908	2.908	3.231									
50	3.029	3.029	3.366									

TABLE 41. Lining Area in Square Feet for 90 Ft Square and 90 Ft Wide Rectangular Reservoirs from 2 to 50 Ft Deep, with Slopes from 0:1 to 3:1

P = 360
2P = 720
3P = 1080
4P = 1440

90

Depth	0:1			1:1			2:1			3:1		
(Ft.)	90	+90	+100	90	+90	+100	90	+90	+100	90	+90	+100
2	9,180	8,640	9,600	9,829	8,968	9,964	9,705	8,906	9,896	9,648	8,878	9,864
4	9,900	9,000	10,000	10,104	9,115	10,128	9,855	8,993	9,992	9,740	8,935	9,928
6	10,620	9,360	10,400	10,367	9,263	10,292	9,989	9,079	10,088	9,816	8,993	9,992
8	11,340	9,720	10,800	10,616	9,410	10,456	10,108	9,166	10,184	9,878	9,050	10,056
10	12,060	10,080	11,200	10,852	9,558	10,620	10,212	9,252	10,280	9,924	9,108	10,120
12	12,780	10,440	11,600	11,075	9,706	10,784	10,300	9,338	10,376	9,955	9,166	10,184
14	13,500	10,800	12,000	11,285	9,853	10,948	10,373	9,425	10,472	9,970	9,223	10,248
16	14,220	11,160	12,400	11,482	10,001	11,112	10,431	9,511	10,568			
18	14,940	11,520	12,800	11,665	10,148	11,276	10,473	9,598	10,664			
20	15,660	11,880	13,200	11,836	10,296	11,440	10,500	9,684	10,760			
22	16,380	12,240	13,600	11,993	10,444	11,604	10,512	9,770	10,856			
24	17,100	12,600	14,000	12,138	10,591	11,768						
26	17,820	12,960	14,400	12,269	10,739	11,932						
28	18,540	13,320	14,800	12,387	10,886	12,096						
30	19,260	13,680	15,200	12,492	11,034	12,260						
32	19,980	14,040	15,600	12,584	11,182	12,424						
34	20,700	14,400	16,000	12,663	11,329	12,588						
36	21,420	14,760	16,400	12,728	11,477	12,752						
38	22,140	15,120	16,800	12,781	11,624	12,916						
40	22,860	15,480	17,200	12,820	11,772	13,080						
42	23,580	15,840	17,600	12,846	11,920	13,244						
44	24,300	16,200	18,000	12,859	12,067	13,408						
46	25,020	16,560	18,400									
48	25,740	16,920	18,800									
50	26,460	17,280	19,200									

TABLE 42. Volume in Millions of Gallons for 100 Ft Square and 100 Ft Wide Rectangular Reservoirs from 2 to 50 Ft Deep, with Slopes from 0:1 to 3:1

Depth	0:1			1:1			2:1			3:1		
(Ft.)	100	+100	+100	100	+100	+100	100	+100	+100	100	+100	+100
2	0.150	0.150	0.150	0.144	0.147	0.147	0.137	0.144	0.144	0.131	0.141	0.141
4	0.299	0.299	0.299	0.276	0.287	0.287	0.257	0.275	0.275	0.234	0.263	0.263
6	0.449	0.449	0.449	0.400	0.422	0.422	0.350	0.395	0.395	0.307	0.368	0.368
8	0.598	0.598	0.598	0.508	0.550	0.550	0.426	0.503	0.503	0.356	0.455	0.455
10	0.748	0.748	0.748	0.608	0.673	0.673	0.490	0.598	0.598	0.390	0.524	0.524
12	0.898	0.898	0.898	0.699	0.790	0.790	0.536	0.682	0.682	0.406	0.574	0.574
14	1.047	1.047	1.047	0.781	0.901	0.901	0.569	0.754	0.754	0.413	0.607	0.607
16	1.197	1.197	1.197	0.855	1.005	1.005	0.595	0.814	0.814	0.416	0.622	0.622
18	1.346	1.346	1.346	0.920	1.104	1.104	0.610	0.862	0.862			
20	1.496	1.496	1.496	0.977	1.197	1.197	0.618	0.898	0.898			
22	1.646	1.646	1.646	1.028	1.284	1.284	0.623	0.922	0.922			
24	1.795	1. 95	1.795	1.071	1.364	1.364	0.623	0.934	0.934			
26	1.945	1.945	1.945	1.108	1.439	1.439						
28	2.094	2.094	2.094	1.141	1.508	1.508						
30	2.244	2.244	2.244	1.167	1.571	1.571						
32	2.394	2.394	2.394	1.188	1.628	1.628						
34	2.543	2.543	2.543	1.207	1.679	1.679						
36	2.693	2.693	2.693	1.219	1.723	1.723						
38	2.842	2.842	2.842	1.229	1.762	1.762						
40	2.992	2.992	2.992	1.237	1.795	1.795						
42	3.142	3.142	3.142	1.242	1.822	1.822						
44	3.291	3.291	3.291	1.244	1.843	1.843						
46	3.441	3.441	3.441	1.247	1.858	1.858						
48	3.590	3.590	3.590	1.247	1.867	1.867						
50	3.740	3.740	3.740									

TABLE 43. Lining Area in Square Feet for 100 Ft Square and 100 Ft Wide Rectangular Reservoirs from 2 to 50 Ft Deep, with Slopes from 0:1 to 3:1

P = 400
2P = 800
3P = 1200
4P = 1600

100

Depth	0:1			1:1			2:1			3:1		
(Ft.)	100	+100	+100	100	+100	+100	100	+100	+100	100	+100	+100
2	11,200	10,600	10,600	11,921	10,964	10,964	11,784	10,896	10,896	11,720	10,864	10,864
4	12,000	11,000	11,000	12,229	11,120	11,120	11,952	10,992	10,992	11,825	10,928	10,928
6	12,800	11,400	11,400	12,525	11,292	11,292	12,107	11,088	11,088	11,915	10,992	10,992
8	13,600	11,800	11,800	12,807	11,456	11,456	12,245	11,184	11,184	11,989	11,056	11,056
10	14,400	12,200	12,200	13,076	11,620	11,620	12,368	11,280	11,280	12,048	11,120	11,120
12	15,200	12,600	12,600	13,332	11,784	11,784	12,476	11,376	11,376	12,092	11,184	11,184
14	16,000	13,000	13,000	13,575	11,948	11,948	12,568	11,472	11,472	12,120	11,248	11,248
16	16,800	13,400	13,400	13,804	12,112	12,112	12,644	11,568	11,568	12,132	11,312	11,312
18	17,600	13,800	13,800	14,021	12,276	12,276	12,706	11,664	11,664			
20	18,400	14,200	14,200	14,224	12,440	12,440	12,752	11,760	11,760			
22	19,200	14,600	14,600	14,414	12,604	12,604	12,783	11,856	11,856			
24	20,000	15,000	15,000	14,591	12,768	12,768	12,798	11,952	11,952			
26	20,800	15,400	15,400	14,755	12,932	12,932						
28	21,600	15,800	15,800	14,906	13,096	13,096						
30	22,400	16,200	16,200	15,044	13,260	13,260						
32	23,200	16,600	16,600	15,169	13,424	13,424						
34	24,000	17,000	17,000	15,280	13,588	13,588						
36	24,800	17,400	17,400	15,379	13,752	13,752						
38	25,600	17,800	17,800	15,464	13,916	13,916						
40	26,400	18,200	18,200	15,536	14,080	14,080						
42	27,200	18,600	18,600	15,595	14,244	14,244						
44	28,000	19,000	19,000	15,605	14,408	14,408						
46	28,800	19,400	19,400	15,674	14,572	14,572						
48	29,600	19,800	19,800	15,693	14,736	14,736						
50	30,400	20,200	20,200									

TABLE 44. Volume in Millions of Gallons for 110 Ft Square and 110 Ft Wide Rectangular Reservoirs from 2 to 50 Ft Deep, with Slopes from 0:1 to 3:1

110

Depth	0:1			1:1			2:1			3:1		
(Ft.)	110	+110	+100	110	+110	+100	110	+110	+100	110	+110	+100
2	0.182	0.182	0.165	0.175	0.178	0.162	0.168	0.175	0.159	0.162	0.171	0.156
4	0.362	0.362	0.329	0.336	0.349	0.317	0.312	0.336	0.305	0.289	0.323	0.293
6	0.543	0.543	0.494	0.486	0.514	0.467	0.433	0.484	0.440	0.385	0.454	0.413
8	0.724	0.724	0.658	0.624	0.671	0.610	0.534	0.619	0.562	0.454	0.566	0.515
10	0.905	0.905	0.823	0.750	0.823	0.748	0.616	0.741	0.673	0.501	0.658	0.598
12	1.086	1.086	0.987	0.866	0.968	0.880	0.681	0.849	0.772	0.530	0.731	0.664
14	1.267	1.267	1.152	0.972	1.106	1.005	0.732	0.945	0.859	0.545	0.783	0.712
16	1.448	1.448	1.316	1.068	1.238	1.125	0.769	1.027	0.934	0.552	0.816	0.742
18	1.629	1.629	1.481	1.154	1.363	1.239	0.795	1.096	0.996	0.553	0.829	0.754
20	1.810	1.810	1.646	1.232	1.481	1.346	0.813	1.152	1.047			
22	1.991	1.991	1.810	1.301	1.593	1.448	0.823	1.195	1.086			
24	2.173	2.173	1.975	1.362	1.698	1.544	0.828	1.224	1.113			
26	2.353	2.353	2.139	1.416	1.797	1.634	0.830	1.241	1.128			
28	2.534	2.534	2.304	1.463	1.889	1.717						
30	2.715	2.715	2.468	1.503	1.975	1.795						
32	2.896	2.896	2.633	1.538	2.054	1.867						
34	3.078	3.078	2.798	1.567	2.126	1.933						
36	3.258	3.258	2.962	1.591	2.192	1.993						
38	3.440	3.440	3.127	1.610	2.251	2.047						
40	3.620	3.620	3.291	1.626	2.304	2.094						
42	3.802	3.802	3.456	1.637	2.350	2.136						
44	3.982	3.982	3.620	1.646	2.389	2.172						
46	4.164	4.164	3.785	1.652	2.422	2.202						
48	4.344	4.344	3.949	1.656	2.449	2.226						
50	4.525	4.525	4.114	1.658	2.468	2.244						

TABLE 45. Lining Area in Square Feet for 110 Ft Square and 110 Ft Wide Rectangular Reservoirs from 2 to 50 Ft Deep, with Slopes from 0:1 to 3:1

P = 440
2P = 880
3P = 1320
4P = 1760

110

Depth	0:1			1:1			2:1			3:1		
(Ft.)	110	+110	+100	110	+110	+100	110	+110	+100	110	+110	+100
2	13,420	12,760	11,600	14,214	13,160	11,964	14,064	13,086	11,896	13,993	13,050	11,864
4	14,300	13,200	12,000	14,555	13,341	12,128	14,252	13,191	11,992	14,111	13,121	11,928
6	15,180	13,640	12,400	14,883	13,521	12,292	14,424	13,297	12,088	14,213	13,191	11,992
8	16,060	14,080	12,800	15,198	13,702	12,456	14,582	13,402	12,184	14,300	13,262	12,056
10	16,940	14,520	13,200	15,500	13,882	12,620	14,724	13,508	12,280	14,372	13,332	12,120
12	17,820	14,960	13,600	15,789	14,062	12,784	14,851	13,614	12,376	14,428	13,402	12,184
14	18,700	15,400	14,000	16,064	14,243	12,948	14,962	13,719	12,472	14,469	13,473	12,248
16	19,580	15,840	14,400	16,327	14,423	13,112	15,058	13,825	12,568	14,495	13,543	12,312
18	20,460	16,280	14,800	16,576	14,604	13,276	15,139	13,930	12,664	14,505	13,614	12,376
20	21,340	16,720	15,200	16,812	14,784	13,440	15,204	14,036	12,760			
22	22,220	17,160	15,600	17,035	14,964	13,604	15,254	14,142	12,856			
24	23,100	17,600	16,000	17,245	15,145	13,768	15,288	14,247	12,952			
26	23,980	18,040	16,400	17,442	15,325	13,932	15,308	14,353	13,048			
28	24,860	18,480	16,800	17,625	15,506	14,096						
30	25,740	18,920	17,200	17,796	15,686	14,260						
32	26,620	19,360	17,600	17,953	15,866	14,424						
34	27,500	19,800	18,000	18,094	16,047	14,588						
36	28,380	20,240	18,400	18,229	16,227	14,752						
38	29,260	20,680	18,800	18,347	16,408	14,916						
40	30,140	21,120	19,200	18,452	16,588	15,080						
42	31,020	21,560	19,600	18,544	16,768	15,244						
44	31,900	22,000	20,000	18,623	16,949	15,408						
46	32,780	22,440	20,400	18,688	17,129	15,572						
48	33,600	22,880	20,800	18,741	17,310	15,736						
50	34,540	23,320	21,200	18,780	17,490	15,900						

TABLE 46. Volume in Millions of Gallons for 120 Ft Square and 120 Ft Wide Rectangular Reservoirs from 2 to 50 Ft Deep, with Slopes from 0:1 to 3:1

120

Depth (Ft.)	0:1			1:1			2:1			3:1		
	120	+120	+100	120	+120	+100	120	+120	+100	120	+120	+100
2	0.215	0.215	0.180	0.206	0.212	0.177	0.199	0.209	0.174	0.193	0.205	0.171
4	0.431	0.431	0.359	0.405	0.416	0.347	0.378	0.402	0.335	0.352	0.388	0.323
6	0.646	0.646	0.538	0.584	0.614	0.512	0.526	0.582	0.485	0.472	0.549	0.458
8	0.862	0.862	0.718	0.750	0.804	0.670	0.651	0.747	0.622	0.562	0.689	0.574
10	1.077	1.077	0.898	0.909	0.987	0.823	0.759	0.898	0.748	0.630	0.808	0.673
12	1.293	1.293	1.077	1.051	1.163	0.969	0.844	1.034	0.862	0.672	0.905	0.754
14	1.508	1.508	1.257	1.182	1.332	1.110	0.912	1.156	0.963	0.698	0.980	0.817
16	1.723	1.723	1.436	1.306	1.494	1.245	0.969	1.264	1.053	0.713	1.034	0.861
18	1.939	1.939	1.616	1.415	1.648	1.373	1.008	1.357	1.131	0.717	1.066	0.889
20	2.154	2.154	1.795	1.514	1.795	1.496	1.036	1.436	1.197			
22	2.370	2.370	1.975	1.608	1.935	1.613	1.058	1.501	1.251			
24	2.585	2.585	2.154	1.686	2.068	1.723	1.069	1.551	1.293			
26	2.801	2.801	2.334	1.761	2.194	1.828	1.074	1.587	1.322			
28	3.016	3.016	2.513	1.829	2.312	1.927	1.078	1.608	1.340			
30	3.231	3.231	2.693	1.885	2.424	2.020						
32	3.442	3.442	2.872	1.934	2.528	2.106						
34	3.662	3.662	3.052	1.980	2.625	2.187						
36	3.878	3.878	3.231	2.016	2.714	2.262						
38	4.093	4.093	3.411	2.047	2.797	2.331						
40	4.308	4.308	3.590	2.075	2.872	2.394						
42	4.524	4.524	3.769	2.096	2.941	2.450						
44	4.739	4.739	3.949	2.112	3.002	2.501						
46	4.955	4.955	4.129	2.128	3.055	2.546						
48	5.170	5.170	4.308	2.137	3.102	2.585						
50	5.386	5.386	4.488	2.143	3.142	2.618						

TABLE 47. Lining Area in Square Feet for 120 Ft Square and 120 Ft Wide Rectangular Reservoirs from 2 to 50 Ft Deep, with Slopes from 0:1 to 3:1

P = 480
2P = 960
3P = 1440
4P = 1920

120

Depth (Ft.)	0:1			1:1			2:1			3:1		
	120	+120	+100	120	+120	+100	120	+120	+100	120	+120	+100
2	15,840	15,120	12,600	16,707	15,557	12,964	16,543	15,475	12,896	16,466	15,437	12,864
4	16,800	15,600	13,000	17,081	15,754	13,128	16,750	15,590	12,992	16,596	15,514	12,928
6	17,760	16,080	13,400	17,442	15,950	13,292	16,942	15,706	13,088	16,711	15,590	12,992
8	18,720	16,560	13,800	17,789	16,147	13,456	17,119	15,821	13,184	16,812	15,667	13,056
10	19,680	17,040	14,200	18,124	16,344	13,620	17,280	15,936	13,280	16,896	15,744	13,120
12	20,640	17,520	14,600	18,445	16,541	13,784	17,426	16,051	13,376	16,965	15,821	13,184
14	21,600	18,000	15,000	18,754	16,738	13,948	17,556	16,166	13,472	17,019	15,898	13,248
16	22,560	18,480	15,400	19,049	16,934	14,112	17,672	16,282	13,568	17,057	15,974	13,312
18	23,520	18,960	15,800	19,331	17,131	14,276	17,772	16,397	13,664	17,080	16,051	13,376
20	24,480	19,440	16,200	19,600	17,328	14,440	17,856	16,512	13,760			
22	25,440	19,920	16,600	19,856	17,525	14,604	17,925	16,627	13,856			
24	26,400	20,400	17,000	20,099	17,722	14,768	17,979	16,742	13,952			
26	27,360	20,880	17,400	20,328	17,918	14,932	18,017	16,858	14,048			
28	28,320	21,360	17,800	20,545	18,115	15,096	18,040	16,973	14,144			
30	29,280	21,840	18,200	20,748	18,312	15,260						
32	30,240	22,320	18,600	20,938	18,509	15,424						
34	31,200	22,800	19,000	21,115	18,706	15,588						
36	32,160	23,280	19,400	21,279	18,902	15,752						
38	33,120	23,760	19,800	21,430	19,099	15,916						
40	34,000	24,240	20,200	21,560	19,296	16,000						
42	25,040	24,720	20,600	21,693	19,493	16,244						
44	36,000	25,200	21,000	21,804	19,690	16,408						
46	36,960	25,680	21,400	21,903	19,886	16,572						
48	37,920	26,160	21,800	21,988	20,083	16,736						
50	38,880	26,640	22,200	22,060	2-,280	16,900						

TABLE 48. Volume in Millions of Gallons for 130 Ft Square and 130 Ft Wide Rectangular Reservoirs from 2 to 50 Ft Deep, with Slopes from 0:1 to 3:1

130

Depth (Ft.)	0:1 130	0:1 +130	0:1 +100	1:1 130	1:1 +130	1:1 +100	2:1 130	2:1 +130	2:1 +100	3:1 130	3:1 +130	3:1 +100
2	0.253	0.253	0.194	0.245	0.249	0.191	0.238	0.245	0.188	0.230	0.241	0.186
4	0.506	0.506	0.389	0.475	0.490	0.377	0.446	0.475	0.365	0.418	0.459	0.353
6	0.758	0.758	0.583	0.691	0.723	0.557	0.627	0.688	0.530	0.568	0.653	0.503
8	1.011	1.011	0.778	0.892	0.949	0.730	0.783	0.887	0.682	0.684	0.825	0.634
10	1.264	1.264	0.972	1.080	1.167	0.898	0.915	1.070	0.823	0.770	0.972	0.748
12	1.517	1.517	1.167	1.254	1.377	1.059	1.026	1.237	0.951	0.832	1.097	0.844
14	1.770	1.770	1.361	1.416	1.579	1.215	1.117	1.389	1.068	0.873	1.198	0.922
16	2.023	2.023	1.556	1.566	1.774	1.364	1.190	1.525	1.173	0.897	1.276	0.981
18	2.275	2.275	1.750	1.703	1.960	1.508	1.248	1.645	1.266	0.909	1.330	1.023
20	2.528	2.528	1.945	1.830	2.139	1.646	1.292	1.750	1.346	0.913	1.361	1.047
22	2.781	2.781	2.139	1.946	2.310	1.777	1.323	1.840	1.415			
24	3.034	3.034	2.334	2.052	2.474	1.903	1.345	1.914	1.472			
26	3.287	3.287	2.528	2.147	2.629	2.023	1.358	1.972	1.517			
28	3.540	3.540	2.723	2.234	2.777	2.136	1.366	2.015	1.550			
30	3.792	3.792	2.917	2.311	2.917	2.244	1.369	2.042	1.571			
32	4.045	4.045	3.112	2.380	3.049	2.346	1.369	2.054	1.580			
34	4.298	4.298	3.306	2.442	3.174	2.441						
36	4.551	4.551	3.501	2.496	3.291	2.531						
38	4.804	4.804	3.365	2.542	3.400	2.615						
40	5.056	5.056	3.890	2.583	3.501	2.693						
42	5.309	5.309	4.084	2.618	3.594	2.765						
44	5.562	5.562	4.279	2.646	3.680	2.830						
46	5.815	5.815	4.473	2.671	3.757	2.890						
48	6.068	6.068	4.668	2.690	3.827	2.944						
50	6.321	6.321	4.862	2.705	3.890	2.992						

TABLE 49. Lining Area in Square Feet for 130 Ft Square and 130 Ft Wide Rectangular Reservoirs from 2 to 50 Ft Deep, with Slopes from 0:1 to 3:1

P = 520
2P = 1040
3P = 1560
4P = 2080

130

Depth (Ft.)	0:1 130	0:1 +130	0:1 +100	1:1 130	1:1 +130	1:1 +100	2:1 130	2:1 +130	2:1 +100	3:1 130	3:1 +130	3:1 +100
2	18,480	17,680	13,600	19,400	18,153	13,964	19,222	18,065	13,896	19,139	18,023	13,864
4	19,520	18,200	14,000	19,807	18,366	14,128	19,448	18,190	13,992	19,282	18,106	13,928
6	20,560	18,720	14,400	20,200	18,580	14,292	19,660	18,314	14,088	19,410	18,190	13,992
8	21,600	19,240	14,800	20,581	18,793	14,456	19,856	18,439	14,184	19,523	18,273	14,056
10	22,640	19,760	15,200	20,948	19,006	14,620	20,036	18,564	14,280	19,620	18,356	14,120
12	23,680	20,280	15,600	21,302	19,219	14,784	20,201	18,689	14,376	19,702	18,439	14,184
14	24,720	20,800	16,000	21,643	19,432	14,948	20,351	18,814	14,472	19,768	18,522	14,248
16	25,760	21,320	16,400	21,971	19,646	15,112	20,485	18,938	14,568	19,820	18,607	14,312
18	26,800	21,840	16,800	20,942	19,859	15,276	20,604	19,063	14,664	19,856	18,689	14,376
20	27,840	22,360	17,200	22,588	20,072	15,440	20,708	19,188	14,760	19,876	18,772	14,440
22	28,880	22,880	17,600	22,877	20,285	15,604	20,796	19,312	14,856			
24	29,920	23,400	18,000	23,152	20,498	15,768	20,869	19,438	14,952			
26	30,960	23,920	18,400	23,415	20,712	15,932	20,927	19,562	15,048			
28	32,000	24,440	18,800	23,664	20,925	16,096	20,969	19,687	15,144			
30	33,040	24,960	19,200	23,900	21,138	16,260	20,996	19,812	15,240			
32	34,080	25,480	19,600	24,123	21,351	16,424	21,008	19,937	15,336			
34	35,120	26,000	20,000	24,333	21,564	16,588						
36	36,160	26,520	20,400	24,530	21,778	16,752						
38	37,200	27,040	20,800	24,713	21,991	16,916						
40	38,240	27,560	21,200	24,884	22,204	17,080						
42	39,280	28,080	21,600	25,041	22,417	17,244						
44	40,320	28,600	22,000	25,186	22,630	17,408						
46	41,360	29,120	22,400	25,317	22,844	17,572						
48	42,400	29,640	22,800	25,435	23,057	17,736						
50	43,440	30,160	23,200	25,540	23,270	17,900						

TABLE 50. Volume in Millions of Gallons for 140 Ft Square and 140 Ft Wide Rectangular Reservoirs from 2 to 50 Ft Deep, with Slopes from 0:1 to 3:1

140

Depth	0:1			1:1			2:1			3:1		
(Ft.)	140	+140	+100	140	+140	+100	140	+140	+100	140	+140	+100
2	0.293	0.293	0.209	0.282	0.289	0.206	0.274	0.285	0.203	0.266	0.281	0.200
4	0.586	0.586	0.419	0.556	0.570	0.407	0.525	0.553	0.395	0.494	0.536	0.383
6	0.880	0.880	0.628	0.806	0.842	0.601	0.737	0.804	0.574	0.673	0.767	0.548
8	1.173	1.173	0.838	1.041	1.106	0.790	0.923	1.039	0.742	0.815	0.972	0.694
10	1.466	1.466	1.047	1.269	1.361	0.972	1.089	1.257	0.898	0.929	1.152	0.823
12	1.759	1.759	1.257	1.475	1.608	1.149	1.225	1.458	1.041	1.010	1.307	0.934
14	2.053	2.053	1.466	1.667	1.847	1.319	1.339	1.642	1.173	1.066	1.437	1.026
16	2.346	2.346	1.676	1.853	2.078	1.484	1.439	1.810	1.293	1.106	1.541	1.101
18	2.639	2.639	1.885	2.019	2.300	1.643	1.514	1.960	1.400	1.127	1.621	1.158
20	2.932	2.932	2.094	2.172	2.513	1.795	1.574	2.094	1.496	1.136	1.676	1.197
22	3.225	3.225	2.303	2.320	2.719	1.942	1.624	2.212	1.580	1.141	1.705	1.218
24	3.519	3.519	2.513	2.450	2.915	2.082	1.657	2.312	1.652			
26	3.812	3.812	2.723	2.569	3.104	2.217	1.680	2.396	1.711			
28	4.105	4.105	2.932	2.684	3.284	2.346	1.698	2.463	1.759			
30	4.398	4.398	3.142	2.783	3.456	2.468	1.705	2.513	1.795			
32	4.691	4.691	3.351	2.872	3.619	2.585	1.708	2.547	1.819			
34	4.985	4.985	3.560	2.957	3.774	2.696	1.711	2.564	1.831			
36	5.278	5.278	3.770	3.029	3.921	2.801						
38	5.571	5.571	3.979	3.092	4.059	2.899						
40	5.864	5.864	4.189	3.153	4.189	2.992						
42	6.158	6.158	4.398	3.202	4.310	3.079						
44	6.451	6.451	4.603	3.244	4.423	3.160						
46	6.744	6.744	4.817	3.284	4.528	3.234						
48	7.037	7.037	5.027	3.315	4.624	3.303						
50	7.330	7.330	5.236	3.340	4.712	3.366						

TABLE 51. Lining Area in Square Feet for 140 Ft Square and 140 Ft Wide Rectangular Reservoirs from 2 to 50 Ft Deep, with Slopes from 0:1 to 3:1

P = 560
2P = 1120
3P = 1680
4P = 2240

140

Depth	0:1			1:1			2:1			3:1		
(Ft.)	140	+140	+100	140	+140	+100	140	+140	+100	140	+140	+100
2	21,280	20,440	14,600	22,293	20,950	14,964	22,101	20,854	14,896	22,012	20,810	14,864
4	22,400	21,000	15,000	22,732	21,179	15,128	22,347	20,989	14,992	22,168	20,899	14,928
6	23,520	21,560	15,400	23,159	21,049	15,292	22,577	21,123	15,088	22,308	20,989	14,992
8	24,640	22,120	15,800	23,572	21,638	15,456	22,792	21,258	15,184	22,440	21,064	15,056
10	25,760	22,680	16,200	23,972	21,868	15,620	22,992	21,392	15,280	22,544	21,168	15,120
12	26,880	23,240	16,600	24,359	22,098	15,784	23,176	21,526	15,376	22,639	21,258	15,184
14	28,000	23,800	17,000	24,733	22,327	15,948	23,345	21,661	15,472	22,718	21,347	15,248
16	29,120	24,360	17,400	25,094	22,557	16,112	23,499	21,795	15,568	22,782	21,437	15,312
18	30,240	24,920	17,800	25,441	22,786	16,276	23,637	21,930	15,664	22,831	21,526	15,376
20	31,360	25,480	18,200	25,776	23,016	16,440	23,760	22,064	15,760	22,864	21,616	15,440
22	32,480	26,040	18,600	26,097	23,246	16,604	23,868	22,198	15,856	22,586	21,706	15,504
24	33,600	26,600	19,000	26,406	23,475	16,768	23,960	22,333	15,952			
26	34,720	27,160	19,400	26,701	23,705	16,932	24,036	22,467	16,048			
28	35,840	27,720	19,800	26,983	23,934	17,096	24,098	22,602	16,144			
30	36,960	28,280	20,200	27,252	24,164	17,260	24,144	22,736	16,240			
32	38,080	28,840	20,600	27,508	24,394	17,424	24,175	22,870	16,336			
34	39,200	29,400	21,000	27,751	24,623	17,588	24,190	23,005	16,432			
36	40,320	29,960	21,400	27,980	24,583	17,752						
38	41,440	30,520	21,800	28,197	25,083	17,916						
40	42,560	31,000	22,200	28,400	25,312	18,080						
42	43,680	31,640	22,600	28,590	25,542	18,244						
44	44,800	32,200	23,000	28,767	25,771	18,408						
46	45,920	32,760	23,400	28,931	26,001	18,572						
48	47,040	33,320	23,800	29,082	26,230	18,736						
50	48,160	33,880	24,200	29,220	26,460	18,900						

TABLE 52. Volume in Millions of Gallons for 150 Ft Square and 150 Ft Wide Rectangular Reservoirs from 2 to 50 Ft Deep, with Slopes from 0:1 to 3:1

150

Depth	0:1			1:1			2:1			3:1		
(Ft.)	150	+150	+100	150	+150	+100	150	+150	+100	150	+150	+100
2	0.337	0.337	0.224	0.328	0.332	0.221	0.319	0.328	0.218	0.310	0.323	0.215
4	0.673	0.673	0.449	0.638	0.655	0.437	0.604	0.637	0.425	0.571	0.619	0.412
6	1.010	1.010	0.673	0.931	0.969	0.646	0.857	0.929	0.619	0.787	0.889	0.592
8	1.346	1.346	0.898	1.208	1.275	0.850	1.080	1.203	0.802	0.961	1.131	0.754
10	1.683	1.683	1.122	1.469	1.571	1.047	1.274	1.459	0.972	1.100	1.346	0.898
12	2.020	2.020	1.346	1.714	1.858	1.239	1.442	1.696	1.131	1.205	1.535	1.023
14	2.356	2.356	1.571	1.944	2.136	1.424	1.586	1.916	1.278	1.283	1.696	1.131
16	2.693	2.693	1.795	2.160	2.406	1.604	1.707	2.118	1.412	1.337	1.831	1.221
18	3.029	3.029	2.020	2.361	2.666	1.777	1.808	2.302	1.535	1.372	1.939	1.293
20	3.366	3.366	2.244	2.548	2.917	1.945	1.890	2.468	1.646	1.391	2.020	1.346
22	3.703	3.703	2.468	2.723	3.160	2.106	1.955	2.617	1.744	1.400	2.073	1.382
24	4.039	4.039	2.693	2.885	3.393	2.262	2.006	2.747	1.831	1.402	2.100	1.400
26	4.376	4.376	2.917	3.034	3.617	2.412	2.043	2.859	1.906			
28	4.712	4.712	3.142	3.172	3.833	2.555	2.070	2.953	1.969			
30	5.049	5.049	3.366	3.299	4.039	2.693	2.087	3.029	2.020			
32	5.386	5.386	3.590	3.415	4.237	2.824	2.097	3.088	2.058			
34	5.722	5.722	3.815	3.520	4.425	2.950	2.102	3.128	2.085			
36	6.059	6.059	4.039	3.616	4.605	3.070	2.104	3.151	2.100			
38	6.395	6.395	4.264	3.702	4.775	3.183						
40	6.732	6.732	4.488	3.780	4.937	3.291						
42	7.069	7.069	4.712	3.849	5.089	3.393						
44	7.405	7.405	4.937	3.910	5.233	3.489						
46	7.742	7.742	5.161	3.964	5.368	3.578						
48	8.078	8.078	5.386	4.011	5.493	3.662						
50	8.415	8.415	5.610	4.052	5.610	3.740						

TABLE 53. Lining Area in Square Feet for 150 Ft Square and 150 Ft Wide Rectangular Reservoirs from 2 to 50 Ft Deep, with Slopes from 0:1 to 3:1

P = 600
2P = 1200
3P = 1800
4P = 2400

150

Depth	0:1			1:1			2:1			3:1		
(Ft.)	150	+150	+100	150	+150	+100	150	+150	+100	150	+150	+100
2	24,300	23,400	15,600	25,385	23,946	15,964	25,180	23,844	15,896	25,084	23,796	15,864
4	25,500	24,000	16,000	25,858	24,192	16,128	25,445	23,988	15,992	25,253	23,892	15,928
6	26,700	24,600	16,400	26,317	24,438	16,292	25,695	24,132	16,088	25,407	23,988	15,992
8	27,900	25,200	16,800	26,763	24,684	16,456	25,929	24,276	16,184	25,545	24,084	16,056
10	29,100	25,800	17,200	27,196	24,930	16,620	26,148	24,420	16,280	25,668	24,180	16,120
12	30,300	26,400	17,600	27,616	25,176	16,784	26,352	24,564	16,376	25,776	24,276	16,184
14	31,500	27,000	18,000	28,023	25,422	16,948	26,540	24,708	16,472	25,868	24,372	16,248
16	32,700	27,600	18,400	28,416	25,668	17,112	26,712	24,852	16,568	25,944	24,468	16,312
18	33,900	28,200	18,800	28,797	25,914	17,276	26,870	24,996	16,664	26,006	24,564	16,376
20	35,100	28,800	19,200	29,164	26,160	17,440	27,012	25,140	16,760	26,052	24,660	16,440
22	36,300	29,400	19,600	29,518	26,406	17,604	27,139	25,284	16,856	26,083	24,756	16,504
24	37,500	30,000	20,000	29,859	26,652	17,768	27,250	25,428	16,952	26,098	24,852	16,568
26	38,700	30,600	20,400	30,187	26,898	17,932	27,346	25,572	17,048			
28	39,900	31,200	20,800	30,502	27,144	18,096	27,427	25,716	17,144			
30	41,100	31,800	21,200	30,804	27,390	18,260	27,492	25,860	17,240			
32	43,200	32,400	21,600	31,093	27,636	18,424	27,542	26,004	17,336			
34	43,500	33,000	22,000	31,368	27,882	18,588	27,576	26,148	17,432			
36	44,700	33,600	22,400	31,631	28,128	18,752	27,596	26,292	17,528			
38	45,900	34,200	22,800	31,880	28,374	18,916						
40	47,100	34,800	23,200	32,116	28,620	19,080						
42	48,300	35,400	23,600	32,339	28,866	19,244						
44	49,500	36,000	24,000	32,549	29,112	19,408						
46	50,700	36,600	24,400	32,746	29,358	19,572						
48	51,900	37,200	24,800	32,929	29,604	19,736						
50	53,100	37,800	25,200	33,100	29,850	19,900						

TABLE 54. Volume in Millions of Gallons for 160 Ft Square and 160 Ft Wide Rectangular Reservoirs from 2 to 50 Ft Deep, with Slopes from 0:1 to 3:1

160

Depth (Ft.)	0:1			1:1			2:1			3:1		
	160	+160	+100	160	+160	+100	160	+160	+100	160	+160	+100
2	0.383	0.383	0.239	0.370	0.378	0.236	0.361	0.373	0.234	0.351	0.369	0.230
4	0.766	0.766	0.479	0.732	0.747	0.467	0.695	0.728	0.455	0.660	0.709	0.443
6	1.149	1.149	0.718	1.065	1.106	0.691	0.985	1.063	0.664	0.910	1.020	0.637
8	1.532	1.532	0.957	1.380	1.455	0.910	1.243	1.379	0.862	1.115	1.302	0.814
10	1.915	1.915	1.197	1.689	1.795	1.122	1.479	1.676	1.047	1.289	1.556	0.972
12	2.298	2.298	1.436	1.970	2.126	1.328	1.677	1.953	1.221	1.419	1.781	1.113
14	2.681	2.681	1.676	2.236	2.446	1.529	1.849	2.212	1.382	1.518	1.977	1.236
16	3.064	3.064	1.915	2.496	2.757	1.723	2.004	2.451	1.532	1.595	2.145	1.340
18	3.447	3.447	2.154	2.729	3.059	1.912	2.128	2.671	1.670	1.644	2.283	1.427
20	3.830	3.830	2.394	2.949	3.351	2.094	2.232	2.872	1.795	1.674	2.394	1.496
22	4.213	4.213	2.633	3.163	3.633	2.271	2.323	3.054	1.909	1.695	2.475	1.547
24	4.596	4.596	2.872	3.355	3.906	2.441	2.390	3.217	2.011	1.700	2.528	1.580
26	4.979	4.979	3.112	3.533	4.170	2.606	2.442	3.361	2.100	1.701	2.552	1.595
28	5.362	5.362	3.351	3.707	4.423	2.765	2.486	3.485	2.178			
30	5.745	5.745	3.590	3.860	4.668	2.917	2.513	3.590	2.244			
32	6.128	6.128	3.830	4.001	4.902	3.064	2.531	3.677	2.298			
34	6.512	6.512	4.069	4.138	5.127	3.204	2.546	3.744	2.340			
36	6.894	6.894	4.308	4.257	5.343	3.339	2.551	3.791	2.370			
38	7.277	7.277	4.548	4.365	5.548	3.468	2.552	3.820	2.388			
40	7.660	7.660	4.787	4.470	5.745	3.590						
42	8.042	8.042	5.027	4.559	5.931	3.707						
44	8.425	8.425	5.266	4.639	6.108	3.818						
46	8.808	8.808	5.505	4.716	6.276	3.923						
48	9.191	9.191	5.745	4.780	6.434	4.021						
50	9.574	9.574	5.984	4.835	6.582	4.114						

TABLE 55. Lining Area in Square Feet for 160 Ft Square and 160 Ft Wide Rectangular Reservoirs from 2 to 50 Ft Deep, with Slopes from 0:1 to 3:1

P = 640
2P = 1280
3P = 1920
4P = 2560

160

Depth (Ft.)	0:1			1:1			2:1			3:1		
	160	+160	+100	160	+160	+100	160	+160	+100	160	+160	+100
2	27,520	26,560	16,600	28,678	27,142	16,964	28,460	27,034	16,896	28,357	26,982	16,864
4	28,800	27,200	17,000	29,183	27,405	17,128	28,744	27,187	16,992	28,539	27,085	16,928
6	30,080	27,840	17,400	29,675	27,667	17,292	29,012	27,341	17,088	28,705	27,187	16,992
8	31,360	28,480	17,800	30,154	27,930	17,456	29,266	27,494	17,184	28,856	27,290	17,056
10	32,640	29,120	18,200	30,620	28,192	17,620	29,504	27,648	17,280	28,992	27,392	17,120
12	33,920	29,760	18,600	31,073	28,454	17,784	29,727	27,802	17,376	29,112	27,494	17,184
14	35,200	30,400	19,000	31,512	28,717	17,948	29,934	27,955	17,472	29,217	27,597	17,248
16	36,480	31,040	19,400	31,939	28,979	18,112	30,126	28,109	17,568	29,307	27,699	17,312
18	37,760	31,680	19,800	32,352	29,242	18,276	30,303	28,262	17,664	29,381	27,802	17,376
20	39,040	32,320	20,200	32,752	29,504	18,440	30,464	28,416	17,760	29,440	27,904	17,440
22	40,320	32,960	20,600	33,139	29,766	18,604	30,610	28,570	17,856	29,484	28,006	17,504
24	41,600	33,600	21,000	33,512	30,029	18,768	30,740	28,723	17,952	29,512	28,109	17,568
26	42,880	34,240	21,400	33,874	30,291	18,932	30,856	28,877	18,048	29,525	28,211	17,632
28	44,160	34,880	21,800	34,221	30,554	19,096	30,956	29,030	18,144			
30	45,440	35,520	22,200	34,556	30,816	19,260	31,040	29,184	18,240			
32	46,720	36,160	22,600	34,877	31,078	19,424	31,109	29,338	18,336			
34	48,000	36,800	23,000	35,186	31,341	19,588	31,163	29,491	18,432			
36	49,280	37,440	23,400	35,481	31,603	19,752	31,201	29,645	18,528			
38	50,560	38,080	23,800	35,763	31,866	19,916	31,224	29,798	18,624			
40	51,840	38,720	24,200	36,032	32,128	20,080						
42	53,120	39,360	24,600	36,288	32,390	20,244						
44	54,400	40,000	25,000	36,531	32,653	20,408						
46	55,680	40,640	25,400	36,760	32,915	20,572						
48	56,960	41,280	25,800	36,977	33,178	20,736						
50	58,240	41,920	26,200	37,180	33,440	20,900						

TABLE 56. Volume in Millions of Gallons for 170 Ft Square and 170 Ft Wide Rectangular Reservoirs from 2 to 50 Ft Deep, with Slopes from 0:1 to 3:1

170

Depth (Ft.)	0:1			1:1			2:1			3:1		
	170	+170	+100	170	+170	+100	170	+170	+100	170	+170	+100
2	0.432	0.432	0.254	0.422	0.427	0.251	0.412	0.422	0.248	0.402	0.417	0.245
4	0.865	0.865	0.509	0.825	0.844	0.497	0.786	0.824	0.485	0.748	0.804	0.473
6	1.297	1.297	0.763	1.208	1.251	0.736	1.123	1.205	0.709	1.042	1.160	0.682
8	1.729	1.729	1.017	1.572	1.648	0.969	1.424	1.567	0.922	1.287	1.485	0.874
10	2.162	2.162	1.272	1.917	2.035	1.197	1.693	1.907	1.122	1.489	1.780	1.047
12	2.594	2.594	1.526	2.245	2.411	1.418	1.931	2.228	1.310	1.651	2.045	1.203
14	3.026	3.026	1.780	2.555	2.777	1.634	2.139	2.528	1.487	1.777	2.279	1.340
16	3.459	3.459	2.035	2.849	3.133	1.843	2.320	2.808	1.652	1.873	2.482	1.460
18	3.891	3.891	2.289	3.125	3.479	2.047	2.476	3.067	1.804	1.943	2.655	1.562
20	4.323	4.323	2.543	3.386	3.815	2.244	2.608	3.306	1.945	1.990	2.798	1.646
22	4.756	4.756	2.798	3.631	4.140	2.435	2.719	3.525	2.073	2.019	2.909	1.711
24	5.188	5.188	3.052	3.861	4.456	2.621	2.810	3.723	2.190	2.034	2.991	1.759
26	5.620	5.620	3.306	4.077	4.761	2.801	2.883	3.901	2.295	2.040	3.042	1.789
28	6.053	6.053	3.560	4.278	5.056	2.974	2.941	4.059	2.388	2.042	3.062	1.801
30	6.485	6.485	3.815	4.466	5.341	3.142	2.985	4.196	2.468			
32	6.918	6.918	4.069	4.640	5.615	3.303	3.016	4.313	2.537			
34	7.350	7.350	4.323	4.802	5.880	3.458	3.038	4.410	2.594			
36	7.782	7.782	4.578	4.952	6.134	3.608	3.051	4.486	2.639			
38	8.215	8.215	4.832	5.089	6.378	3.752	3.059	4.542	2.672			
40	8.647	8.647	5.086	5.216	6.612	3.890	3.062	4.578	2.693			
42	9.079	9.079	5.341	5.332	6.836	4.021	3.062	4.593	2.702			
44	9.512	9.512	5.595	5.437	7.050	4.147						
46	9.944	9.944	5.849	5.533	7.253	4.267						
48	10.38	10.38	6.104	5.620	7.446	4.380						
50	10.81	10.81	6.358	5.697	7.630	4.488						

TABLE 57. Lining Area in Square Feet for 170 Ft Square and 170 Ft Wide Rectangular Reservoirs from 2 to 50 Ft Deep, with Slopes from 0:1 to 3:1

P = 680
2P = 1360
3P = 2040
4P = 2720

170

Depth (Ft.)	0:1			1:1			2:1			3:1		
	170	+170	+100	170	+170	+100	170	+170	+100	170	+170	+100
2	30,940	29,920	17,600	32,171	30,539	17,964	31,939	30,423	17,964	31,830	30,369	17,864
4	32,300	30,600	18,000	32,709	30,818	18,128	32,242	30,586	17,992	32,024	30,478	17,928
6	33,660	31,280	18,400	33,234	31,096	18,292	32,530	30,750	18,088	32,204	30,586	17,992
8	35,020	31,960	18,800	33,745	31,375	18,456	32,803	30,913	18,184	32,368	30,695	18,056
10	36,380	32,640	19,200	34,244	31,654	18,620	33,060	31,076	18,280	32,516	30,804	18,120
12	37,740	33,320	19,600	34,729	31,933	18,784	33,302	31,239	18,376	32,649	30,913	18,184
14	39,100	34,000	20,000	35,202	32,212	18,948	33,528	31,402	18,472	32,767	31,022	18,248
16	40,460	34,680	20,400	35,661	32,490	19,112	33,740	31,566	18,568	32,869	31,130	18,312
18	41,820	35,360	20,800	36,107	32,769	19,276	33,936	31,729	18,664	32,856	31,239	18,376
20	43,180	36,040	21,200	36,540	32,048	19,440	34,116	31,892	18,760	33,028	31,348	18,440
22	44,540	36,720	21,600	36,960	33,327	19,604	34,281	32,055	18,856	33,084	31,457	18,504
24	45,900	37,400	22,000	37,367	33,606	19,768	34,431	32,218	18,952	33,125	31,566	18,568
26	47,260	38,080	22,400	37,760	33,884	19,932	34,565	32,382	19,048	33,151	31,674	18,632
28	48,620	38,760	22,800	38,141	34,163	20,096	34,684	32,545	19,144	33,161	31,783	18,696
30	49,980	39,440	23,200	38,508	34,442	20,260	34,788	32,708	19,240			
32	51,340	40,120	23,600	38,862	34,721	20,424	34,876	32,871	19,336			
34	52,700	40,800	24,000	39,203	35,000	20,588	34,949	33,034	19,432			
36	54,060	41,480	24,400	39,531	35,278	20,752	35,007	33,198	19,528			
38	55,420	42,160	24,800	39,846	35,557	20,916	35,049	33,361	19,624			
40	56,780	42,840	25,200	40,148	35,836	21,080	35,076	33,524	19,720			
42	58,140	43,520	25,600	40,437	36,115	21,244	35,088	33,687	19,816			
44	59,500	44,200	26,000	40,712	36,384	21,408						
46	60,860	44,880	26,400	40,975	36,672	21,572						
48	62,220	45,560	26,800	41,224	36,951	21,736						
50	63,580	46,240	27,200	41,270	37,230	21,900						

TABLE 58. Volume in Millions of Gallons for 180 Ft Square and 180 Ft Wide Rectangular Reservoirs from 2 to 50 Ft Deep, with Slopes from 0:1 to 3:1

180

Depth (Ft.)	0:1 180	0:1 +180	0:1 +100	1:1 180	1:1 +180	1:1 +100	2:1 180	2:1 +180	2:1 +100	3:1 180	3:1 +180	3:1 +100
2	0.484	0.484	0.26	0.469	0.479	0.266	0.459	0.474	0.263	0.449	0.469	0.260
4	0.969	0.969	0.539	0.932	0.948	0.527	0.890	0.926	0.515	0.850	0.905	0.503
6	1.454	1.454	0.808	1.359	1.406	0.781	1.269	1.357	0.754	1.183	1.309	0.727
8	1.939	1.939	1.077	1.767	1.853	1.029	1.611	1.766	0.981	1.464	1.680	0.934
10	2.424	2.424	1.346	2.169	2.289	1.272	1.929	2.154	1.197	1.709	2.020	1.122
12	2.908	2.908	1.616	2.538	2.714	1.508	2.202	2.520	1.400	1.900	2.327	1.293
14	3.393	3.393	1.885	2.888	3.129	1.738	2.443	2.866	1.592	2.052	2.601	1.445
16	3.878	3.878	2.154	3.233	3.533	1.963	2.666	3.188	1.771	2.180	2.844	1.580
18	4.362	4.362	2.424	3.548	3.926	2.181	2.850	3.490	1.939	2.268	3.054	1.696
20	4.847	4.847	2.693	3.846	4.308	2.394	3.009	3.770	2.094	2.331	3.231	1.795
22	5.332	5.332	2.962	4.138	4.680	2.600	3.153	4.028	2.238	2.380	3.377	1.876
24	5.816	5.816	3.231	4.403	5.041	2.801	3.266	4.265	2.370	2.404	3.490	1.939
26	6.301	6.301	3.501	4.653	5.391	2.995	3.359	4.481	2.489	2.416	3.571	1.984
28	6.786	6.786	3.770	4.897	5.730	3.183	3.442	4.675	2.597	2.425	3.619	2.011
30	7.271	7.271	4.039	5.116	6.059	3.366	3.501	4.847	2.693			
32	7.755	7.755	4.308	5.321	6.377	3.543	3.545	4.998	2.777			
34	8.240	8.240	4.578	5.522	6.684	3.713	3.584	5.127	2.848			
36	8.725	8.725	4.847	5.700	6.980	3.878	3.606	5.235	2.908			
38	9.209	9.209	5.116	5.865	7.265	4.036	3.620	5.321	2.956			
40	9.694	9.694	5.386	6.027	7.540	4.189	3.632	5.386	2.992			
42	10.18	10.18	5.655	6.168	7.804	4.335	3.634	5.429	3.016			
44	10.66	10.66	5.924	6.297	8.057	4.476	3.634	5.450	3.028			
46	11.15	11.15	6.193	6.424	8.299	4.611						
48	11.63	11.63	6.463	6.532	8.531	4.739						
50	12.12	12.12	6.732	6.630	8.752	4.862						

TABLE 59. Lining Area in Square Feet for 180 Ft Square and 180 Ft Wide Rectangular Reservoirs from 2 to 50 Ft Deep, with Slopes from 0:1 to 3:1

P = 720
2P = 1440
3P = 2160
4P = 2880

180

Depth (Ft.)	0:1 180	0:1 +180	0:1 +180	1:1 180	1:1 +180	1:1 +100	2:1 180	2:1 +180	2:1 +100	3:1 180	3:1 +180	3:1 +100
2	34,560	33,480	18,600	35,864	34,135	18,964	35,618	34,013	18,896	35,503	33,955	18,864
4	36,000	34,200	19,000	36,435	34,430	19,128	35,940	34,186	18,992	35,710	34,070	18,928
6	37,440	34,920	19,400	36,992	34,726	19,292	36,248	34,358	19,088	35,902	34,186	18,992
8	38,880	35,640	19,800	37,457	35,021	19,456	36,540	34,531	19,184	36,079	34,301	19,056
10	40,320	36,360	20,200	38,068	35,316	19,620	36,816	34,704	19,280	36,240	34,416	19,120
12	41,760	37,080	20,600	38,586	35,611	19,784	37,077	34,877	19,376	36,386	34,531	19,184
14	43,200	37,800	21,000	39,091	35,906	19,948	37,323	35,050	19,248	36,516	34,646	19,248
16	44,640	38,520	21,400	39,583	36,202	20,112	37,553	35,222	19,568	36,632	34,762	19,312
18	46,080	39,240	21,800	40,062	36,497	20,276	37,768	35,395	19,664	36,732	34,877	19,376
20	47,520	39,960	22,200	40,528	36,792	20,440	37,968	35,568	19,760	36,816	34,992	19,440
22	48,960	40,680	22,600	40,981	37,087	20,604	38,152	35,741	19,856	36,885	35,107	19,504
24	50,400	41,400	23,000	41,420	37,382	20,768	38,321	35,914	19,952	36,939	35,222	19,568
26	51,840	42,120	23,400	41,847	37,678	20,932	38,475	36,086	20,048	36,977	35,338	19,632
28	53,280	42,840	23,800	42,260	37,973	21,096	38,613	36,259	20,144	37,000	35,453	19,696
30	54,720	43,560	24,200	42,664	38,268	21,260	38,736	36,432	20,240			
32	56,160	44,280	24,600	43,047	38,563	21,424	38,844	36,605	20,336			
34	57,600	45,000	25,000	43,421	38,858	21,588	38,936	36,778	20,432			
36	59,040	45,720	25,400	43,782	39,154	21,752	39,012	36,950	20,528			
38	60,480	46,440	25,800	44,109	39,449	21,916	39,074	37,123	20,624			
40	61,920	47,160	26,200	44,464	39,744	22,080	39,120	37,296	20,720			
42	63,360	47,880	26,600	44,785	40,039	22,244	39,151	37,469	20,816			
44	64,800	48,600	27,000	45,094	40,334	22,408	39,166	37,642	20,912			
46	66,240	49,320	27,400	45,389	40,630	22,572						
48	67,680	50,040	27,800	45,671	40,925	22,736						
50	69,120	50,760	28,200	45,940	41,220	22,900						

TABLE 60. Volume in Millions of Gallons for 190 Ft Square and 190 Ft Wide Rectangular Reservoirs from 2 to 50 Ft Deep, with Slopes from 0:1 to 3:1

190

Depth (Ft.)	0:1			1:1			2:1			3:1		
	190	+190	+100	190	+190	+100	190	+190	+100	190	+190	+100
2	0.540	0.540	0.284	0.529	0.534	0.281	0.518	0.529	0.278	0.507	0.523	0.275
4	1.080	1.080	0.568	1.035	1.057	0.557	0.992	1.035	0.545	0.949	1.012	0.533
6	1.620	1.620	0.852	1.520	1.569	0.826	1.424	1.518	0.799	1.333	1.467	0.772
8	2.160	2.160	1.137	1.983	2.069	1.089	1.817	1.978	1.041	1.660	1.887	0.993
10	2.700	2.700	1.421	2.426	2.558	1.346	2.172	2.416	1.272	1.937	2.274	1.197
12	3.240	3.240	1.705	2.848	3.036	1.598	2.491	2.831	1.490	2.168	2.626	1.382
14	3.780	3.780	1.990	3.251	3.502	1.843	2.776	3.223	1.696	2.355	2.945	1.550
16	4.320	4.320	2.274	3.634	3.957	2.082	3.029	3.593	1.891	2.505	3.229	1.699
18	4.861	4.861	2.558	3.998	4.400	2.316	3.251	3.940	2.073	2.621	3.480	1.831
20	5.401	5.401	2.842	4.343	4.832	2.543	3.446	4.264	2.244	2.708	3.695	1.945
22	5.941	5.941	3.127	4.671	5.253	2.765	3.614	4.565	2.403	2.769	3.877	2.041
24	6.481	6.481	3.411	4.981	5.662	2.980	3.758	4.843	2.549	2.810	4.025	2.118
26	7.021	7.021	3.695	5.275	6.060	3.189	3.879	5.099	2.684	2.834	4.139	2.178
28	7.561	7.561	3.979	5.551	6.447	3.393	3.980	5.332	2.806	2.846	4.218	2.220
30	8.101	8.101	4.264	5.812	6.822	3.590	4.062	5.543	2.917	2.850	4.264	2.244
32	8.641	8.641	4.548	6.057	7.168	3.782	4.127	5.730	3.016			
34	9.181	9.181	4.832	6.287	7.538	3.967	4.177	5.895	3.103			
36	9.721	9.721	5.116	6.503	7.879	4.147	4.215	6.037	3.178			
38	10.26	10.26	5.401	6.704	8.209	4.320	4.241	6.157	3.240			
40	10.80	10.80	5.685	6.892	8.527	4.488	4.259	6.253	3.291			
42	11.34	11.34	5.969	7.066	8.834	4.650	4.269	6.327	3.330			
44	11.88	11.88	6.253	7.228	9.130	4.805	4.274	6.378	3.357			
46	12.42	12.42	6.538	7.378	9.414	4.955	4.275	6.407	3.372			
48	12.96	12.96	6.822	7.515	9.687	5.098						
50	13.50	13.50	7.106	7.642	9.948	5.236						

TABLE 61. Lining Area in Square Feet for 190 Ft Square and 190 Ft Wide Rectangular Reservoirs from 2 to 50 Ft Deep, with Slopes from 0:1 to 3:1

P = 760
2P = 1520
3P = 2280
4P = 3040

190

Depth (Ft.)	0:1			1:1			2:1			3:1		
	190	+190	+100	190	+190	+100	190	+190	+100	190	+190	+100
2	38,380	37,240	19,600	39,757	37,932	19,964	39,497	37,802	19,896	39,376	37,742	19,864
4	39,900	38,000	20,000	40,360	38,243	20,128	39,839	37,985	19,992	39,596	37,863	19,928
6	41,420	38,760	20,400	40,951	38,555	20,292	40,165	38,167	20,088	39,800	37,985	19,992
8	42,940	39,520	20,800	41,528	38,866	20,456	40,476	38,350	20,184	39,990	38,106	20,056
10	44,460	40,280	21,200	42,092	39,178	20,620	40,772	38,532	20,280	40,164	38,228	20,120
12	45,980	41,040	21,600	42,643	39,490	20,784	41,052	38,714	20,376	40,323	38,350	20,184
14	47,500	41,800	22,000	43,181	39,801	20,948	41,317	38,897	20,472	40,466	38,471	20,248
16	49,020	42,560	22,400	43,706	40,113	21,112	41,567	39,079	20,568	40,594	38,593	20,312
18	55,540	43,320	22,800	44,217	40,424	21,276	41,801	39,262	20,664	40,707	38,714	20,376
20	52,060	44,080	23,200	44,716	40,736	21,440	42,020	39,444	20,760	40,804	38,836	20,440
22	53,580	44,840	23,600	45,201	41,048	21,604	42,224	39,626	20,856	40,886	38,958	20,504
24	55,100	45,600	24,000	45,674	41,359	21,768	42,412	39,809	20,952	40,952	39,064	20,560
26	56,620	46,360	24,400	46,133	41,671	21,932	42,584	39,991	21,048	41,004	39,201	20,632
28	58,140	47,120	24,800	46,579	41,982	22,096	42,742	40,174	21,144	41,040	39,322	20,696
30	59,660	47,880	25,200	47,012	42,294	22,260	42,884	40,356	21,240	41,060	39,444	20,760
32	61,180	48,640	25,600	47,432	42,606	22,424	43,011	40,538	21,336			
34	62,700	49,400	26,000	47,839	42,917	22,588	43,122	40,721	21,432			
36	64,220	50,160	26,400	48,232	43,229	22,752	43,218	40,903	21,528			
38	65,740	50,920	26,800	48,613	43,540	22,916	43,299	41,086	21,624			
40	67,260	51,680	27,200	48,980	43,952	23,080	43,364	41,268	21,720			
42	68,780	52,440	27,600	49,334	44,126	23,224	43,414	41,450	21,816			
44	70,300	53,200	28,000	49,675	44,475	23,408	43,448	41,633	21,912			
46	71,820	53,960	28,400	50,003	44,787	23,572	43,468	41,815	22,008			
48	73,340	54,720	28,800	50,318	45,098	23,736						
50	74,860	55,480	29,200	50,620	45,410	23,900						

TABLE 62. Volume in Millions of Gallons for 200 Ft Square and 200 Ft Wide Rectangular Reservoirs from 2 to 50 Ft Deep, with Slopes from 0:1 to 3:1

Depth (Ft.)	0:1			1:1			2:1			3:1		
	200	+200	+100	200	+200	+100	200	+200	+100	200	+200	+100
2	0.598	0.598	0.299	0.580	0.592	0.296	0.569	0.586	0.293	0.557	0.580	0.290
4	1.197	1.197	0.598	1.155	1.173	0.586	1.109	1.149	0.574	1.064	1.125	0.562
6	1.795	1.795	0.898	1.690	1.741	0.871	1.588	1.687	0.844	1.491	1.634	0.817
8	2.394	2.394	1.197	2.202	2.298	1.149	2.026	2.202	1.101	1.860	2.106	1.053
10	2.992	2.992	1.496	2.708	2.842	1.421	2.434	2.693	1.346	2.189	2.543	1.272
12	3.590	3.590	1.795	3.177	3.375	1.687	2.798	3.160	1.580	2.453	2.944	1.472
14	4.189	4.189	2.094	3.625	3.896	1.948	3.125	3.602	1.801	2.672	3.309	1.655
16	4.787	4.787	2.394	4.067	4.404	2.202	3.423	4.021	2.011	2.861	3.638	1.819
18	5.386	5.386	2.693	4.474	4.901	2.450	3.679	4.416	2.208	3.001	3.931	1.966
20	5.984	5.984	2.992	4.862	5.386	2.693	3.906	4.787	2.394	3.109	4.189	2.094
22	6.582	6.582	3.291	5.245	5.858	2.929	4.115	5.134	2.567	3.197	4.410	2.205
24	7.181	7.181	3.590	5.595	6.319	3.160	4.286	5.457	2.729	3.251	4.596	2.298
26	7.779	7.779	3.890	5.927	6.768	3.384	4.432	7.757	2.878	3.287	4.745	2.373
28	8.378	8.378	4.189	6.255	7.205	3.602	4.565	6.032	3.016	3.313	4.859	2.430
30	8.796	8.796	4.488	6.552	7.630	3.815	4.668	6.283	3.142	3.321	4.937	2.468
32	9.574	9.574	4.787	6.833	8.042	4.021	4.751	6.511	3.255	3.322	4.979	2.489
34	10.17	10.17	5.086	7.110	8.443	4.222	4.826	6.714	3.357			
36	10.77	10.77	4.386	7.359	8.832	4.416	4.877	6.894	3.447			
38	11.37	11.37	5.684	7.592	9.209	4.605	4.915	7.049	3.525			
40	11.97	11.97	5.984	7.823	9.574	4.787	4.949	7.181	3.590			
42	12.57	12.57	6.283	8.027	9.927	4.964	4.966	7.288	3.644			
44	13.17	13.17	6.582	8.218	10.27	5.134	4.976	7.372	3.686			
46	13.76	13.76	6.882	8.407	10.60	5.299	4.986	7.432	3.716			
48	14.36	14.36	7.181	8.571	10.92	5.457	4.986	7.468	3.734			
50	14.96	14.96	7.480	8.723	11.22	5.610						

TABLE 63. Lining Area in Square Feet for 200 Ft Square and 200 Ft Wide Rectangular Reservoirs from 2 to 50 Ft Deep, with Slopes from 0:1 to 3:1

P = 800
2P = 1600
3P = 2400
4P = 3200

200

Depth (Ft.)	0:1			1:1			2:1			3:1		
	200	+200	+100	200	+200	+100	200	+200	+100	200	+200	+100
2	42,400	41,200	20,600	43,849	41,928	20,964	43,576	41,792	20,896	43,438	41,728	20,864
4	44,000	42,000	21,000	44,486	42,256	21,128	43,937	41,984	20,992	43,681	41,856	20,928
6	45,600	42,800	21,400	45,099	42,584	21,292	44,283	42,176	21,088	43,899	41,984	20,992
8	47,200	43,600	21,800	45,719	42,912	21,456	44,613	42,368	21,184	44,101	42,104	21,052
10	48,800	44,400	22,200	46,316	43,240	21,620	44,928	42,560	21,280	44,288	42,240	21,120
12	50,400	45,200	22,600	46,900	43,568	21,784	45,228	42,752	21,376	44,460	42,368	21,184
14	52,000	46,000	23,000	47,471	43,896	21,948	45,512	42,944	21,472	44,616	42,496	21,248
16	53,600	46,800	23,400	48,026	44,224	22,112	45,780	43,136	21,568	44,756	42,624	21,312
18	55,200	47,600	23,800	48,573	44,552	22,276	46,034	43,328	21,664	44,882	42,752	21,376
20	56,800	48,400	24,200	49,104	44,880	22,440	46,272	43,520	21,760	44,992	42,880	21,440
22	58,400	49,200	24,600	49,662	45,208	22,604	46,495	43,712	21,856	45,087	43,008	21,504
24	60,000	50,000	25,000	50,127	45,536	22,768	46,702	43,904	21,952	45,166	43,136	21,568
26	61,600	50,800	25,400	50,619	45,864	22,932	46,984	44,096	22,048	45,230	43,264	21,632
28	63,200	51,600	25,800	51,098	46,192	23,096	47,071	44,288	22,144	45,279	43,392	21,696
30	64,800	52,400	26,200	51,564	46,520	23,260	47,232	44,480	22,240	45,312	43,520	21,760
32	66,400	53,200	26,600	52,017	46,848	23,424	47,378	44,672	22,336	45,329	43,648	21,824
34	68,000	54,000	27,000	52,456	47,176	23,588	47,508	44,864	22,432			
36	69,600	54,800	27,400	52,883	47,504	23,752	47,624	45,056	22,528			
38	71,200	55,600	27,800	53,296	47,832	23,916	47,724	45,248	22,624			
40	72,800	56,400	28,200	53,696	48,160	24,080	47,808	45,440	22,720			
42	74,400	57,200	28,600	54,083	48,488	24,244	47,877	45,632	22,816			
44	76,000	58,000	29,000	54,457	48,816	24,408	47,931	45,824	22,912			
46	77,600	58,800	29,400	54,818	49,144	24,572	47,969	46,016	23,008			
48	79,200	59,600	29,800	55,165	49,476	24,738	47,992	46,208	23,104			
50	80,800	60,400	30,200	55,500	49,800	24,900						

TABLE 64. Volume in Millions of Gallons for 220 Ft Square and 220 Ft Wide Rectangular Reservoirs from 2 to 50 Ft Deep, with Slopes from 0:1 to 3:1

220

Depth (Ft.)	0:1			1:1			2:1			3:1		
	220	+220	+100	220	+220	+100	220	+220	+100	220	+220	+100
2	0.724	0.724	0.329	0.704	0.717	0.326	0.691	0.711	0.323	0.678	0.704	0.320
4	1.448	1.448	0.658	1.403	1.422	0.646	1.352	1.395	0.634	1.302	1.369	0.622
6	2.172	2.172	0.987	2.056	2.113	0.960	1.944	2.054	0.934	1.829	1.994	0.907
8	2.896	2.896	1.316	2.684	2.791	1.269	2.489	2.686	1.221	2.305	2.580	1.173
10	3.620	3.620	1.646	3.308	3.456	1.571	3.008	3.291	1.496	2.728	3.127	1.421
12	4.344	4.344	1.975	3.888	4.107	1.867	3.465	3.864	1.756	3.078	3.633	1.652
14	5.068	5.068	2.304	4.444	4.746	2.157	3.882	4.423	2.011	3.375	4.101	1.864
16	5.793	5.793	2.633	4.997	5.371	2.441	4.276	4.950	2.250	3.637	4.529	2.058
18	6.517	6.517	2.962	5.508	5.983	2.720	4.617	5.450	2.477	3.841	4.917	2.235
20	7.241	7.241	3.291	5.998	6.582	2.992	4.922	5.924	2.693	4.005	5.266	2.394
22	7.965	7.965	3.620	6.484	7.168	3.258	5.208	6.372	2.896	4.145	5.575	2.534
24	8.689	8.689	3.949	6.931	7.741	3.519	5.449	6.793	3.088	4.242	5.845	2.657
26	9.413	9.413	4.279	7.358	8.300	3.773	5.660	7.188	3.267	4.313	6.076	2.762
28	10.14	10.14	4.608	7.781	8.847	4.021	5.856	7.557	3.435	4.370	6.266	2.848
30	10.86	10.86	4.937	8.168	9.380	4.264	6.014	7.899	3.590	4.398	6.418	2.917
32	11.59	11.59	5.266	8.536	9.900	4.500	6.148	8.215	3.734	4.413	6.530	2.969
34	12.31	12.31	5.595	8.902	10.41	4.730	6.271	8.504	3.866	4.423	6.602	3.001
36	13.03	13.03	5.924	9.233	10.90	4.955	6.364	8.768	3.985	4.425	6.635	3.016
38	13.76	13.76	6.253	9.547	11.38	5.173	6.438	9.005	4.093			
40	14.48	14.48	6.582	9.859	11.85	5.386	6.506	9.215	4.189			
42	15.21	15.21	6.912	10.14	12.30	5.592	6.550	9.400	4.273			
44	15.93	15.93	7.241	10.40	12.74	5.793	6.581	9.558	4.344			
46	16.65	16.65	7.570	10.67	13.17	5.987	6.611	9.689	4.404			
48	17.38	17.38	7.899	10.90	13.59	6.175	6.624	9.795	4.452			
50	18.10	18.10	8.228	11.12	13.99	6.358	6.630	9.874	4.488			

TABLE 65. Lining Area in Square Feet for 220 Ft Square and 220 Ft Wide Rectangular Reservoirs from 2 to 50 Ft Deep, with Slopes from 0:1 to 3:1

P = 880
2P = 1760
3P = 2640
4P = 3520

220

Depth (Ft.)	0:1			1:1			2:1			3:1		
	220	+220	+100	220	+220	+100	220	+220	+100	220	+220	+100
2	51,040	49,720	22,600	52,635	50,521	22,964	52,335	50,371	22,896	52,194	50,301	22,864
4	52,800	50,600	23,000	53,337	50,882	23,128	52,734	50,582	22,992	52,452	50,442	22,928
6	54,560	51,480	23,400	54,026	51,242	23,292	53,118	50,794	23,088	52,696	50,582	22,992
8	56,320	52,360	23,800	54,701	51,603	23,456	53,487	51,005	23,184	52,924	50,723	23,056
10	58,080	53,240	24,200	55,364	51,964	23,620	53,840	51,216	23,280	53,136	50,864	23,120
12	59,840	54,120	24,600	56,013	52,325	23,784	54,178	51,427	23,376	53,333	51,005	23,184
14	61,600	55,000	25,000	56,650	52,686	23,948	54,500	51,638	23,472	53,515	51,146	23,248
16	63,360	55,880	25,400	57,273	53,046	24,112	54,808	51,850	23,568	53,681	51,286	23,312
18	65,120	56,760	25,800	57,883	53,407	24,276	55,100	52,061	23,664	53,832	51,427	23,376
20	66,880	57,640	26,200	58,480	53,768	24,440	55,376	52,272	23,760	53,968	51,568	23,440
22	68,640	58,520	26,600	59,064	54,129	24,604	55,637	52,483	23,856	54,088	51,709	23,504
24	70,400	59,400	27,000	59,635	54,490	24,768	55,883	52,694	23,952	54,193	51,850	23,568
26	72,160	60,280	27,400	60,192	54,850	24,932	56,113	52,906	24,048	54,283	51,990	23,632
28	73,920	61,160	27,800	60,737	55,211	25,096	56,328	53,117	24,144	54,357	52,131	23,696
30	75,680	62,040	28,200	61,268	55,572	25,260	56,528	53,328	24,240	54,416	52,272	23,760
32	77,440	62,920	28,600	61,786	55,933	25,424	56,708	53,539	24,336	54,460	52,413	23,824
34	79,200	63,800	29,000	62,291	56,294	25,588	56,881	53,750	24,432	54,488	52,554	23,888
36	80,960	64,680	29,400	62,783	56,654	25,752	57,035	53,962	24,528	54,500	52,694	23,952
38	82,720	65,560	29,800	63,262	57,015	25,916	57,173	54,173	24,624			
40	84,480	66,440	30,200	63,728	57,376	26,080	57,296	54,384	24,720			
42	86,240	67,320	30,600	64,181	57,737	26,244	57,404	54,595	24,816			
44	88,000	68,200	31,000	64,620	58,098	26,408	57,495	54,806	24,912			
46	89,760	69,080	31,400	65,047	58,458	26,572	57,572	55,018	25,008			
48	91,520	69,960	31,800	65,460	58,819	26,736	57,634	55,229	25,104			
50	93,280	70,840	32,200	65,860	59,180	26,900	57,680	55,440	25,200			

TABLE 66. Volume in Millions of Gallons for 240 Ft Square and 240 Ft Wide Rectangular Reservoirs from 2 to 50 Ft Deep, with Slopes from 0:1 to 3:1

240

Depth (Ft.)	0:1			1:1			2:1			3:1		
	240	+240	+100	240	+240	+100	240	+240	+100	240	+240	+100
2	0.362	0.362	0.359	0.339	0.855	0.356	0.325	0.347	0.353	0.311	0.340	0.350
4	1.723	1.723	0.718	1.675	1.695	0.706	1.619	1.666	0.694	1.565	1.637	0.682
6	2.585	2.585	1.077	2.458	2.520	1.050	2.335	2.456	1.023	2.217	2.391	0.996
8	3.447	3.447	1.436	3.214	3.332	1.388	3.001	3.217	1.340	2.796	3.102	1.293
10	4.308	4.308	1.795	3.967	4.129	1.720	3.638	3.949	1.646	3.321	3.770	1.571
12	5.170	5.170	2.154	4.670	4.912	2.047	4.205	4.653	1.939	3.774	4.395	1.831
14	6.032	6.032	2.513	5.348	5.680	2.367	4.727	5.328	2.220	4.167	4.976	2.073
16	6.894	6.894	2.872	6.023	6.434	2.681	5.225	5.974	2.489	4.504	5.515	2.298
18	7.755	7.755	3.231	6.650	7.174	2.989	5.661	6.592	2.747	4.789	6.010	2.504
20	8.617	8.617	3.590	7.253	7.899	3.291	6.058	7.181	2.992	5.027	6.463	2.693
22	9.479	9.479	3.949	7.854	8.610	3.587	6.434	7.741	3.225	5.226	6.872	2.863
24	10.34	10.34	4.308	8.410	9.306	3.878	6.756	8.272	3.447	5.377	7.238	3.016
26	11.20	11.20	4.668	8.943	9.988	4.162	7.044	8.775	3.656	5.494	7.561	3.151
28	12.06	12.06	5.027	9.475	10.66	4.440	7.315	9.249	3.854	5.590	7.481	3.267
30	12.93	12.93	5.386	9.963	11.31	4.712	7.540	9.694	4.039	5.655	8.078	3.366
32	13.79	13.79	5.745	10.43	11.95	4.979	7.736	10.11	4.213	5.699	8.272	3.447
34	14.65	14.65	6.104	10.90	12.57	5.239	7.920	10.50	4.374	5.729	8.423	3.510
36	15.51	15.51	6.462	11.32	13.18	5.493	8.065	10.86	4.524	5.739	8.531	3.554
38	16.37	16.37	6.822	11.73	13.78	5.742	8.188	11.19	4.662	5.744	8.595	3.581
40	17.23	17.23	7.181	12.13	14.36	5.984	8.302	11.49	4.787			
42	18.10	18.10	7.540	12.50	14.93	6.220	8.384	11.76	4.910			
44	18.96	18.96	7.899	12.85	15.48	6.451	8.450	12.01	5.003			
46	19.82	19.82	8.258	13.20	16.02	6.675	8.508	12.22	5.092			
48	20.68	20.68	8.617	13.51	16.55	6.894	8.548	12.41	5.170			
50	21.54	21.54	8.976	13.81	17.05	7.106	8.577	12.57	5.236			

TABLE 67. Lining Area in Square Feet for 240 Ft Square and 240 Ft Wide Rectangular Reservoirs from 2 to 50 Ft Deep, with Slopes from 0:1 to 3:1

P = 960
2P = 1920
3P = 2880
4P = 3840

240

Depth (Ft.)	0:1			1:1			2:1			3:1		
	240	+240	+100	240	+240	+100	240	+240	+100	240	+240	+100
2	60,480	59,040	24,600	62,221	59,914	24,964	61,893	59,750	24,896	61,740	59,674	24,864
4	62,400	60,000	25,000	62,988	60,307	25,128	62,331	59,981	24,992	62,024	59,827	24,928
6	64,320	60,960	25,400	63,743	60,701	25,292	62,753	60,211	25,088	62,292	59,981	24,992
8	66,240	61,920	25,800	64,484	61,094	25,456	63,160	60,442	25,184	62,546	60,134	25,056
10	68,160	62,880	26,200	65,212	61,488	25,620	63,552	60,672	25,280	62,784	60,288	25,120
12	70,080	63,840	26,600	65,927	61,882	25,784	63,928	60,902	25,376	63,007	60,442	25,184
14	72,000	64,800	27,000	66,629	62,275	25,948	64,289	61,133	25,472	63,214	60,595	25,248
16	73,920	65,760	27,400	67,318	62,669	26,112	64,635	61,363	25,568	63,406	60,749	25,312
18	75,840	66,720	27,800	67,993	63,062	26,276	64,965	61,594	25,664	63,583	60,902	25,376
20	77,760	67,680	28,200	68,656	63,456	26,440	65,280	61,824	25,760	63,744	61,056	25,440
22	79,680	68,640	28,600	69,305	63,850	26,604	65,580	62,054	25,856	63,890	61,210	25,504
24	81,600	69,600	29,000	69,942	64,243	26,768	65,864	62,285	25,952	64,020	61,363	25,568
26	83,520	70,560	29,400	70,565	64,637	26,932	66,132	62,515	26,048	64,136	61,517	25,632
28	85,440	71,520	29,800	71,175	65,030	27,096	66,386	62,746	26,144	64,236	61,670	25,696
30	87,360	72,480	30,200	71,772	65,424	27,260	66,624	62,876	26,240	64,320	61,824	25,760
32	89,280	73,440	30,600	72,356	65,818	27,424	66,847	63,206	26,336	64,389	61,978	25,824
34	91,200	74,400	31,000	72,927	66,211	27,588	67,054	63,437	26,432	64,443	62,131	25,888
36	93,120	75,360	31,400	73,484	66,605	27,752	67,246	63,667	26,528	64,481	62,285	25,952
38	95,040	76,320	31,800	74,029	66,998	27,916	67,423	63,898	26,624	64,504	62,438	26,016
40	96,960	77,280	32,200	74,560	67,392	28,080	67,584	64,128	26,720			
42	98,880	78,240	32,600	75,078	67,786	28,244	67,730	64,358	26,816			
44	100,800	79,200	33,000	75,583	68,179	28,408	67,860	64,589	26,912			
46	102,720	80,160	33,400	76,075	68,573	28,572	67,976	64,819	27,008			
48	104,640	81,120	33,800	76,554	68,966	28,736	68,076	65,050	27,104			
50	106,560	82,080	34,200	77,020	69,360	28,900	68,160	65,280	27,200			

TABLE 68. Volume in Millions of Gallons for 260 Ft Square and 260 Ft Ft Wide Rectangular Reservoirs from 2 to 50 Ft Deep, with Slopes from 0:1 to 3:1

260

Depth (Ft.)	0:1 260	0:1 +260	0:1 +100	1:1 260	1:1 +260	1:1 +100	2:1 260	2:1 +260	2:1 +100	3:1 260	3:1 +260	3:1 +100
2	1.011	1.011	0.389	0.995	1.004	0.386	0.980	0.996	0.383	0.964	0.988	0.380
4	2.023	2.023	0.778	1.962	1.991	0.766	1.901	1.960	0.754	1.843	1.929	0.742
6	3.034	3.034	1.167	2.896	2.963	1.140	2.762	2.894	1.113	2.633	2.824	1.086
8	4.045	4.045	1.556	3.800	3.921	1.508	3.567	3/796	1.460	3.344	3.672	1.412
10	5.056	5.056	1.945	4.678	4.862	1.870	4.319	4.668	1.795	3.980	4.473	1.720
12	6.068	6.068	2.334	5.525	3.788	2.226	5.017	5.508	2.118	4.543	5.228	2.011
14	7.070	7.070	2.723	6.343	6.698	2.576	5.664	6.317	2.430	5.038	5.936	2.283
16	8.090	8.090	3.112	7.135	7.592	2.920	6.263	7.095	2.729	5.471	6.597	2.537
18	9.102	9.102	3.501	7.900	8.472	3.258	6.814	7.841	3.016	5.844	7.211	2.774
20	10.11	10.11	3.890	8.637	9.335	3.590	7.320	8.557	3.291	6.163	7.779	2.992
22	11.12	11.12	4.279	9.348	10.18	3.917	7.784	9.242	3.554	6.433	8.300	3.192
24	12.14	12.14	4.668	10.03	11.02	4.237	8.206	9.895	3.806	6.655	8.775	3.375
26	13.15	13.15	5.056	10.69	11.83	4.551	8.589	10.52	4.045	6.831	9.203	3.540
28	14.16	14.16	5.445	11.33	12.63	4.859	8.936	11.11	4.273	6.981	9.584	3.686
30	15.17	15.17	5.834	11.94	13.42	5.161	9.245	11.67	4.488	7.091	9.918	3.815
32	16.18	16.18	6.223	12.53	14.19	5.457	9.521	12.20	4.691	7.173	10.21	3.926
34	17.19	17.19	6.612	13.09	14.94	5.748	9.768	12.70	4.883	7.231	10.45	4.018
36	18.20	18.20	7.001	13.63	15.68	6.032	9.983	13.16	5.062	7.268	10.64	4.093
38	19.22	19.22	7.390	14.15	16.41	6.310	10.17	13.60	5.230	7.290	10.79	4.150
40	20.23	20.23	7.779	14.64	17.11	6.582	10.33	14.00	5.386	7.301	10.89	4.189
42	21.24	21.24	8.168	15.12	17.81	6.849	10.47	14.38	5.529	7.304	10.95	4.210
44	22.25	22.25	8.557	15.56	18.48	7.109	10.59	14.72	5.661			
46	23.26	23.26	8.946	16.00	19.15	7.363	10.68	15.03	5.781			
48	24.27	24.27	9.335	16.41	19.79	7.612	10.76	15.31	5.888			
50	25.28	25.28	9.724	16.81	20.42	7.854	10.82	15.56	5.984			

TABLE 69. Lining Area in Square Feet for 260 Ft Square and 260 Ft Wide Rectangular Reservoirs from 2 to 50 Ft Deep, with Slopes from 0:1 to 3:1

P = 1040
2P = 2080
3P = 3120
4P = 4160

260

Depth (Ft.)	0:1 260	0:1 +260	0:1 +100	1:1 260	1:1 +260	1:1 +100	2:1 260	2:1 +260	2:1 +100	3:1 260	3:1 +260	3:1 +100
2	70,720	69,160	26,600	72,606	70,106	26,964	72,252	69,930	26,896	72,085	69,346	26,864
4	72,800	70,200	27,000	73,439	70,533	27,128	72,728	70,179	26,992	72,395	70,013	26,928
6	74,880	72,140	27,400	74,259	70,959	27,292	73,188	70,429	27,088	72,689	70,179	26,992
8	76,960	72,280	27,800	75,066	71,386	27,456	73,634	70,678	27,184	72,968	70,346	27,056
10	79,040	73,320	28,200	75,860	71,812	27,620	74,064	70,928	27,280	73,232	70,512	27,120
12	81,120	74,360	28,600	76,641	72,238	27,784	74,478	71,178	27,376	73,480	70,678	27,184
14	83,200	75,400	29,000	77,408	72,665	27,948	74,878	71,427	27,472	73,713	70,845	27,248
16	85,280	76,440	29,400	78,163	73,091	28,112	75,262	71,677	27,568	73,930	71,011	27,312
18	87,360	77,480	29,800	78,904	73,518	28,276	75,631	71,926	27,664	74,133	71,178	27,376
20	89,440	78,520	30,200	79,632	73,944	28,440	75,984	72,176	27,760	74,320	71,344	27,440
22	91,520	79,560	30,600	80,347	74,370	28,604	76,322	72,426	27,856	74,492	71,510	27,504
24	93,600	80,600	31,000	81,049	74,797	28,768	76,644	72,675	27,952	74,648	71,677	27,568
26	95,680	81,640	31,400	81,738	75,223	28,932	76,952	72,925	28,048	74,788	71,843	27,632
28	97,760	82,680	31,800	82,413	75,650	29,096	77,244	73,174	28,144	74,914	72,010	27,696
30	99,840	83,720	32,200	83,076	76,076	29,260	77,520	73,424	28,240	75,024	72,176	27,760
32	101,920	84,760	32,600	83,725	76,502	29,424	77,781	73,674	28,336	75,119	72,342	27,824
34	104,000	85,800	33,000	84,362	76,929	29,588	78,027	73,923	28,432	75,198	72,509	27,888
36	106,080	86,840	33,400	84,985	77,355	29,752	78,257	74,173	28,528	75,262	72,675	27,952
38	108,160	87,880	33,800	85,595	77,782	29,916	78,472	74,422	28,624	75,311	72,842	28,016
40	110,240	88,920	34,200	86,192	78,208	30,080	78,672	74,672	28,720	75,344	73,008	28,080
42	112,320	89,960	34,600	86,776	78,634	30,244	78,856	74,922	28,816	75,362	73,174	28,144
44	114,400	91,000	35,000	87,347	79,061	30,408	79,025	75,171	28,912			
46	116,480	92,040	35,400	87,904	79,487	30,572	79,179	75,421	29,008			
48	118,560	93,080	35,800	88,449	79,914	30,736	79,317	75,670	29,104			
50	120,640	94,120	36,200	88,980	80,340	30,900	79,440	75,920	29,200			

TABLE 70. Volume in Millions of Gallons for 280 Ft Square and 280 Ft Wide Rectangular Reservoirs from 2 to 50 Ft Deep, with Slopes from 0:1 to 3:1

280

Depth (Ft.)	0:1			1:1			2:1			3:1		
	280	+280	+100	280	+280	+100	280	+280	+100	280	+280	+100
2	1.173	1.173	0.419	1.155	1.164	0.416	1.139	1.156	0.413	1.122	1.148	0.410
4	2.346	2.346	0.838	2.280	2.312	0.826	2.215	2.279	0.814	2.151	2.245	0.802
6	3.519	3.519	1.257	3.370	3.443	1.230	3.224	3.368	1.203	3.086	3.292	1.176
8	4.691	4.691	1.675	4.427	4.557	1.628	4.175	4.423	1.580	3.923	4.289	1.532
10	5.864	5.864	2.094	5.466	5.655	2.020	5.077	5.445	1.945	4.707	5.236	1.870
12	7.037	7.037	2.513	6.451	6.736	2.406	5.900	6.434	2.298	5.383	6.132	2.190
14	8.210	8.210	2.932	7.406	7.800	2.786	6.677	7.389	2.634	5.992	6.979	2.492
16	9.393	9.393	3.351	8.352	8.847	3.160	7.403	8.311	2.968	6.542	7.774	2.777
18	10.56	10.56	3.770	9.257	9.877	3.523	8.074	9.199	3.285	7.008	8.520	3.043
20	11.73	11.73	4.189	10.13	10.89	3.890	8.696	10.05	3.590	7.419	9.215	3.291
22	12.90	12.90	4.608	10.98	11.89	4.246	9.272	10.87	3.884	7.776	9.860	3.522
24	14.07	14.07	5.027	11.80	12.87	4.596	9.800	11.66	4.165	8.077	10.46	3.734
26	15.25	15.25	5.445	12.59	13.83	4.490	10.28	12.42	4.434	8.329	11.00	3.928
28	16.42	16.42	5.864	13.36	14.78	5.278	10.73	13.14	4.691	8.539	11.49	4.105
30	17.59	17.59	6.283	14.09	15.71	5.610	11.13	13.82	4.937	8.707	11.94	4.264
32	18.77	18.77	6.702	14.80	16.62	5.936	11.49	14.48	5.170	8.839	12.33	4.404
34	19.94	19.94	7.121	15.50	17.52	6.256	11.82	15.10	5.392	8.940	12.68	4.527
36	21.12	21.12	7.540	16.15	18.40	6.570	12.12	15.68	5.601	9.013	12.97	4.632
38	22.28	22.28	7.959	16.78	19.26	6.879	12.38	16.24	5.798	9.063	13.21	4.718
40	23.46	23.46	8.378	17.39	20.11	7.181	12.61	16.76	5.984	9.096	13.40	4.787
42	24.63	24.63	8.796	17.98	20.94	7.477	12.81	17.24	6.158	9.113	13.55	4.838
44	25.80	25.80	9.215	18.54	21.75	7.767	12.98	17.69	6.319	9.120	13.64	4.871
46	26.98	26.98	9.634	19.08	22.54	8.051	13.14	18.11	6.469	9.126	13.68	4.886
48	28.15	28.15	10.05	19.60	23.32	8.330	13.26	18.49	6.606			
50	29.32	29.32	10.47	20.10	24.09	8.602	13.36	18.85	6.732			

TABLE 71. Lining Area in Square Feet for 280 Ft Square and 280 Ft Wide Rectangular Reservoirs from 2 to 50 Ft Deep, with Slopes from 0:1 to 3:1

P = 1120
2P = 2240
3P = 3360
4P = 4480

280

Depth (Ft.)	0:1			1:1			2:1			3:1		
	280	+280	+100	280	+280	+100	280	+280	+100	280	+280	+100
2	81,760	80,080	28,600	83,792	81,099	28,964	83,410	80,909	28,896	83,231	80,819	28,864
4	84,000	81,200	29,000	84,691	81,558	29,128	83,924	81,178	28,992	83,566	80,998	28,928
6	86,240	82,320	29,400	85,576	82,018	29,292	84,424	81,446	29,088	83,886	81,178	28,992
8	88,480	83,440	29,800	86,448	82,477	29,456	84,908	81,715	29,184	84,191	81,357	29,056
10	90,720	84,560	30,200	87,308	82,936	29,620	85,376	81,984	29,280	84,480	81,536	29,120
12	92,600	85,680	30,600	88,154	83,395	29,784	85,829	82,253	20,376	84,754	81,715	29,184
14	95,200	86,800	31,000	88,987	83,854	29,948	86,267	82,522	29,472	85,012	81,894	29,248
16	97,440	87,920	31,400	89,807	84,314	30,112	86,689	82,790	29,568	85,256	82,074	29,312
18	99,680	89,040	31,800	90,614	84,773	30,276	87,096	83,059	29,664	85,484	82,253	29,376
20	101,920	90,160	32,200	91,408	85,232	30,440	87,488	83,328	29,760	85,696	82,432	29,440
22	104,160	91,280	32,600	92,189	85,691	30,604	87,864	83,597	29,856	85,893	82,611	29,504
24	106,400	92,400	33,000	92,956	86,150	30,768	88,225	83,866	29,952	86,075	82,790	29,568
26	108,640	93,520	33,400	93,761	86,610	30,932	88,571	84,134	30,048	86,241	82,970	29,632
28	110,880	94,640	33,800	94,452	87,069	31,096	88,901	84,403	30,144	86,392	83,149	29,696
30	113,120	95,760	34,200	95,180	87,528	31,260	89,216	84,672	30,240	86,528	83,328	29,760
32	115,360	96,880	34,600	95,895	87,987	31,424	89,516	84,941	30,336	86,648	83,507	29,824
34	117,600	98,000	35,000	96,597	88,446	31,588	89,800	85,210	30,432	86,753	83,686	29,888
36	119,840	99,120	35,400	97,286	88,906	31,752	90,068	85,478	30,528	86,843	83,866	29,952
38	122,080	100,240	35,800	97,961	89,365	31,916	[illegible]	85,747	30,624	86,917	84,045	30,016
40	124,320	101,360	36,200	98,624	89,824	32,080	90,560	86,016	30,720	86,976	84,224	30,080
42	126,560	102,480	36,600	99,273	90,283	32,244	90,782	86,285	30,816	87,020	84,403	30,144
44	128,800	103,600	37,000	99,909	90,742	32,408	90,990	86,554	30,912	87,048	84,582	30,208
46	131,040	104,720	37,400	100,533	91,202	32,572	91,182	86,822	31,008	87,060	84,762	30,272
48	133,280	105,840	37,800	101,143	91,661	32,736	91,359	87,091	31,104			
50	135,520	106,960	38,200	101,740	92,120	32,900	91,520	87,360	31,200			

TABLE 72. Volume in Millions of Gallons for 300 Ft Square and 300 Ft Wide Rectangular Reservoirs from 2 to 50 Ft Deep, with Slopes from 0:1 to 3:1

300

Depth	0:1			1:1			2:1			3:1		
(Ft.)	300	+300	+100	300	+300	+100	300	+300	+100	300	+300	+100
2	1.346	1.346	0.449	1.327	1.337	0.446	1.310	1.328	0.443	1.292	1.319	0.440
4	2.693	2.693	0.898	2.623	2.657	0.886	2.553	2.621	0.874	2.484	2.585	0.862
6	4.039	4.039	1.346	3.880	3.958	1.319	3.725	3.878	1.293	3.574	3.797	1.266
8	5.386	5.386	1.795	5.102	5.242	1.747	4.830	5.098	1.699	4.569	4.955	1.652
10	6.732	6.732	2.244	6.306	6.508	2.169	5.875	6.283	2.094	5.476	6.059	2.020
12	8.078	8.078	2.693	7.449	7.755	2.585	6.855	7.432	2.477	6.295	7.109	2.370
14	9.425	9.425	3.142	8.560	8.985	2.995	7.774	8.545	2.848	7.031	8.105	2.702
16	10.77	10.77	3.590	9.664	10.20	3.399	8.638	9.622	3.207	7.702	9.048	3.016
18	12.12	12.12	4.039	10.72	11.39	3.797	9.442	10.66	3.554	8.279	9.936	3.312
20	13.46	13.46	4.488	11.75	12.48	4.159	10.19	11.67	3.890	8.796	10.77	3.590
22	14.81	14.81	4.937	12.76	13.72	4.575	10.89	12.64	4.213	9.250	11.55	3.851
24	16.16	16.16	5.386	13.71	14.86	4.955	11.54	13.57	4.524	9.642	12.28	4.093
26	17.50	17.50	5.834	14.64	15.99	5.329	12.14	14.47	4.823	9.978	12.95	4.317
28	18.85	18.85	6.283	15.55	17.09	5.697	12.69	15.33	5.110	10.27	13.57	4.524
30	20.20	20.20	6.732	16.43	18.18	6.059	13.20	16.16	5.386	10.50	14.14	4.712
32	21.54	21.54	7.181	17.27	19.25	6.415	13.66	16.95	5.649	10.70	14.65	4.883
34	22.89	22.89	7.630	18.09	20.30	6.765	14.08	17.70	5.900	10.85	15.11	5.036
36	24.24	24.24	8.078	18.88	21.33	7.109	14.46	18.42	6.140	10.97	15.51	5.170
38	25.58	25.58	8.527	19.65	22.34	7.447	14.81	19.10	6.367	11.06	15.86	5.287
40	26.93	26.93	8.976	20.39	23.34	7.779	15.12	19.75	6.582	11.13	16.16	5.386
42	28.27	28.27	9.425	21.10	24.32	8.105	15.40	20.36	6.786	11.17	16.40	5.466
44	29.62	29.62	9.874	21.78	25.28	8.425	15.64	20.93	6.977	11.20	16.59	5.529
46	30.97	30.97	10.32	22.44	26.22	8.740	15.86	21.47	7.157	11.22	16.72	5.574
48	32.31	32.31	10.77	23.08	27.14	9.048	16.05	21.97	7.324	11.22	16.80	5.601
50	33.66	33.66	11.22	23.69	28.05	9.350	16.21	22.44	7.480			

TABLE 73. Lining Area in Square Feet for 300 Ft Square and 300 Ft Wide Rectangular Reservoirs from 2 to 50 Ft Deep, with Slopes from 0:1 to 3:1

P = 1200
2P = 2400
3P = 3600
4P = 4800

300

Depth	0:1			1:1			2:1			3:1		
(Ft.)	300	+300	+100	300	+300	+100	300	+300	+100	300	+300	+100
2	93,600	91,800	30,600	95,777	92,892	30,964	95,368	92,688	30,896	95,176	92,592	30,864
4	96,000	93,000	31,000	96,742	93,384	31,128	95,920	92,976	30,992	95,537	92,784	30,928
6	98,400	94,200	31,400	97,693	93,876	31,292	96,459	93,264	31,088	95,883	92,976	30,992
8	100,800	95,400	31,800	98,630	94,368	31,456	96,981	93,552	31,184	96,213	93,168	31,056
10	103,200	96,600	32,200	99,556	94,860	31,620	97,488	93,840	31,280	96,528	93,360	31,120
12	105,600	97,800	32,600	100,468	95,352	31,784	97,980	94,128	31,376	96,828	93,552	31,184
14	108,000	99,000	33,000	101,367	95,844	31,948	98,456	94,416	31,472	97,112	93,744	31,248
16	110,400	100,200	33,400	102,252	96,336	32,112	98,916	94,704	31,568	97,380	93,936	31,312
18	112,800	101,400	33,800	103,125	96,828	32,276	99,362	94,992	31,664	97,634	94,128	31,376
20	115,200	102,600	34,200	103,984	97,320	32,440	99,792	95,280	31,760	97,872	94,320	31,440
22	117,600	103,800	34,600	104,830	97,812	32,604	100,207	95,568	31,856	98,095	94,512	31,504
24	120,000	105,000	35,000	105,663	98,304	32,768	100,606	95,856	31,952	98,302	94,680	31,560
26	122,400	106,200	35,400	106,483	98,796	32,932	100,990	96,144	32,048	98,494	94,896	31,632
28	124,800	107,400	35,800	107,290	99,288	33,096	101,359	96,432	32,144	98,671	95,088	31,696
30	127,200	108,600	36,200	108,084	99,780	33,260	101,712	96,720	32,240	98,832	95,280	31,760
32	129,600	109,800	36,600	108,865	100,272	33,424	102,050	97,008	32,336	98,978	95,472	31,824
34	132,000	111,000	37,000	109,632	100,764	33,588	102,372	97,296	32,432	99,108	95,664	31,888
36	134,400	112,200	37,400	110,387	101,256	33,752	102,680	97,584	32,528	99,224	95,856	31,952
38	136,800	113,400	37,800	111,128	101,748	33,916	102,972	97,872	32,624	99,324	96,048	32,016
40	139,200	114,600	38,200	111,856	102,240	34,080	103,248	98,160	32,720	99,408	96,240	32,080
42	141,600	115,800	38,600	112,571	102,732	34,244	103,509	98,448	32,816	99,477	96,432	32,144
44	144,000	117,000	39,000	113,273	103,224	34,408	103,755	98,736	32,912	99,531	96,624	32,208
46	146,400	118,200	39,400	113,962	103,716	34,572	103,985	99,024	33,008	99,569	96,816	32,272
48	148,800	119,400	39,800	114,637	104,208	34,736	104,200	99,312	33,104	99,592	97,008	32,336
50	151,200	120,600	40,200	115,300	104,700	34,900	104,400	99,600	33,200			

TABLE 74. Volume in Millions of Gallons for 320 Ft Square and 320 Ft Wide Rectangular Reservoirs from 2 to 50 Ft Deep, with Slopes from 0:1 to 3:1

320

Depth	0:1			1:1			2:1			3:1		
(Ft.)	320	+320	+100	320	+320	+100	320	+320	+100	320	+320	+100
2	1.532	1.532	0.479	1.511	1.522	0.476	1.492	1.513	0.473	1.474	1.503	0.470
4	3.064	3.064	0.957	2.989	3.026	0.945	2.915	2.987	0.934	2.841	2.949	0.922
6	4.596	4.596	1.436	4.426	4.510	1.409	4.260	4.423	1.382	4.098	4.337	1.355
8	6.128	6.128	1.915	5.825	5.974	1.867	5.534	5.821	1.819	5.253	5.668	1.771
10	7.660	7.660	2.394	7.192	7.420	2.319	6.743	7.181	2.244	6.314	6.941	2.169
12	9.191	9.191	2.872	8.519	8.847	2.765	7.882	8.502	2.657	7.278	8.157	2.549
14	10.72	10.72	3.351	9.811	10.25	3.204	8.955	9.785	3.058	8.154	9.316	2.911
16	12.26	12.26	3.830	11.07	11.64	3.638	9.969	11.03	3.447	8.947	10.42	3.255
18	13.79	13.79	4.308	12.29	13.01	4.066	10.92	12.24	3.824	0.657	11.46	3.581
20	15.32	15.32	4.787	13.48	14.36	4.488	11.81	13.40	4.189	10.29	12.45	3.890
22	16.85	16.85	5.266	14.64	15.69	4.904	12.64	14.53	4.542	10.86	13.38	4.180
24	18.38	18.38	5.745	15.76	17.00	5.314	13.42	15.63	4.883	11.35	14.25	4.452
26	19.92	19.92	6.223	16.85	18.30	5.718	14.14	16.68	5.212	11.78	15.06	4.706
28	21.45	21.45	6.702	17.91	19.57	6.116	14.82	17.69	5.529	12.16	15.82	4.943
30	22.98	22.98	7.181	18.94	20.82	15.44	15.44	18.67	5.834	12.48	16.52	5.161
32	24.51	24.51	7.660	19.93	22.06	6.894	16.01	19.61	6.128	12.75	17.16	5.362
34	26.04	26.04	8.138	20.90	23.28	7.274	16.54	20.51	6.409	12.97	17.74	5.544
36	27.57	27.57	8.617	21.84	24.47	7.648	17.03	21.37	6.678	13.11	18.27	5.709
38	29.11	29.11	9.096	22.74	25.65	8.016	17.47	22.19	6.935	13.29	18.74	5.855
40	30.64	30.64	9.574	23.62	26.81	8.378	17.87	22.98	7.181	13.41	19.15	5.984
42	32.17	32.17	10.05	24.64	27.95	8.734	18.24	23.73	7.414	13.49	19.50	6.095
44	33.70	33.70	10.53	25.28	29.07	9.084	18.56	24.43	7.636	13.54	19.80	6.187
46	35.23	35.23	11.01	26.08	30.17	9.428	18.86	25.10	7.845	13.58	20.04	6.262
48	36.77	36.77	11.49	26.84	31.25	9.766	19.12	25.74	8.042	13.60	20.22	6.319
50	38.30	38.30	11.97	27.58	32.31	10.10	19.35	26.33	8.228	13.61	20.35	6.358

TABLE 75. Lining Area in Square Feet for 320 Ft Square and 320 Ft Wide Rectangular Reservoirs from 2 to 50 Ft Deep, with Slopes from 0:1 to 3:1

P = 1280
2P = 2560
3P = 3840
4P = 5120

320

Depth	0:1			1:1			2:1			3:1		
(Ft.)	320	+320	+100	320	+320	+100	320	+320	+100	320	+320	+100
2	106,240	104,320	32,600	108,563	105,485	32,964	108,127	105,267	32,896	107,922	105,165	32,864
4	108,800	105,600	33,000	109,593	106,010	33,128	108,718	105,574	32,992	108,308	105,370	32,928
6	111,360	106,880	33,400	110,610	106,534	33,292	109,294	105,882	33,088	108,679	105,574	32,992
8	113,920	108,160	33,800	111,613	107,059	33,456	109,855	106,189	33,184	109,036	105,779	33,056
10	116,480	109,440	34,200	112,604	107,584	33,620	110,400	106,496	33,280	109,376	105,984	33,120
12	119,040	110,720	34,600	113,581	108,109	33,784	110,930	106,790	33,372	109,701	106,189	33,184
14	121,600	112,000	35,000	114,546	106,634	33,948	111,444	107,110	33,472	110,011	106,394	33,248
16	124,160	113,280	35,400	115,497	109,158	34,112	111,944	107,418	33,568	110,305	106,598	33,312
18	126,720	114,560	35,800	116,434	109,683	34,276	112,428	107,724	33,664	110,584	106,803	33,376
20	129,280	115,840	36,200	117,360	110,208	34,440	112,896	108,032	33,760	110,848	107,008	33,440
22	131,840	117,120	36,600	118,272	110,733	34,604	113,349	108,339	33,856	111,096	107,213	33,504
24	134,400	118,400	37,000	119,171	111,258	34,768	113,787	108,646	33,952	111,329	107,418	33,568
26	136,960	119,680	37,400	120,056	111,782	34,932	114,209	108,954	34,048	111,547	107,622	33,632
28	139,520	120,960	37,800	120,929	112,307	35,096	114,616	109,261	34,144	111,749	107,827	33,696
30	142,080	122,240	38,200	121,788	112,832	35,260	115,008	109,568	34,240	111,936	108,032	33,760
32	144,640	123,520	38,600	122,634	113,357	35,424	115,384	109,875	34,336	112,108	108,237	33,824
34	147,200	124,800	39,000	123,467	113,882	35,588	115,745	110,182	34,432	112,264	108,442	33,888
36	149,760	126,080	39,400	124,287	114,406	35,752	116,091	110,490	34,528	112,404	108,646	33,952
38	152,320	127,360	39,800	125,094	114,931	35,916	116,421	110,797	34,624	112,530	108,851	34,016
40	154,880	128,640	40,200	125,888	115,456	36,080	116,736	111,104	34,720	112,640	109,056	34,080
42	157,440	129,920	40,600	126,669	115,981	36,244	117,036	111,412	34,816	112,735	109,261	34,144
44	160,000	131,200	41,000	127,436	116,506	36,408	117,320	111,718	34,912	112,814	109,466	34,208
46	162,560	132,480	41,400	128,191	117,030	36,572	117,588	112,026	35,008	112,878	109,670	34,272
48	165,120	133,760	41,800	128,932	117,555	36,736	117,842	112,333	35,104	112,926	109,875	34,336
50	167,680	135,040	42,200	129,660	118,080	36,900	118,080	112,640	35,200	112,960	110,080	34,400

TABLE 76. Volume in Millions of Gallons for 340 Ft Square and 340 Ft Wide Rectangular Reservoirs from 2 to 50 Ft Deep, with Slopes from 0:1 to 3:1

340

Depth	0:1			1:1			2:1			3:1		
(Ft.)	340	+340	+100	340	+340	+100	340	+340	+100	340	+340	+100
2	1.729	1.729	0.509	1.707	1.719	0.506	1.687	1.709	0.503	1.667	1.699	0.500
4	3.459	3.459	1.017	3.380	3.418	1.005	3.300	3.377	0.993	3.222	3.337	0.981
6	5.188	5.188	1.526	5.007	5.097	1.499	4.831	5.005	1.472	4.658	4.913	1.445
8	6.918	6.918	2.035	6.595	6.755	1.987	6.285	6.592	1.939	5.985	6.429	1.891
10	8.647	8.647	2.543	8.150	8.392	2.468	7.671	8.138	2.394	7.212	7.884	2.319
12	10.38	10.38	3.052	9.661	10.01	2.944	8.980	9.644	2.836	8.334	9.278	2.729
14	12.11	12.11	3.560	11.13	11.61	3.414	10.22	11.11	3.267	9.307	10.61	3.121
16	13.84	13.84	4.069	12.58	13.18	3.878	11.40	12.53	3.686	10.30	11.88	3.495
18	15.56	15.56	4.578	13.98	14.74	4.335	12.50	13.92	4.093	11.14	13.09	3.851
20	17.29	17.29	5.086	15.34	16.28	4.787	13.54	15.26	4.488	11.91	14.24	4.189
22	19.02	19.02	5.595	16.67	17.79	5.233	14.53	16.56	4.871	12.60	15.33	4.509
24	20.75	20.75	6.104	17.96	19.29	5.673	15.44	17.82	5.242	13.20	16.36	4.811
26	22.48	22.48	6.612	19.22	20.76	6.107	16.31	19.04	5.601	13.74	17.32	5.095
28	24.21	24.21	7.121	20.44	22.22	6.535	17.11	20.22	5.948	14.22	18.23	5.362
30	25.94	25.94	7.630	21.63	23.65	6.956	17.86	21.36	6.283	14.63	19.07	5.610
32	27.67	27.67	8.138	22.79	25.07	7.372	18.56	22.46	6.606	14.99	19.86	5.840
34	29.40	29.40	8.647	23.91	26.46	7.782	19.21	23.52	6.918	15.29	20.58	5.053
36	31.13	31.13	9.156	25.00	27.83	8.186	19.81	24.54	7.217	15.54	21.24	6.247
38	32.86	32.86	9.664	26.06	29.19	8.584	20.36	25.51	7.503	15.75	21.84	6.424
40	34.59	34.59	10.17	27.09	30.52	8.976	20.87	26.45	7.779	15.92	22.38	6.582
42	36.32	36.32	10.68	27.93	31.83	9.362	21.33	27.34	8.042	16.05	22.86	6.723
44	38.05	38.05	11.19	29.05	33.12	9.742	21.75	28.20	8.294	16.15	23.28	6.846
46	39.78	39.78	11.70	29.99	34.39	10.12	22.13	29.01	8.533	16.22	23.63	6.950
48	41.51	41.51	12.21	30.89	35.65	10.48	22.48	29.79	8.761	16.28	23.93	7.04
50	43.23	43.23	12.72	31.76	36.88	10.85	22.79	30.52	8.976	16.31	24.16	7.106

TABLE 77. Lining Area in Square Feet for 340 Ft Square and 340 Ft Wide Rectangular Reservoirs from 2 to 50 Ft Deep, with Slopes from 0:1 to 3:1

P = 1360
2P = 2720
3P = 4080
4P = 5440

340

Depth	0:1			1:1			2:1			3:1		
(Ft.)	340	+340	+100	340	+340	+100	340	+340	+100	340	+340	+100
2	119,680	117,640	34,600	122,149	118,878	34,964	121,685	118,646	34,896	121,468	118,538	34,864
4	122,400	119,000	35,000	123,244	119,435	35,128	122,315	118,973	34,992	121,880	118,755	34,928
6	125,120	120,360	35,400	124,327	119,993	35,292	122,929	119,299	35,088	122,276	118,973	34,992
8	127,840	121,720	35,800	125,396	120,550	35,456	123,528	119,626	35,184	122,657	119,190	35,056
10	130,560	123,080	36,200	126,452	121,108	35,620	124,112	119,952	35,280	123,024	119,408	35,120
12	133,280	124,440	36, 00	127,495	121,666	35,784	124,680	120,278	35,376	123,375	119,626	35,184
14	136,000	125,800	37,000	128,525	122,223	35,948	125,232	120,605	35,472	123,670	119,843	35,248
16	138,720	127,160	37,400	129,554	122,781	36,112	125,771	120,931	35,568	123,990	120,061	35,312
18	141,440	128,520	37,800	130,545	123,338	36,276	126,293	121,258	35,664	124,335	120,278	35,376
20	144,160	129,880	38,200	131,496	123,896	36,440	126,800	121,584	35,760	124,624	120,496	35,440
22	146,880	131,240	38,600	132,513	124,454	36,604	127,292	121,910	35,856	124,898	120,714	35,504
24	149,600	132,600	39,000	133,478	125,011	36,768	127,768	122,237	35,952	125,156	120,931	35,568
26	152,320	133,960	39,400	134,429	125,569	36,932	128,228	122,563	36,048	125,400	121,149	35,632
28	155,040	135,320	39,800	135,366	126,126	37,096	128,673	122,890	36,144	125,628	121,366	35,696
30	157,560	136,680	40,200	136,292	126,684	37,260	129,104	123,216	36,240	125,840	121,584	35,760
32	160,480	138,040	40,600	137,204	127,242	37,424	129,519	123,542	36,336	126,037	121,802	35,824
34	163,200	139,400	41,000	138,103	127,799	37,588	129,918	123,869	36,432	126,219	122,019	35,888
36	165,920	140,760	41,400	138,988	128,357	37,752	130,302	124,195	36,528	126,385	122,237	35,952
38	168,640	142,120	41,800	139,861	128,914	37,916	130,671	124,522	36,624	126,536	122,454	36,016
40	171,360	143,480	42,200	140,720	129,472	38,080	131,024	124,848	36,720	126,672	122,672	36,080
42	174,080	144,840	42,600	141,566	130,030	38,244	131,362	125,174	36,816	126,792	122,890	36,144
44	176,800	146,200	43,000	142,399	130,587	38,408	131,684	125,501	36,912	126,897	123,107	36,208
46	179,520	147,560	43,400	143,219	131,145	38,572	131,992	125,827	37,008	126,987	123,325	36,272
48	182,240	148,920	43,800	144,026	131,702	38,736	132,284	126,154	37,104	127,061	123,542	36,336
50	184,960	150,280	44,200	144,820	132,260	38,900	132,560	126,480	37,200	127,120	123,760	36,400

TABLE 78. Volume in Millions of Gallons for 360 Ft Square and 360 Ft Wide Rectangular Reservoirs from 2 to 50 Ft Deep, with Slopes from 0:1 to 3:1

360

Depth (Ft.)	0:1 360	0:1 +360	0:1 +100	1:1 360	1:1 +360	1:1 +100	2:1 360	2:1 +360	2:1 +100	3:1 360	3:1 +360	3:1 +100
2	1.939	1.939	0.539	1.915	1.928	0.536	1.894	1.917	0.533	1.873	1.907	0.530
4	3.878	3.878	1.077	3.794	3.835	1.065	3.710	3.791	1.053	3.627	3.748	1.041
6	5.816	5.816	1.616	5.625	5.720	1.589	5.437	5.623	1.562	5.254	5.526	1.535
8	7.755	7.755	2.154	7.414	7.583	2.106	7.085	7.411	2.058	6.765	7.238	2.011
10	9.964	9.964	2.693	9.167	9.425	2.618	8.659	9.156	2.543	8.170	8.886	2.468
12	11.63	11.63	3.231	10.88	11.25	3.124	10.15	10.86	3.016	9.461	10.47	2.908
14	13.57	13.57	3.770	12.54	13.04	3.623	11.57	12.52	3.477	10.65	11.99	3.330
16	15.51	15.51	4.308	14.17	14.82	4.117	12.92	14.13	3.926	11.74	13.44	3.734
18	17.45	17.45	4.847	15.76	16.58	4.605	14.19	15.70	4.362	12.74	14.83	4.120
20	19.39	19.39	5.386	17.31	18.31	5.086	15.40	17.23	4.787	13.64	16.16	4.488
22	21.33	21.33	5.924	18.29	20.02	5.562	16.54	18.72	5.200	14.64	17.42	4.838
24	23.27	23.27	6.463	20.30	21.72	6.032	17.61	20.16	5.601	15.20	18.61	5.170
26	25.21	25.21	7.001	21.74	23.38	6.496	18.62	21.56	5.990	15.86	19.74	5.484
28	27.14	27.14	7.540	23.14	25.03	6.953	19.58	22.92	6.367	16.45	20.81	5.781
30	29.08	29.08	8.078	24.50	26.66	7.405	20.47	24.24	6.732	16.97	21.81	6.059
32	31.02	31.02	8.617	25.83	28.26	7.851	21.30	25.51	7.085	17.42	22.75	6.319
34	32.96	32.96	9.156	27.13	29.85	8.291	22.08	26.73	7.426	17.81	23.62	6.561
36	34.90	34.90	9.694	28.38	31.41	8.725	22.80	27.92	7.755	18.15	24.43	6.786
38	36.84	36.84	10.23	29.61	32.95	9.153	23.47	29.06	8.072	18.43	25.17	6.992
40	38.78	38.78	10.77	30.80	34.47	9.574	24.10	30.16	8.378	18.67	25.85	7.181
42	40.72	40.72	11.31	31.95	35.97	9.990	24.67	31.22	8.671	18.87	26.47	7.351
44	42.65	42.65	11.85	33.08	37.44	10.40	25.20	32.23	8.952	19.02	27.01	7.504
46	44.59	44.59	12.39	34.17	38.90	10.80	25.69	33.20	9.221	19.14	27.50	7.639
48	46.53	46.53	12.93	35.23	40.33	11.20	26.13	34.12	9.479	19.23	27.92	7.755
50	48.47	48.47	13.46	36.25	41.74	11.59	26.53	35.01	9.724	19.30	28.27	7.854

TABLE 79. Lining Area in Square Feet for 360 Ft Square and 360 Ft Wide Rectangular Reservoirs from 2 to 50 Ft Deep, with Slopes from 0:1 to 3:1

P = 1440
2P = 2880
3P = 4320
4P = 5760

360

Depth (Ft.)	0:1 360	0:1 +360	0:1 +100	1:1 360	1:1 +360	1:1 +100	2:1 360	2:1 +360	2:1 +100	3:1 360	3:1 +360	3:1 +100
2	133,920	131,760	36,600	136,534	133,070	36,964	136,044	132,826	36,896	135,813	132,710	36,864
4	136,800	133,200	37,000	137,695	133,661	37,128	136,712	133,171	36,992	136,251	132,941	36,928
6	139,680	134,640	37,400	138,843	134,251	37,292	137,364	133,517	37,088	136,673	133,171	36,992
8	142,560	136,080	37,800	139,978	134,842	37,456	138,002	133,862	37,184	137,080	133,402	37,056
10	145,440	137,520	38,200	141,100	135,432	37,620	138,624	134,208	37,280	137,472	133,632	37,120
12	148,320	138,960	38,600	142,209	136,022	37,784	139,231	134,554	37,376	137,848	133,862	37,184
14	151,200	140,400	39,000	143,304	136,613	37,948	139,822	134,899	37,472	138,209	134,093	37,248
16	154,080	141,840	39,400	144,387	137,203	38,112	140,398	135,245	37,568	138,555	134,323	37,312
18	156,960	143,280	39,800	145,456	137,794	38,276	140,959	135,590	37,664	138,885	134,554	37,376
20	159,840	144,720	40,200	146,512	138,384	38,440	141,504	135,936	37,760	139,200	134,784	37,440
22	162,570	146,160	40,600	147,555	138,974	38,604	142,034	136,282	37,856	139,500	135,014	37,504
24	165,600	147,600	41,000	148,585	139,536	38,760	142,548	136,627	37,952	139,784	135,245	37,568
26	168,480	149,040	41,400	149,602	140,155	38,932	143,048	136,973	38,048	140,052	135,475	37,632
28	171,360	150,480	41,800	150,605	140,746	39,096	143,532	137,318	38,144	140,306	135,706	37,696
30	174,240	151,920	42,200	151,596	141,336	39,260	144,000	137,664	38,240	140,544	135,936	37,760
32	177,120	153,360	42,600	152,573	141,926	39,424	144,453	138,010	38,336	140,767	136,166	37,824
34	180,000	154,800	43,000	153,538	142,488	39,580	144,891	138,355	38,432	140,974	136,368	37,880
36	182,880	156,240	43,400	154,489	143,107	39,752	145,313	138,701	38,528	141,166	136,627	37,952
38	185,760	157,680	43,800	155,426	143,698	39,916	145,720	139,046	38,624	141,343	136,858	38,016
40	188,640	159,120	44,200	156,363	144,288	40,000	146,110	139,392	38,720	141,504	137,088	38,080
42	191,520	160,560	44,600	157,264	144,878	40,244	146,488	139,738	38,816	141,650	137,318	38,144
44	194,400	162,000	45,000	158,163	145,469	40,408	146,849	140,083	38,912	141,780	137,549	38,208
46	197,280	163,440	45,400	159,048	146,059	40,572	147,195	140,429	39,008	141,896	137,779	38,272
48	200,160	164,880	45,800	159,921	146,650	40,736	147,525	140,774	39,104	141,996	138,010	38,336
50	203,040	166,320	46,200	160,780	147,240	40,900	147,840	141,120	39,200	142,080	138,240	38,400

TABLE 80. Volume in Millions of Gallons for 380 Ft Square and 380 Ft Wide Rectangular Reservoirs from 2 to 50 Ft Deep, with Slopes from 0:1 to 3:1

380

Depth (Ft.)	0:1			1:1			2:1			3:1		
	380	+380	+100	380	+380	+100	380	+380	+100	380	+380	+100
2	2.160	2.160	0.568	2.135	2.149	0.565	2.113	2.137	0.562	2.091	2.126	0.560
4	4.320	4.320	1.137	4.232	4.275	1.125	4.143	4.229	1.113	4.055	4.184	1.101
6	6.481	6.481	1.705	6.278	6.378	1.679	6.080	6.276	1.652	5.886	6.174	1.625
8	8.641	8.641	2.273	8.280	8.459	2.226	7.932	8.277	2.178	7.593	8.095	2.130
10	10.80	10.80	2.842	10.25	10.52	2.768	9.706	10.23	2.693	9.187	9.948	2.618
12	12.96	12.96	3.411	12.16	12.55	3.303	11.39	12.14	3.195	10.66	11.73	3.088
14	15.12	15.12	3.979	14.03	14.56	3.833	13.00	14.01	3.686	12.02	13.45	3.540
16	17.28	17.28	4.548	15.87	16.55	4.356	14.54	15.83	4.165	13.29	15.10	3.973
18	19.44	19.44	5.116	17.66	18.52	4.874	15.99	17.60	4.632	14.44	16.68	4.389
20	21.60	21.60	5.685	19.41	20.47	5.386	17.37	19.33	5.086	15.50	18.19	4.787
22	23.76	23.76	6.253	21.12	22.39	5.891	18.69	21.01	5.529	16.47	19.64	5.167
24	25.92	25.92	6.822	22.79	24.29	6.391	19.93	22.65	5.960	17.34	21.01	5.529
26	28.08	28.08	7.390	24.41	26.16	6.885	21.10	24.24	6.379	18.13	22.32	5.873
28	30.24	30.24	7.959	26.01	28.02	7.372	22.21	25.79	6.786	18.84	23.56	6.199
30	32.40	32.40	8.527	27.56	29.85	7.854	23.25	27.29	7.181	19.48	24.73	6.508
32	34.56	34.56	9.096	29.07	31.65	8.330	24.23	28.74	7.564	20.04	25.83	6.798
34	36.72	36.72	9.664	30.55	33.44	8.799	25.15	30.15	7.935	20.54	26.87	7.070
36	38.88	38.88	10.23	31.98	35.20	9.263	26.01	31.52	8.294	20.97	27.83	7.324
38	41.04	41.04	10.80	33.38	36.94	9.721	26.81	32.84	8.641	21.34	28.73	7.561
40	43.20	43.20	11.37	34.75	38.66	10.17	27.57	34.11	8.976	21.66	29.56	7.779
42	45.37	45.37	11.94	36.08	40.35	10.62	28.26	35.34	9.299	21.93	30.32	7.980
44	47.53	47.53	12.51	37.37	42.02	11.06	28.91	36.52	9.610	22.15	31.02	8.162
46	49.69	49.69	13.08	38.63	43.67	11.49	29.51	37.66	9.910	22.34	31.64	8.327
48	51.85	51.85	13.64	39.85	45.30	10.20	30.06	38.75	10.20	22.48	32.20	8.473
50	54.01	54.01	14.21	41.04	46.90	12.34	30.57	39.79	10.47	22.59	32.69	8.602

TABLE 81. Lining Area in Square Feet for 380 Ft Square and 380 Ft Wide Rectangular Reservoirs from 2 to 50 Ft Deep, with Slopes from 0:1 to 3:1

P = 1520
2P = 3040
3P = 4560
4P = 6080

380

Depth (Ft.)	0:1			1:1			2:1			3:1		
	380	+380	+100	380	+380	+100	380	+380	+100	380	+380	+100
2	148,960	146,680	38,600	151,720	148,063	38,964	151,202	147,805	38,896	150,959	147,683	38,864
4	152,000	148,200	39,000	152,947	148,686	39,128	151,908	148,170	38,992	151,422	147,896	38,928
6	155,040	149,720	39,400	154,160	149,310	39,292	152,600	148,534	39,088	151,870	148,170	38,992
8	158,080	151,240	39,800	155,361	149,933	39,456	153,276	148,899	39,184	152,303	148,413	39,056
10	161,120	152,760	40,200	156,548	150,556	39,620	153,936	149,264	39,280	152,720	148,656	39,120
12	164,160	154,280	40,600	157,722	151,179	39,784	154,580	149,629	39,376	153,122	148,899	39,184
14	167,200	155,800	41,000	158,883	151,802	39,948	155,211	149,994	39,472	153,508	149,142	39,248
16	170,240	157,320	41,400	160,031	152,426	40,112	155,825	150,358	39,568	153,879	149,386	39,312
18	173,280	158,840	41,800	161,166	153,049	40,276	156,424	150,723	39,664	154,236	149,629	39,376
20	176,320	160,360	42,200	162,288	153,672	40,440	157,008	151,088	39,760	154,576	149,872	39,440
22	179,360	161,8 0	42,600	163,397	154,295	40,604	157,576	151,453	39,856	154,901	150,115	39,504
24	182,400	163,400	43,000	164,492	154,918	40,768	158,138	151,818	39,952	155,211	150,358	39,568
26	185,440	164,920	43,400	165,575	155,542	40,932	158,667	152,182	40,048	155,505	150,602	39,632
28	188,480	166,440	43,800	166,644	156,165	41,096	159,189	152,547	40,144	155,784	150,845	39,696
30	191,520	167,960	44,200	167,700	156,788	41,260	159,696	152,912	40,240	156,048	151,088	39,760
32	194,560	169,480	44,600	168,743	157,411	41,424	160,188	153,277	40,336	156,296	151,331	39,824
34	197,600	171,000	45,000	169,773	158,034	41,588	160,664	153,642	40,432	156,529	151,574	39,888
36	200,640	172,520	45,400	170,790	158,658	41,752	161,124	154,006	40,528	156,747	151,818	39,952
38	203,680	174,040	45,800	171,793	159,281	41,916	161,570	154,371	40,624	156,949	152,061	40,016
40	206,720	175,560	46,200	172,784	159,904	42,080	162,000	154,736	40,720	157,136	152,304	40,080
42	209,760	177,080	46,600	173,761	160,527	42,244	162,415	155,101	40,816	157,308	152,547	40,144
44	212,800	178,600	47,000	174,726	161,150	42,408	162,814	155,466	40,912	157,464	152,790	40,208
46	215,840	180,120	47,400	175,677	161,774	42,572	163,198	155,830	41,008	157,604	153,034	40,272
48	218,880	181,640	47,800	176,614	162,397	42,736	163,567	156,195	41,104	157,730	153,277	40,336
50	221,920	183,160	48,200	177,540	163,020	42,900	163,920	156,560	41,200	157,840	153,520	40,400

TABLE 82. Volume in Millions of Gallons for 400 Ft Square and 400 Ft Wide Rectangular Reservoirs from 2 to 50 Ft Deep, with Slopes from 0:1 to 3:1

Depth	0:1			1:1			2:1			3:1		
(Ft.)	400	+400	+100	400	+400	+100	400	+400	+100	400	+400	+100
2	2.394	2.394	0.598	2.367	2.382	0.595	2.344	2.370	0.592	2.320	2.358	0.589
4	4.787	4.787	1.197	4.694	4.739	1.185	4.601	4.691	1.173	4.508	4.644	1.161
6	7.181	7.181	1.795	6.968	7.073	1.768	6.759	6.965	1.741	6.554	6.858	1.714
8	9.574	9.574	2.394	9.194	9.383	2.346	8.827	9.633	2.408	8.469	9.000	2.250
10	11.97	11.97	2.992	11.38	11.67	2.917	10.81	11.37	2.842	10.27	11.07	2.768
12	14.36	14.36	3.590	13.52	13.93	3.483	12.71	13.50	3.375	11.93	13.07	3.267
14	16.76	16.76	4.189	15.61	16.17	4.042	14.52	15.58	3.896	13.48	15.00	3.749
16	19.15	19.15	4.787	17.66	18.38	4.596	16.25	17.62	4.404	14.92	16.85	4.213
18	21.54	21.54	5.386	19.66	20.57	5.143	17.90	19.60	4.901	16.25	18.63	4.659
20	23.94	23.94	5.894	21.62	22.74	5.685	19.47	21.54	5.386	17.47	20.35	5.086
22	26.33	26.33	6.582	23.54	24.88	6.220	20.96	23.43	5.858	18.60	21.99	5.496
24	28.72	28.72	7.181	25.41	27.00	6.750	22.38	25.28	6.319	19.62	23.55	5.888
26	31.12	31.12	7.779	27.25	29.09	7.274	23.73	27.07	6.768	20.56	25.05	6.262
28	33.51	33.51	8.378	29.04	31.17	7.791	25.01	28.82	7.205	21.41	26.47	6.618
30	35.90	35.90	8.976	30.79	33.21	8.303	26.21	30.52	7.630	22.17	27.83	6.956
32	38.30	38.30	9.574	32.50	35.23	8.808	27.35	32.17	8.042	22.86	29.11	7.277
34	40.69	40.69	10.17	34.17	37.23	9.308	28.43	33.77	8.443	23.47	30.32	7.579
36	43.09	43.09	10.77	35.80	39.21	9.802	29.44	35.33	8.832	24.01	31.45	7.863
38	45.48	45.48	11.37	37.38	41.16	10.29	30.38	36.84	9.210	24.48	32.52	8.129
40	47.87	47.87	11.97	36.94	43.09	10.77	31.28	38.30	9.574	24.90	33.51	8.378
42	50.27	50.27	12.57	40.45	44.99	11.25	32.11	39.71	9.927	25.25	34.43	8.608
44	52.66	52.66	13.17	41.92	46.87	11.72	32.89	41.07	10.27	25.55	35.28	8.820
46	55.05	55.05	13.76	43.36	48.72	12.18	33.61	42.39	10.60	25.80	36.06	9.015
48	57.45	57.45	14.36	44.76	50.55	12.64	34.28	43.66	10.92	26.01	36.77	9.191
50	59.84	59.84	14.96	46.13	52.36	13.09	34.91	44.88	11.22	26.18	37.40	9.350

TABLE 83. Lining Area in Square Feet for 400 Ft Square and 400 Ft Wide Rectangular Reservoirs from 2 to 50 Ft Deep, with Slopes from 0:1 to 3:1

P = 1600
2P = 3200
3P = 4800
4P = 6400

Depth	0:1			1:1			2:1			3:1		
(Ft.)	400	+400	+100	400	+400	+100	400	+400	+100	400	+400	+100
2	164,800	162,400	40,600	167,705	163,856	40,964	167,160	163,584	40,896	166,904	163,456	40,864
4	168,000	164,000	41,000	168,998	164,512	41,128	167,905	163,968	40,992	167,393	163,712	40,928
6	171,200	165,600	41,400	170,277	165,168	41,292	168,635	164,352	41,088	167,867	163,968	40,992
8	174,400	167,200	41,800	171,543	165,824	41,456	169,349	164,736	41,184	168,325	164,224	41,056
10	177,600	168,800	42,200	172,796	166,480	41,620	170,048	165,120	41,280	168,768	164,480	41,120
12	180,800	170,400	42,600	174,036	167,136	41,784	170,732	165,504	41,376	169,196	164,736	41,184
14	184,000	172,000	43,000	175,262	167,936	41,948	171,400	165,888	41,472	169,608	164,992	41,248
16	187,200	173,600	43,400	176,476	168,448	42,112	172,052	166,272	41,568	170,004	165,248	41,312
18	190,400	175,200	43,800	177,677	169,104	42,276	172,690	166,656	41,664	170,386	165,504	41,376
20	193,600	176,800	44,200	178,864	169,760	42,440	173,312	167,040	41,760	170,752	165,760	41,440
22	196,800	178,400	44,600	180,038	170,416	42,604	173,919	167,424	41,856	171,103	166,016	41,504
24	200,000	180,000	45,000	181,199	171,072	42,768	174,510	167,808	41,952	171,438	166,272	41,568
26	203,200	181,600	45,400	182,347	171,728	42,932	175,086	168,192	42,048	171,758	166,528	41,632
28	206,400	183,200	45,800	183,482	172,384	43,096	175,647	168,576	42,144	172,063	166,784	41,696
30	209,600	184,800	46,200	184,604	173,040	43,260	176,192	168,960	42,240	172,352	167,040	41,760
32	212,800	186,400	46,600	185,713	173,696	43,424	176,722	169,344	42,336	172,626	167,296	41,824
34	216,000	188,000	47,000	186,808	174,352	43,588	177,236	169,728	42,432	172,884	167,552	41,888
36	219,200	189,600	47,400	187,891	175,008	43,752	177,736	170,112	42,528	173,128	167,808	41,952
38	222,400	191,200	47,800	188,960	175,664	43,916	178,220	170,496	42,624	173,356	168,064	42,016
40	225,600	192,800	48,200	190,916	176,320	44,080	178,688	170,880	42,720	173,568	168,320	42,080
42	228,800	194,400	48,600	191,059	176,976	44,244	179,141	171,264	42,816	173,765	168,576	42,144
44	232,000	196,000	49,000	192,089	177,632	44,408	179,579	171,648	42,912	173,947	168,832	42,208
46	235,200	197,600	49,400	193,106	182,288	45,572	180,001	172,032	43,008	174,113	169,088	42,272
48	238,400	199,200	49,800	194,109	178,944	44,736	180,408	172,416	43,104	174,264	169,344	42,336
50	241,600	200,800	50,200	195,100	179,600	44,900	180,800	172,800	43,200	174,400	169,600	42,400

TABLE 84. Volume in Millions of Gallons for 420 Ft Square and 420 Ft Wide Rectangular Reservoirs from 2 to 50 Ft Deep, with Slopes from 0:1 to 3:1

420

Depth (Ft.)	0:1 420	0:1 +420	0:1 +100	1:1 420	1:1 +420	1:1 +100	2:1 420	2:1 +420	2:1 +100	3:1 420	3:1 +420	3:1 +100
2	2.639	2.639	0.628	2.614	2.625	0.625	2.589	2.612	0.622	2.564	2.600	0.619
4	5.278	5.278	1.257	5.178	5.229	1.245	5.079	5.179	1.233	4.982	5.128	1.221
6	7.917	7.917	1.885	7.693	7.804	1.858	7.473	7.690	1.831	7.258	7.577	1.804
8	10.56	10.56	2.513	10.16	10.35	2.465	9.771	10.16	2.418	9.395	9.954	2.370
10	13.19	13.19	3.142	12.58	12.88	3.067	11.98	12.57	2.992	11.40	12.25	2.917
12	15.83	15.83	3.770	14.95	15.38	3.662	14.09	14.93	3.554	13.27	14.48	3.447
14	18.47	18.47	4.398	17.27	17.86	4.252	16.12	17.24	4.105	15.02	16.62	3.958
16	21.11	21.11	5.027	19.54	20.31	4.835	18.06	19.50	4.644	16.65	18.70	4.452
18	23.75	23.75	5.654	21.77	22.73	5.413	19.91	21.71	5.170	18.17	20.70	4.928
20	26.39	26.39	6.283	23.96	25.13	5.984	21.68	23.88	5.685	19.57	22.62	5.386
22	29.03	29.03	6.912	26.09	27.51	6.549	23.37	25.99	6.187	20.86	24.47	5.825
24	31.67	31.67	7.540	28.19	29.86	7.109	24.98	28.05	6.678	22.11	26.24	6.247
26	34.31	34.31	8.168	30.23	32.18	7.662	26.51	30.06	7.157	23.14	27.93	6.651
28	36.95	36.95	8.796	32.24	34.48	8.210	27.97	32.02	7.624	24.13	29.56	7.037
30	39.58	39.58	9.425	34.20	36.76	8.752	29.35	36.57	8.078	25.04	31.10	7.405
32	42.22	42.22	10.05	36.12	39.01	9.287	30.66	35.79	8.521	25.86	32.57	7.755
34	44.86	44.86	10.68	37.99	41.23	9.817	31.90	37.60	8.952	26.60	33.97	8.087
36	47.50	47.50	11.31	39.82	43.43	10.34	33.08	39.36	9.371	27.26	35.29	8.402
38	50.14	50.14	11.94	41.61	45.61	10.86	34.18	41.07	9.778	27.85	36.53	8.698
40	52.78	52.78	12.57	43.36	47.75	11.37	35.23	42.71	10.17	28.36	37.70	8.976
42	55.42	55.42	13.19	45.07	49.90	11.88	36.21	44.35	10.56	28.82	38.79	9.236
44	58.06	58.06	13.82	46.74	51.95	12.37	37.13	45.91	10.93	29.21	39.81	9.479
46	60.70	60.70	14.45	48.37	54.05	12.87	37.99	47.42	11.29	29.55	40.75	9.703
48	63.33	63.33	15.08	49.96	56.11	13.36	38.79	48.85	11.63	29.83	41.62	9.910
50	65.97	65.97	15.71	51.51	58.13	13.84	39.54	50.27	11.97	30.01	42.42	10.10

TABLE 85. Lining Area in Square Feet for 420 Ft Square and 420 Ft Wide Rectangular Reservoirs from 2 to 50 Ft Deep, with Slopes from 0:1 to 3:1

P = 1680
2P = 3360
3P = 5040
4P = 6720

420

Depth (Ft.)	0:1 420	0:1 +420	0:1 +100	1:1 420	1:1 +420	1:1 +100	2:1 420	2:1 +420	2:1 +100	3:1 420	3:1 +420	3:1 +100
2	181,440	178,920	42,600	184,491	180,449	42,964	183,919	180,163	42,896	183,650	180,029	42,864
4	184,800	180,600	43,000	185,849	181,378	43,128	184,702	180,566	42,992	184,164	180,298	42,928
6	188,160	182,280	43,400	187,194	181,826	43,292	185,470	180,970	43,088	184,664	180,566	42,992
8	191,520	183,960	43,800	188,525	182,515	43,456	186,222	181,373	43,184	185,148	180,835	43,056
10	194,880	185,640	44,200	189,844	183,204	43,620	186,960	181,776	43,280	185,616	181,104	43,120
12	198,240	187,320	44,600	191,149	183,893	43,784	187,681	182,179	43,376	186,069	181,373	43,184
14	201,600	189,000	45,000	192,442	184,582	43,948	188,388	182,582	43,472	186,507	181,642	43,248
16	204,960	190,680	45,400	193,721	185,270	44,112	189,080	182,986	43,568	186,929	181,910	43,312
18	208,320	192,360	45,800	194,987	185,959	44,276	189,755	183,389	43,664	187,336	182,179	43,376
20	211,680	194,040	46,200	196,240	186,648	44,440	190,416	183,792	43,760	187,728	181,448	43,440
22	215,040	195,720	46,600	197,479	187,337	44,604	191,061	184,195	43,856	188,104	182,717	43,504
24	218,400	197,400	47,000	198,707	188,026	44,768	191,691	184,598	43,952	188,465	182,986	43,568
26	221,760	199,080	47,400	199,920	188,714	44,932	192,305	185,002	44,048	188,811	183,254	43,632
28	225,120	200,760	47,800	201,120	189,403	45,096	192,904	185,405	44,144	189,141	183,523	43,696
30	228,480	202,440	48,200	202,308	190,092	45,260	193,488	185,808	44,240	189,456	183,792	43,760
32	231,840	204,120	48,600	203,482	190,781	45,424	194,056	186,211	44,336	189,756	184,061	43,824
34	235,200	205,800	49,000	204,643	191,470	45,588	194,609	186,614	44,432	190,040	184,330	43,888
36	238,560	207,480	49,400	205,791	192,158	45,752	195,147	187,018	44,528	190,308	184,598	43,952
38	241,920	209,160	49,800	206,926	192,847	45,916	195,669	187,421	44,624	190,562	184,867	44,016
40	245,280	210,840	50,200	208,048	193,536	46,080	196,176	187,824	44,720	190,800	185,136	44,080
42	248,640	212,520	50,600	209,157	194,225	46,244	196,668	188,227	44,816	191,023	185,405	44,144
44	252,000	214,200	51,000	210,252	194,914	46,408	197,144	188,630	44,912	191,230	185,674	44,208
46	255,360	215,880	51,400	211,335	195,602	46,572	197,604	189,034	45,008	191,422	185,942	44,272
48	258,720	217,560	51,800	212,403	196,291	46,736	198,050	189,437	45,104	191,599	186,211	44,336
50	262,080	219,240	52,200	213,460	196,980	46,900	198,480	189,840	45,200	191,760	186,480	44,400

TABLE 86. Volume in Millions of Gallons for 440 Ft Square and 440 Ft Wide Rectangular Reservoirs from 2 to 50 Ft Deep, with Slopes from 0:1 to 3:1

440

Depth (Ft.)	0:1 440	0:1 +440	0:1 +100	1:1 440	1:1 +440	1:1 +100	2:1 440	2:1 +440	2:1 +100	3:1 440	3:1 +440	3:1 +100
2	2.896	2.896	0.658	2.870	2.882	0.655	2.844	2.869	0.652	2.818	2.856	0.649
4	5.793	5.793	1.316	5.688	5.742	1.305	5.584	5.689	1.293	5.482	5.636	1.281
6	8.689	8.689	1.945	8.454	8.571	1.948	8.223	8.452	1.921	7.997	8.334	1.894
8	11.59	11.59	2.633	11.17	11.37	2.585	10.76	11.16	2.537	10.37	10.95	2.489
10	14.48	14.48	3.291	13.83	14.15	3.216	13.20	13.82	3.142	12.60	13.50	3.069
12	17.38	17.38	3.949	16.45	16.90	3.842	15.55	16.43	3.734	14.69	15.95	3.626
14	20.27	20.27	4.608	19.01	19.63	4.461	17.80	18.98	4.314	16.65	18.34	4.168
16	23.17	23.17	5.266	21.53	23.33	5.074	19.96	21.49	4.883	18.48	20.64	4.691
18	26.07	26.07	5.924	23.99	25.00	5.682	22.03	23.93	5.439	20.19	22.87	5.197
20	28.96	28.96	6.582	26.41	27.65	6.283	24.02	26.33	5.984	21.76	25.01	5.685
22	31.86	31.86	7.241	28.78	30.27	6.879	25.91	28.67	6.517	23.26	27.08	6.155
24	34.76	34.76	7.899	31.10	32.86	7.468	27.72	30.96	7.037	24.62	29.07	6.606
26	37.65	37.65	8.557	33.38	35.42	8.051	29.45	33.20	7.546	25.88	30.98	7.040
28	40.55	40.55	9.215	35.61	37.97	8.629	31.10	35.38	8.042	27.04	32.81	7.456
30	43.44	43.44	9.874	37.79	40.48	9.200	32.67	37.52	8.527	28.09	34.56	7.854
32	46.34	46.34	10.53	39.93	42.97	9.766	34.17	39.60	9.000	29.06	36.23	8.234
34	49.24	49.24	11.19	42.02	45.45	10.33	35.59	41.63	9.461	29.94	37.82	8.596
36	52.13	52.13	11.85	44.07	47.87	10.88	36.93	43.60	9.910	30.73	39.34	8.940
38	55.03	55.03	12.51	46.07	50.29	11.43	38.21	45.54	10.35	31.44	40.77	9.266
40	57.93	57.93	13.16	48.03	52.67	11.97	39.41	47.39	10.77	32.07	42.13	9.574
42	60.82	60.82	13.82	49.95	55.00	12.50	40.55	49.19	11.18	32.64	43.41	9.865
44	63.72	63.72	14.48	51.82	57.33	13.03	41.63	51.00	11.59	33.13	44.62	10.14
46	66.61	66.61	15.14	53.66	59.66	13.56	42.64	52.66	11.97	33.57	45.72	10.39
48	69.51	69.51	15.80	55.45	61.91	14.07	43.59	54.34	12.35	33.94	46.77	10.63
50	72.41	72.41	16.46	57.20	64.20	14.59	44.48	55.97	12.72	34.26	47.74	10.85

TABLE 87. Lining Area in Square Feet for 440 Ft Square and 440 Ft Wide Rectangular Reservoirs from 2 to 50 Ft Deep, with Slopes from 0:1 to 3:1

P = 1760
2P = 3520
3P = 5280
4P = 7040

440

Depth (Ft.)	0:1 440	0:1 +440	0:1 +100	1:1 440	1:1 +440	1:1 +100	2:1 440	2:1 +440	2:1 +100	3:1 440	3:1 +440	3:1 +100
2	198,880	196,240	44,600	202,077	197,842	44,864	201,477	197,542	44,896	201,196	197,402	44,864
4	202,400	198,000	45,000	203,500	198,563	45,128	202,299	197,965	44,992	201,736	197,683	44,928
6	205,920	199,760	45,400	204,911	199,285	45,292	203,105	198,387	45,088	202,260	197,965	44,992
8	209,440	201,520	45,800	206,308	200,006	45,456	203,896	198,810	45,184	202,770	198,246	45,056
10	212,960	203,280	46,200	207,692	200,728	45,620	204,672	199,232	45,280	203,264	198,528	45,120
12	216,480	205,040	46,600	209,063	201,450	45,784	205,423	199,654	45,376	203,743	198,810	45,184
14	220,000	206,800	47,000	210,421	202,171	45,948	206,177	200,077	45,472	204,206	199,091	45,248
16	223,520	208,560	47,400	211,766	202,893	46,112	206,907	200,499	45,568	204,654	199,373	45,312
18	227,040	210,320	47,800	213,097	203,614	46,276	207,621	200,922	45,664	205,087	199,654	45,376
20	230,560	212,080	48,200	214,416	204,336	46,440	208,320	201,334	45,760	205,504	199,936	45,440
22	234,080	213,840	48,600	215,721	205,058	46,604	209,004	201,766	45,856	205,906	200,218	45,504
24	237,600	215,600	49,000	217,014	205,779	46,768	209,672	202,189	45,952	206,292	200,499	45,568
26	241,120	217,360	49,400	218,293	206,501	46,932	210,324	202,611	46,048	206,664	200,781	45,632
28	244,640	219,120	49,800	219,559	207,222	47,096	210,962	203,034	46,144	207,020	201,062	45,696
30	248,160	220,880	50,200	220,812	207,944	47,260	211,584	203,456	46,240	207,360	201,344	45,760
32	251,680	222,640	50,600	222,051	208,666	47,424	212,191	203,878	46,336	207,685	201,626	45,824
34	255,200	224,400	51,000	223,295	209,387	47,588	212,782	204,301	46,432	207,995	201,907	45,888
36	258,720	226,160	51,400	224,492	210,109	47,752	213,358	204,723	46,528	208,289	202,189	45,952
38	262,240	227,920	51,800	225,690	210,830	47,916	213,919	205,146	46,624	208,560	202,470	46,016
40	265,760	229,680	52,200	226,880	211,552	48,080	214,464	205,568	46,720	208,832	202,752	46,080
42	269,280	231,440	52,600	228,054	212,274	48,244	214,994	205,990	46,816	209,080	203,034	46,144
44	272,800	233,200	53,000	229,215	212,995	48,408	215,508	206,413	46,912	209,313	203,315	46,208
46	276,320	234,960	53,400	230,363	213,717	48,572	216,008	206,835	47,008	209,531	203,597	46,272
48	279,840	236,720	53,800	231,498	214,438	48,736	216,491	207,258	47,104	209,733	203,878	46,336
50	283,360	238,480	54,200	232,620	215,160	48,900	216,960	207,680	47,200	209,920	204,160	46,400

TABLE 88. Volume in Millions of Gallons for 460 Ft Square and 460 Ft Wide Rectangular Reservoirs from 2 to 50 Ft Deep, with Slopes from 0:1 to 3:1

460

Depth (Ft.)	0:1 460	0:1 +460	0:1 +100	1:1 460	1:1 +460	1:1 +100	2:1 460	2:1 +460	2:1 +100	3:1 460	3:1 +460	3:1 +100
2	3.166	3.166	0.688	3.138	3.151	0.685	3.111	3.137	0.682	3.084	3.123	0.670
4	6.331	6.331	1.376	6.222	6.274	1.364	6.113	6.219	1.352	6.006	6.164	1.340
6	9.497	9.497	2.064	9.251	9.375	2.038	9.010	9.251	2.011	8.773	9.126	1.984
8	12.66	12.66	2.753	12.23	12.44	2.705	11.80	12.22	2.657	11.39	12.00	2.609
10	15.83	15.83	3.441	15.15	15.48	3.366	14.49	15.14	3.291	13.85	14.79	3.216
12	18.99	18.99	4.129	18.02	18.50	4.021	17.08	18.00	3.914	16.18	17.51	3.806
14	22.16	22.16	4.817	20.84	21.49	4.671	19.57	20.81	4.524	18.36	20.13	4.377
16	25.32	25.32	5.505	23.60	24.44	5.314	21.96	23.56	5.122	20.41	22.68	4.931
18	28.49	28.49	6.193	26.32	27.37	5.951	24.26	26.26	5.709	22.32	25.14	5.466
20	31.66	31.66	6.882	28.98	30.28	6.582	26.47	28.90	6.283	24.12	27.53	24.12
22	34.82	34.82	7.570	31.60	33.16	7.208	28.58	31.49	6.846	25.78	29.83	6.484
24	37.99	37.99	8.258	34.16	36.00	7.827	30.61	34.02	7.396	27.34	32.04	6.965
26	41.15	41.15	8.496	36.68	38.82	8.440	32.55	36.50	7.935	28.77	31.36	6.818
28	44.32	44.32	9.634	39.14	41.62	9.048	34.40	38.92	8.461	30.10	36.23	7.875
30	47.48	47.48	10.32	41.56	44.39	9.649	36.17	41.29	8.976	31.33	38.19	8.303
32	50.65	50.65	11.01	43.93	47.10	10.24	37.86	43.60	9.479	32.45	40.08	8.713
34	53.81	53.81	11.70	46.25	49.82	10.83	39.47	45.86	9.969	33.48	41.88	9.105
36	56.98	56.98	12.39	48.53	52.53	11.42	41.00	48.07	10.45	34.41	43.60	9.479
38	60.15	60.15	13.08	50.76	55.15	11.99	42.46	50.19	10.91	35.26	45.24	9.835
40	63.31	63.31	13.76	52.94	57.82	12.57	43.84	52.30	11.37	36.02	46.78	10.17
42	66.48	66.48	14.45	55.08	60.40	13.13	45.15	54.33	11.81	36.71	48.25	10.49
44	69.64	69.64	15.14	57.17	62.97	13.69	46.39	56.30	12.24	37.32	49.68	10.80
46	72.81	72.81	15.83	59.21	65.50	14.24	47.57	58.37	12.69	37.86	50.97	11.08
48	75.97	75.97	16.52	61.22	68.03	14.79	48.67	60.12	13.07	38.33	52.21	11.35
50	79.14	79.14	17.20	63.18	70.52	15.33	49.72	61.92	13.46	38.75	53.31	11.59

TABLE 89. Lining Area in Square Feet for 460 Ft Square and 460 Ft Wide Rectangular Reservoirs from 2 to 50 Ft Deep, with Slopes from 0:1 to 3:1

P = 1840
2P = 3680
3P = 5520
4P = 7360

460

Depth (Ft.)	0:1 460	0:1 +460	0:1 +100	1:1 460	1:1 +460	1:1 +100	2:1 460	2:1 +460	2:1 +100	3:1 460	3:1 +460	3:1 +100
2	217,120	214,360	46,600	220,462	216,034	46,964	219,836	215,722	46,896	219,541	215,574	46,864
4	220,800	216,200	47,000	221,951	216,789	47,128	220,696	216,163	46,992	220,107	215,869	46,928
6	224,480	218,040	47,400	223,427	217,543	47,292	221,540	216,605	47,088	220,657	216,163	46,992
8	228,160	219,880	47,800	224,890	218,298	47,456	222,370	217,046	47,184	221,192	216,458	47,056
10	231,840	221,720	48,200	226,340	219,052	47,620	223,184	217,488	47,280	221,712	216,752	47,120
12	235,520	223,560	48,600	227,777	219,806	47,784	223,983	217,930	47,376	222,216	217,046	47,184
14	239,200	225,400	49,000	229,200	220,561	47,948	224,766	218,371	47,472	222,705	217,341	47,248
16	242,880	227,240	49,400	230,611	221,315	48,112	225,534	218,813	47,568	223,179	217,635	47,312
18	246,560	229,080	49,800	232,008	222,070	48,276	226,287	219,254	47,664	223,637	217,930	47,376
20	250,240	230,920	50,200	233,392	222,824	48,440	227,024	219,696	47,760	224,080	218,224	47,440
22	253,920	232,760	50,600	234,763	223,578	48,604	227,746	220,138	47,856	224,508	218,518	47,504
24	257,600	234,600	51,000	236,121	224,333	48,768	228,452	220,579	47,952	224,920	218,813	47,568
26	261,280	236,440	51,400	237,466	225,087	48,932	229,144	221,021	48,048	225,316	219,107	47,632
28	264,960	238,280	51,800	238,797	225,842	49,096	229,820	221,462	48,144	225,698	219,402	47,696
30	268,640	240,120	52,200	240,116	226,596	49,260	230,480	221,904	48,240	226,064	219,696	47,760
32	272,320	241,960	52,600	241,421	227,350	49,424	231,125	222,346	48,336	226,415	219,990	47,824
34	276,000	243,800	53,000	242,714	228,105	49,588	231,755	222,787	48,432	226,750	220,285	47,888
36	279,680	245,640	53,400	243,993	228,859	49,752	232,369	223,229	48,528	227,070	220,579	47,952
38	283,360	247,480	53,800	245,259	229,614	49,916	232,968	223,670	48,624	227,375	220,874	48,016
40	287,040	249,320	54,200	246,512	230,368	50,080	233,552	224,112	48,720	227,664	221,168	48,080
42	290,720	251,160	54,600	247,752	231,122	50,244	234,120	224,554	48,816	227,938	221,462	48,144
44	294,400	253,000	55,000	248,979	231,877	50,408	234,623	224,995	48,912	228,196	221,757	48,208
46	298,080	254,840	55,400	250,192	232,631	50,572	235,211	225,437	49,008	228,440	222,051	48,272
48	301,760	256,680	55,800	251,393	233,386	50,736	235,733	225,878	49,104	228,668	222,346	48,336
50	305,440	258,520	56,200	252,580	234,140	50,900	236,240	226,320	49,200	228,880	222,640	48,400

TABLE 90. Volume in Millions of Gallons for 480 Ft Square and 480 Ft Wide Rectangular Reservoirs from 2 to 50 Ft Deep, with Slopes from 0:1 to 3:1

480

Depth (Ft.)	0:1 480	0:1 +480	0:1 +100	1:1 480	1:1 +480	1:1 +100	2:1 480	2:1 +480	2:1 +100	3:1 480	3:1 +480	3:1 +100
2	3.447	3.447	0.718	3.418	3.432	0.715	3.390	3.418	0.712	3.361	3.403	0.709
4	6.894	6.894	1.436	6.779	6.835	1.424	6.667	6.778	1.412	6.555	6.720	1.400
6	10.34	10.34	2.154	10.08	10.21	2.127	9.832	10.08	2.100	9.584	9.950	2.073
8	13.79	13.79	2.872	13.33	13.56	2.824	12.89	13.33	2.777	12.45	13.10	2.729
10	17.23	17.23	3.590	16.53	16.88	3.516	15.84	16.52	3.441	15.17	16.16	3.366
12	20.68	20.68	4.308	19.66	20.16	4.201	18.68	19.65	4.093	17.73	19.13	3.985
14	24.13	24.13	5.027	22.75	23.42	4.880	21.42	22.72	4.733	20.15	22.02	4.587
16	27.57	27.57	5.745	25.78	26.65	5.553	24.06	25.74	5.362	22.43	24.82	5.170
18	31.02	31.02	6.463	28.75	29.86	6.220	26.60	28.69	5.978	24.56	27.53	5.736
20	34.47	34.47	7.181	31.68	33.03	6.882	29.04	31.59	6.582	26.57	30.16	6.283
22	37.91	37.91	7.899	34.54	36.18	7.537	31.39	34.44	7.175	28.44	32.70	6.813
24	41.36	41.36	8.617	37.36	39.29	8.186	33.64	37.22	7.755	30.19	35.16	7.324
26	44.81	44.81	9.335	40.13	42.38	8.829	35.80	39.96	8.324	31.82	37.53	7.818
28	48.25	48.25	10.05	42.84	45.44	9.467	37.87	42.62	8.880	33.34	39.81	8.294
30	51.70	51.70	10.77	45.51	48.48	10.10	39.85	45.24	9.425	34.74	42.01	8.752
32	55.15	55.15	11.49	48.12	51.47	10.72	41.75	47.79	9.957	36.03	44.12	9.191
34	58.60	58.60	12.21	50.69	54.43	11.34	43.56	50.30	10.48	37.22	46.14	9.613
36	62.04	62.04	12.93	53.20	57.41	11.96	45.29	52.75	10.99	38.31	48.10	10.02
38	65.49	65.49	13.64	55.67	60.29	12.56	46.94	55.10	11.48	39.31	49.92	10.40
40	68.94	68.94	14.36	58.08	63.17	13.16	48.51	57.46	11.97	40.21	51.70	10.77
42	72.38	72.38	15.08	60.45	66.05	13.76	50.00	59.71	12.44	41.03	53.38	11.12
44	75.83	75.83	15.80	62.78	68.88	14.35	51.42	61.92	12.90	41.77	54.96	11.45
46	79.28	79.28	16.52	65.05	71.66	14.93	52.77	64.08	13.35	42.43	56.50	11.77
48	82.72	82.72	17.23	67.28	74.45	15.51	54.05	66.19	13.79	43.02	57.89	12.06
50	86.17	86.17	17.95	69.46	77.18	16.08	55.25	68.21	14.21	43.53	59.23	12.34

TABLE 91. Lining Area in Square Feet for 480 Ft Square and 480 Ft Wide Rectangular Reservoirs from 2 to 50 Ft Deep, with Slopes from 0:1 to 3:1

P = 1920
2P = 3840
3P = 5760
4P = 7680

480

Depth (Ft.)	0:1 480	0:1 +480	0:1 +100	1:1 480	1:1 +480	1:1 +100	2:1 480	2:1 +480	2:1 +100	3:1 480	3:1 +480	3:1 +100
2	236,160	233,280	48,600	239,647	235,027	48,964	238,994	234,701	48,896	238,687	234,547	48,864
4	240,000	235,200	49,000	241,203	235,814	49,128	239,892	235,162	48,992	239,278	234,854	48,928
6	243,840	237,120	49,400	242,744	236,602	49,292	240,776	235,622	49,088	239,854	235,162	48,992
8	247,680	239,040	49,800	244,273	237,389	49,456	241,644	236,083	49,184	240,414	235,469	49,056
10	251,520	240,960	50,200	245,788	238,176	49,620	242,496	236,544	49,280	240,960	235,776	49,120
12	255,360	242,880	50,600	247,290	238,963	49,784	243,333	237,005	49,376	241,490	236,083	49,184
14	259,200	244,800	51,000	248,779	239,750	49,948	244,155	237,466	49,472	242,004	236,390	49,248
16	263,040	246,720	51,400	250,255	240,538	50,112	244,961	237,926	49,568	242,504	236,698	49,312
18	266,880	248,640	51,800	251,718	241,325	50,276	245,752	238,387	49,664	243,005	237,005	49,376
20	270,720	250,560	52,200	253,168	242,112	50,440	246,528	238,848	49,760	243,456	237,312	49,440
22	274,560	252,480	52,600	254,605	242,899	50,604	247,288	239,309	49,856	243,909	237,619	49,504
24	278,400	254,400	53,000	256,028	243,686	50,768	248,033	239,770	49,952	244,347	237,926	49,568
26	282,240	256,320	43,400	257,439	244,474	50,932	248,763	240,230	50,048	244,769	238,234	49,632
28	286,080	258,240	53,800	258,836	245,261	51,096	249,477	240,691	50,144	245,176	238,541	49,696
30	289,920	260,160	54,200	260,220	246,048	51,260	250,176	241,152	50,240	245,568	238,848	49,760
32	293,760	262,080	54,600	261,591	246,835	51,424	250,860	241,613	50,336	245,944	239,155	49,824
34	297,600	264,000	55,000	262,949	247,622	51,588	251,528	242,074	50,432	246,305	239,462	49,888
36	301,440	265,920	55,400	264,294	248,410	51,752	252,180	242,534	50,528	246,651	239,770	49,952
38	305,280	267,840	55,800	265,626	249,197	51,916	252,810	242,995	50,624	246,901	240,077	50,016
40	309,120	269,760	56,200	266,944	249,984	52,080	253,440	243,456	50,720	247,296	240,384	50,080
42	312,960	271,680	56,600	268,249	250,771	52,244	254,047	243,917	50,816	247,596	240,691	50,144
44	316,800	273,600	57,000	269,541	251,558	52,408	254,638	244,378	50,912	247,880	240,998	50,208
46	320,640	275,520	57,400	270,821	252,346	52,572	255,214	244,838	51,008	248,148	241,306	50,272
48	324,480	277,440	57,800	272,087	253,133	52,736	255,774	245,299	51,104	248,402	241,613	50,336
50	328,320	279,360	58,200	273,340	253,920	52,900	256,320	245,760	51,200	248,640	241,920	50,400

TABLE 92. Volume in Millions of Gallons for 500 Ft Square and 500 Ft Wide Rectangular Reservoirs from 2 to 50 Ft Deep, with Slopes from 0:1 to 3:1

Depth	0:1			1:1			2:1			3:1		
(Ft.)	500	+500	+100	500	+500	+100	500	+500	+100	500	+500	+100
2	3.740	3.740	0.748	3.710	3.725	0.745	3.680	3.710	0.742	3.651	3.695	0.739
4	7.480	7.480	7.361	7.361	7.420	1.484	7.243	7.360	1.472	7.127	7.300	1.460
6	11.22	11.22	2.244	10.95	11.09	2.217	10.69	10.95	2.190	10.43	10.82	2.163
8	14.96	14.96	2.992	14.49	14.72	2.944	14.02	14.48	2.896	13.57	14.24	2.848
10	18.70	18.70	3.744	17.96	18.33	3.665	17.24	17.95	3.590	16.55	17.58	3.516
12	22.44	22.44	4.488	21.38	21.90	4.380	20.35	21.37	4.273	19.36	20.83	4.165
14	26.18	26.18	5.236	24.74	25.45	5.089	23.36	24.72	4.943	22.03	23.98	4.796
16	29.92	29.92	5.984	28.05	28.97	5.793	26.25	28.01	5.601	24.54	27.05	5.410
18	33.66	33.66	6.732	31.31	32.45	6.490	29.05	31.24	6.247	26.91	30.03	6.005
20	37.40	37.40	7.480	34.49	35.91	7.181	31.74	34.41	6.882	29.14	32.91	6.582
22	41.14	41.14	8.228	37.63	39.33	7.866	34.32	37.52	7.504	31.23	35.71	7.142
24	44.88	44.88	8.976	40.71	42.73	8.545	36.81	40.57	8.114	33.20	38.42	7.683
26	48.62	48.62	9.724	43.74	46.09	9.218	39.20	43.57	8.713	35.02	41.04	8.207
28	52.36	52.36	10.47	46.71	49.43	9.886	41.51	46.50	9.299	36.74	43.57	8.713
30	56.10	56.10	11.22	49.64	52.75	10.55	43.71	49.37	9.874	38.33	46.00	9.200
32	59.84	59.84	11.97	52.51	56.00	11.20	45.83	52.20	10.44	39.80	48.35	9.670
34	63.58	63.58	12.72	55.33	59.25	11.85	47.85	54.95	10.99	41.17	50.60	10.12
36	67.32	67.32	13.46	58.09	62.45	12.49	49.79	57.65	11.53	42.43	52.80	10.56
38	71.06	71.06	14.21	60.81	65.65	13.13	51.65	60.25	12.05	43.58	54.85	10.97
40	74.80	74.80	14.96	63.47	68.80	13.76	53.42	62.85	12.57	44.74	56.85	11.37
42	78.54	78.54	15.71	66.08	71.95	14.39	55.11	65.35	13.07	45.61	58.75	11.75
44	82.28	82.28	16.46	68.65	75.05	15.01	56.72	67.80	13.56	46.48	60.55	12.11
46	86.02	86.02	17.20	71.16	78.10	15.62	58.25	70.20	14.04	47.27	62.30	12.46
48	89.76	89.76	17.95	73.63	81.15	16.23	59.70	72.55	14.51	47.98	63.90	12.78
50	93.50	93.50	18.70	76.05	84.15	16.83	61.09	74.80	14.96	48.62	65.45	13.09

TABLE 93. Lining Area in Square Feet for 500 Ft Square and 500 Ft Wide Rectangular Reservoirs from 2 to 50 Ft Deep, with Slopes from 0:1 to 3:1

P = 2000
2P = 4000
3P = 6000
4P = 8000

Depth	0:1			1:1			2:1			3:1		
(Ft.)	500	+500	+100	500	+500	+100	500	+500	+100	500	+500	+100
2	256,000	253,000	50,600	259,633	254,820	50,964	258,952	254,480	50,896	258,632	254,320	50,864
4	260,000	255,000	51,000	261,254	255,640	51,128	259,889	254,960	50,992	259,249	254,640	50,929
6	264,000	257,000	51,400	262,361	256,460	51,292	260,811	255,440	51,088	259,851	254,960	50,992
8	268,000	259,000	51,800	264,455	257,280	51,456	261,717	255,920	51,184	260,437	255,280	51,056
10	272,000	261,000	52,200	266,036	258,100	51,620	262,608	256,400	51,280	261,008	255,600	51,120
12	276,000	263,000	52,600	267,604	258,920	51,784	263,484	256,880	51,376	261,564	255,920	51,184
14	280,000	265,000	53,000	269,159	259,740	51,948	264,344	257,360	51,472	262,104	256,240	51,248
16	284,000	267,000	53,400	270,700	260,560	52,112	265,188	257,840	51,568	262,628	256,560	51,312
18	288,000	269,000	53,800	272,229	261,380	52,276	266,018	258,320	51,664	263,138	256,880	51,376
20	292,000	271,000	54,200	273,744	262,200	52,440	266,832	258,800	51,760	263,632	257,200	51,440
22	296,000	273,000	54,600	275,246	263,020	52,604	267,631	259,280	51,856	264,111	257,520	51,504
24	300,000	275,000	55,000	276,735	263,840	52,768	268,414	259,760	51,952	264,574	257,840	51,568
26	304,000	277,000	55,400	278,211	264,660	52,932	269,182	260,240	52,048	265,022	258,160	51,632
28	308,000	279,000	55,800	279,674	265,480	63,096	269,935	260,720	52,144	265,455	258,480	51,696
30	312,000	281,000	56,200	281,124	266,300	53,260	270,672	261,200	52,240	265,872	258,800	51,760
32	316,000	283,000	56,600	282,561	267,120	53,424	271,394	261,680	52,336	266,274	259,120	51,824
34	320,000	285,000	57,000	283,984	267,940	53,588	272,100	262,160	52,432	266,660	259,440	51,888
36	324,000	287,000	57,400	285,395	268,760	53,752	272,792	262,640	52,528	267,032	259,760	51,952
38	328,000	289,000	57,800	286,792	269,580	53,916	273,468	263,120	52,624	267,388	260,080	52,016
40	332,000	291,000	58,200	288,176	270,400	54,080	274,128	263,600	52,720	267,728	260,400	52,080
42	336,000	293,000	58,600	289,547	271,220	54,244	274,773	264,080	52,816	268,053	260,720	52,144
44	340,000	295,000	59,000	290,905	272,040	54,408	275,403	264,560	52,912	268,363	261,040	52,208
46	344,000	297,000	59,400	292,250	272,860	54,572	276,017	265,040	53,008	268,567	261,370	52,272
48	348,000	299,000	59,800	293,581	273,680	54,736	276,616	265,520	53,104	268,936	261,680	52,336
50	352,000	301,000	60,200	294,940	274,500	54,900	277,200	266,000	53,200	269,200	262,000	52,400

Glossary

aquifer: an underground natural reservoir.

berm: the level portion of the reservoir at the top of the slope. A roadway is often built on the berm.

bonded lining: a lining system that is adhesively bonded 100% to the substrate, which is usually trowel finished concrete.

CPE: abbreviation for chlorinated polyethylene membrane.

count: short term for fabric count, the number of threads per inch in the fabric, expressed as two numbers separated by an "x." The first number is the number of warp yarns per inch, and the second is the number of fill yarns (picks) per inch.

critical reservoir: a reservoir so located as to cause possible loss of life or extensive property damage if it failed. Sometimes used to designate a body of water, functional or decorative, that sits over a potable water reservoir. Also, a body of water, functional or decorative, that sits above a normally occupied area where leakage could cause inconvenience to persons or equipment below.

cured membranes: materials that cure, usually by the chemically cross-linking of molecules to yield a compound that is not softened upon subsequent exposure to heat.

cut-and-fill reservoir: a reservoir constructed by scooping out a place in the earth and utilizing the excavated material to increase the depth of the cut. Even though the earth from the hole is not used in the embankment, the facility is still referred to as a cut-and-fill reservoir.

denier: a measure of synthetic yarn size. A unit of weight equal to 0.05 g/450 m of yarn.

dewatering: the process of emptying a reservoir.

drawdown: the lowering of the water level in a reservoir.

elastomers: the general name given to butyl rubber, neoprene, EPDM, and EPT.

electronic bonding: use of high-frequency radio circuitry to effect thermal bonding of two pieces of PVC membranes. CPE and some Hypalon formulations may be seamed by this technique.

EDPM: a tripolymer consisting of ethylene, propylene, and an unidentified diene monomer. "EPDM" is often used interchangeably with "butyl rubber."

EPT: abbreviation for ethylene propylene terpolymer, a tripolymer consisting of ethylene, propylene, and an unidentified third polymer. "EPT" is often used interchangeably with "EPDM," designating members of the butyl rubber family.

fish mouth: a seam defect in which one member of the seam, because of wrinkles or unequal stretching, is longer than the mating piece. It is most common in EPDM or butyl family lining work; when such materials are joined, the longer piece may "pop up," opening up a passage through which water may flow.

freeboard: the part of the reservoir that is above the maximum waterline of the reservoir. The freeboard distance is expressed in feet, measured vertically.

Gunite: a trademark used for a mixture of cement and sand, pumped dry through a flexible hose. At the mixing head, water is introduced and the mixture sprayed as a wet mix onto the surface to be coated.

Hypalon: chlorosulfonated PE resin or membrane lining material.

leakage tolerance: the maximum value, expressed as gallons per minute, that will be tolerated for acceptance of a reservoir for service under a full head of water.

leno weave: a plain woven fabric in which two warp yarns weave in a crisscross pattern, locking onto each fill yarn. This gives loosely woven, lightweight fabrics more stability in handling.

lining: the portion of a reservoir responsible for the first line of defense against seepage, that is, the part immediately adjacent to the liquid being held.

membrane: a lining, especially one of the continuous, flexible, impervious types.

MG: million gallons.

mgd: million gallons per day.

mil: one-thousandth of an inch.

mm: millimeter; 1 mm = 1/25.4 in.

MVT: moisture vapor transmission, expressed as grams/100 sq in./24 hr/mil of thickness. Measurement is done at 90% relative humidity and 100° F.

panel: a single unit or panel of a factory fabricated, flexible lining piece. Also, a short term for prefabricated asphalt panel.

panel lining: linings made of prefabricated asphalt panels. Also, a single panel of a factory fabricated, flexible lining piece.

PE: abbreviation for polyethylene membrane.

perm: permeance of a sample; the MVT rate expressed as grains/sq ft/hr/in. of mercury moisture vapor pressure difference.

plasticizer: an oily base material, which under heat and pressure causes PVC resin to go into solution during processing in internal mixers. This is the volatile component of the mix, which is slowly sublimed by heat and leached out by water. The plasticizer loss is largely responsible for the stiffening of PVC membranes upon aging.

popcorn asphalt: an open-graded asphalt concrete blanket used in the construction of continuous underdrain blankets and as a slope protection against wave erosion.

Preload tank: generic name for a prestressed wire wrapped tank, from the name of the firm that did much to develop this type of holding facility.

prime lining: a lining or a lining system that is engineered to prevent the intrusion of water into the substrate. Generally, a prime lining system is used in conjunction with an engineered underdrain system. Normally, PVC and PE films are not classed as prime lining materials.

PVC: abbreviation for polyvinyl chloride membrane or resin.

radio-frequency bonding: technical name for electronic bonding of plastic lining materials.

reverse hydrostatic: the uplift pressures sometimes developed beneath the lining by water trapped there when the reservoir is being dewatered.

rubber: in this book, the meaning is "a synthetic rubber," such as butyl, EPT, EPDM, or neoprene.

scrim: loosely woven fabric with spaces between yarns easily visible. Total fabric weights for scrims for lining reinforcement are usually on the order of 2.0 oz/sq yd.

second-feet: cubic feet per second.

shotcrete: a form of sprayed concrete similar in final appearance to Gunite.

sink hole: a depression in the substrate, usually caused by settlement or substrate particle removal by water migrating behind the lining. The hole is deep in comparison to its diameter.

slope: the side slope of a cut-and-fill reservoir is stated as the ratio of the horizontal distance to the vertical distance. Vertical slopes have zero slope.

sterilant a chemical added to the soil just before application of the lining to control weed growths above the waterline.

thermoplastics: plastic materials, that, upon heating, rapidly lose their strength properties.

toe of slope: the line formed by the intersection of bottom and sloping side walls of the cut-and-fill reservoir.

tongue-and-groove joint: a joint common to butyl and EPDM linings, whereby one edge of the lining fits into a groove of the adjacent piece and both flaps of the groove are sealed down tight. Flaps are usually about 6 in. wide.

top of slope: the line formed by the intersection of the berm and sloping side wall of the cut-and fill reservoir.

underdrain system: a system built into the reservoir substrate whose sole purpose is to collect any stray water that gets into the substrate for any reason and lead it therefrom without removing any soil particles in the process.

Visqueen: generic term for PE film lining.

Bibliography

American Society of Civil Engineers, "Review of Slope Protection Methods," *Proceedings,* Vol. 74, June 1948, Subcommittee on Slope Protection, Soil Mechanics and Foundations Division.

Asphalt Institute, *Asphalt in Hydraulics,* 5th printing, College Park, Md., March 1965.

Association of American Portland Cement Manufacturers, *Concrete Review,* Vol. 4, No. 2, Bull. No. 23, Philadelphia, 1910.

Baumeister, T. and Marks, L. S., *Standard Handbook for Mechanical Engineers,* 7th edition, McGraw-Hill Book Co., New York, 1967 p. 9-198.

California, State of *Investigation of the Failure of Baldwin Hills Reservoir,* Resources Agency, Department of Water Resources, April 1964, 64 pp.

Carr, Donald E., *Death of the Sweet Waters,* W. W. Norton Co., New York, 1966, 231 pp.

Casagrande, A., "Classification and Identification of Soils," *Transactions of the American Society of Civil Engineers,* 1948.

Cedergren, Harry R., "Control of Seepage in Earth Dams," *Proceedings of the Second Seepage Symposium,* Phoenix, Ariz., Mar. 25–27, 1968.

Colorado Department of Health, "1976 Drinking Water Standards," Denver.

Cook, J. Gordon, *The World of Water,* Dial Press, New York, 1957, 192 pp.

Davis, Charles F., *The Law of Irrigation,* Publishers Press Room & Bindery Co., Denver, Colo., 1915.

Davis, C. V., and Sorensen, K. V., *Handbook of Applied Hydraulics,* 3rd edition, McGraw-Hill Book Co., New York, 1969, pp. 4–20.

Davy, H., *Philosophical Transactions of the Royal Society of London,* Vol. 114, 1824.

Decker, Ray S., "Sealing Small Reservoirs with Chemical Dispersants," *Proceedings of First Seepage Symposium,* Phoenix, Ariz., Feb. 19–21, 1963.

Fisher, Douglas Alan, *The Epic of Steel,* Harper & Row, New York, 1963, 344 pp.

Galanopoulos, A. G. and Bacon, Edward, *Atlantis: The Truth behind the Legend.* Bobbs-Merrill Co., New York, 1969, 199 pp.

Hough, B. K., *Basic Soils Engineering,* Ronald Press, New York, 1957, 416 pp.

Huffman, Roy E., *Irrigation Development and Public Water Supplies,* Ronald Press, New York, 1953, 336 pp.

Kenmir, Russell C., "Concrete Reservoir Design," *Journal of the American Water Works Association,* Vol. 60, No. 10, October 1968, pp. 1188–1189.

Kinney, Clessen S., *A Treatise on the Law of Irrigation,* W. H. Lowdermilk & Co., Washington, D. C., 1894.

LaQue, F. L. and Copson, H. R., *Corrosion Resistance of Metals and Alloys,* American Chemical Society Nomograph Series, Reinhold Publishing Corp., New York, 1965, 685 pp.

Lesley, Robert W., *History of the Portland Cement Industry in the United States,* International Trade Press, Chicago, Ill., 1924.

Marron, Hal E., "Steel Liner for Earthen Reservoir," Paper presented before the American Water Works Association Convention, Southern California Section, Santa Monica, Calif., Oct. 25, 1962.

Miller, David W., Geraghty, James S., and Collins, Robert S., *Water Atlas of the United States,* Water Information Center, Port Washington, N.Y., 1962, 86 pp.

Outland, Charles, *Man-Made Disaster—The Story of St. Francis Dam,* Arthur H. Clark Co., Glendale, Calif., 1963, 249 pp.

Overman, Michael, *Water,* Doubleday & Co., Garden City, N. Y., 1969, 192 pp.

Payne, Robert, *The Canal Builders,* Macmillan Co., New York, 1959, 278 pp.

Portland Cement Association, *Concrete Lined Reservoirs,* Chicago, 1955.

Potter, James H., *Handbook of Engineering Sciences,* Vol. 1, D. Van Nostrand & Co., Princeton, N. J., 1967, p. 996.

Proctor, R. R., "Fundamental Principles of Soil Compaction," *Engineering News-Record,* Aug. 31, Sept. 7, Sept. 21, and Sept. 28, 1933.

"Railway Roadside Water Tanks for Locomotive Supply," *Railway Age Gzaette,* Nov. 19, 1915, p. 955.

Singer, Charles, Holmyard, E. J., Hall, A. R., and Williams, Trevor I., *A History of Technology,* Clarendon Press, Oxford, England, 1958, 888 pp.

Sowers, George B., and Sowers, George F., *Introductory Soil Mechanics and Foundations,* Macmillan Co., New York, 1965, 378 pp.

Tschebotarioff, Gregory P., *Soil Mechanics, Foundations, and Earth Structures,* McGraw-Hill Book Co., New York, 1951, 329 pp.

U. S. Army Transportation Research Command, *Heliocopter Downwash Blast Effects Study,* Vicksburg, Miss., 1964.

U. S. Bureau of Reclamation, *Linings for Irrigation Canals,* U. S. Government Printing Office, Washington, D. C., 1963.

U. S. Bureau of Reclamation, *Design of Small Dams,* U. S. Government Printing Office, Washington, D. C., 1960, pp. 201–204.

U. S. Waterways Experiment Station, "Unified Soil Classification System," Technical Memorandum No. 3-357, Vicksburg, Miss., 1953.

Ven Te Chow, *Open Channel Hydraulics,* McGraw-Hill Book Co., New York, 1959, 680 pp.

Water Information Center, *Water Newsletter,* Port Washington, N. Y., Vol. 18, No. 15, Aug. 9, 1976.

Westerman, W. L., "The Development of the Irrigation System of Egypt," *Classical Philology,* Vol. XIV, No. 2, April 1919, pp. 158–164

Winiarski, Lawrence D., and Bryam, Kenneth V., *"Reflective Cooling Ponds,"* Unpublished Paper, National Pollution Research Program, U.S. Department of the Interior, Federal Water Quality Administration, Pacific Northwest Water Laboratory, Corvallis, Ore., 1970.

Index